20th International Conference on the Application of High Magnetic Fields in Semiconductor Physics

(HMF-20)

AA002826

Journal of Physics: Conference Series Volume 456

Chamonix, France
22-27 July 2012

ISBN: 978-1-62993-094-7
ISSN: 1742-6588

Printed from e-media with permission by:

Curran Associates, Inc.
57 Morehouse Lane
Red Hook, NY 12571

Some format issues inherent in the e-media version may also appear in this print version.

Printed by Curran Associates, Inc. (2013)

For permission requests, please contact the Institute of Physics
at the address below.

Institute of Physics
Dirac House, Temple Back
Bristol BS1 6BE UK

Phone: 44 1 17 929 7481
Fax: 44 1 17 920 0979

techtracking@iop.org

Additional copies of this publication are available from:

Curran Associates, Inc.
57 Morehouse Lane
Red Hook, NY 12571 USA
Phone: 845-758-0400
Fax: 845-758-2634
Email: curran@proceedings.com
Web: www.proceedings.com

TABLE OF CONTENTS

Journal of Physics: Conference Series **456** (2013) 011001 doi:10.1088/1742-6596/456/1/011001

Preface

The 20th International Conference on 'High Magnetic Fields in Semiconductor Physics' (HMF-20) was held on 22–27 July 2012, in Chamonix Mont Blanc, France, as a satellite meeting to the 31st International Conference on the Physics of Semiconductors.

HMF-20 followed a series of biennial conferences, initiated by Gottfried Landwehr, in Wurzburg, Germany, in 1972. Primarily focused on 'semiconductors' and 'magnetic fields', the main topics of the conference have evolved with time and are now dominated by current themes related to the physics of low dimensional systems in conjunction with the application of magnetic fields.

The list of HMF-20 topics included: quantum Hall effect phenomena, graphene and carbon nanotubes, quantum wells, dots and wires, bulk semiconductors, topological insulators and organic conductors, magneto-transport and magneto-spectroscopy, electron correlations and magnetic field driven phases, spin-dependent phenomena and non-equilibrium effects, as well as novel phenomena and new techniques in high magnetic fields.

The HMF-20 conference gathered 200 participants from 23 different countries. It was organized by the Laboratoire National des Champs Magnétiques Intenses, Grenoble, France, and greatly sponsored by the European High Magnetic Field Laboratory under the EC-FP7 framework. The 21st edition of the HMF conference series will take place during the summer of 2014 in Florida, USA.

We thank the participants who, through their presentations, convivial discussions, and the papers presented here, contributed to the success of HMF-20 and advancements in the physics related to the applications of high magnetic fields.

Clément Faugeras, Milan Orlita, Benjamin Piot and Marek Potemski

Laboratoire National des Champs Magnétiques Intenses,
CNRS/UJF/UPS/INSA Grenoble,
France

Journal of Physics: Conference Series **456** (2013) 011002 doi:10.1088/1742-6596/456/1/011002

Peer review statement

All papers published in this volume of *Journal of Physics: Conference Series* have been peer reviewed through processes administered by the proceedings Editors. Reviews were conducted by expert referees to the professional and scientific standards expected of a proceedings journal published by IOP Publishing.

Quantum, normal and anomalous Hall effect in 2D ferromagnetic structures: GaAs/InGaAs/GaAs quantum well with remote Mn delta-layer

B. Aronzon,[1] A. Davydov,[1] M. Goiran,[2] B. Raquet,[2] A. Lashkul[3] and E. Lahderanta[3]

[1]RRC "Kurchatov institute", Kurchtov sq. 1, Moscow, 123182, Russia

[2] LNCMI, CNRS-UPR 3228, INSA, UJF, UPS, 143 Avenue de Rangueil, 31400 Toulouse, France and 25 rue des Martyrs, 38042 Grenoble, France

[3]Lappeenranta University of Technology, Lappeenranta, P.O. box 20, 53850, Finland

E-mail: aronzon@mail.ru

Abstract. We study experimentally the electronic transport and magnetism of GaAs/InGaAs/GaAs quantum wells (QW) with remote Mn δ-layer. The 2D energy spectrum of the carriers is revealed by quantum Hall Effect measurements. The ferromagnetic ordering is evidenced by anomalous Hall Effect and direct magnetization measurements. A signature of this state is also observed on the temperature dependence of the sample resistance. The dependence of the Curie temperature, T_C, on manganese content, quantum well depth, and on the spacer thickness between QW and Mn δ-layer shed light on the mechanisms of exchange interaction in such structures.

1. Introduction

Due to fundamental limits for further miniaturization of transistors in microprocessors and memory cells, electronic devices based on new physical principles are demanded. One of the most promising proposal is to transfer and process information not only through the charge but also via the electrons spin. This domain of research is called spintronics. In this context, Dilute Magnetic Semiconductors (DMS), mostly based on $A^{III}B^{V}$-type semiconductors, are currently the object of a widespread attention. However, two-dimensional (2D) structures, which proved to be suitable for electronics, are much less studied in the context of spintronics, and only recently some papers appeared, see for example [1, 2, 3]. For such structures, the possibility to achieve a spatial separation of magnetic dopants and charge carriers seems to be crucial and promising. Having this in mind, we have studied heterostructures GaAs/Mn δ-layer/GaAs/InGaAs/GaAs quantum wells with a remote Mn δ-layer introduced into one of the GaAs layers. The Mn content (Mn layer effective thickness) ΔMn varied in the range of 0.25 -1.2 ML, the spacer thickness d, between the Mn layer and the QW was fixed from 1 to 5 nm, the thickness of the QW was 10 nm and its depth U_0 in the range from 80 to 160 meV was controlled by In concentration. For detailed description of the structure see [3]. In these samples we observed ferromagnetic (FM) ordering and different manifestations of the spin polarization of the charge carriers.

Journal of Physics: Conference Series **456** (2013) 012001 doi:10.1088/1742-6596/456/1/012001

2. Results and discussion

Unlike previous studies of $A^{III}B^{V}$-type DMS with Mn δ-layers, in our case the Mn layer and the conductive channel (the quantum well filled with charge carriers) are well separated from each other. This results in a relatively high mobility of the carriers, two orders of magnitude higher as compared to the DMS-based structures studied earlier [4, 5]. The hole spectrum is unambiguously 2D, as it is confirmed by the observation of the quantum Hall effect shown in Fig. 1. At low magnetic fields we have observed the anomalous Hall effect (AHE) up to relatively high temperatures (140 K) [3, 6] (see Fig.2) unveiling the spin polarization of the 2D hole gas. The magnetic ordering in the structure under study was also detected by the presence of a hump in the temperature dependence of the electrical resistance $R(T)$ presented in Fig. 3, which is a signature of FM ordering, while samples doped with nonmagnetic acceptor do not show such a hump [3, 6]. The position of this hump is often assumed as Curie temperature T_c [7].

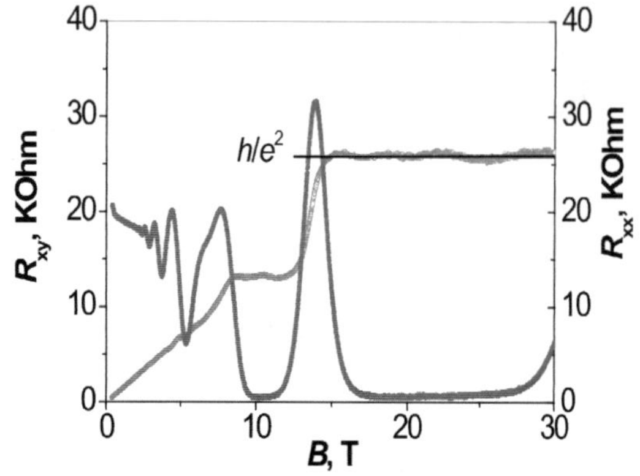

Fig. 1. Magnetic field dependence of the Hall and longitudinal resistance of the structure with ΔMn = 0.3 ML and d = 3nm.

The polarization of photo- and electroluminescence, proving the spin polarization of the carriers, was recently observed in similar structures [8]. Direct magnetization measurements were also performed. At high temperatures, the magnetization curves correspond to a magnetic cluster glass, while at low temperatures and at low Mn content a hysteresis loop was observed [9]. For structures with high Mn content exchange bias was also observed [10]. This is due to spatial disorder of Mn distribution, which results in the existence of antiferromagnetic regions that could pin the magnetic moment of FM areas.

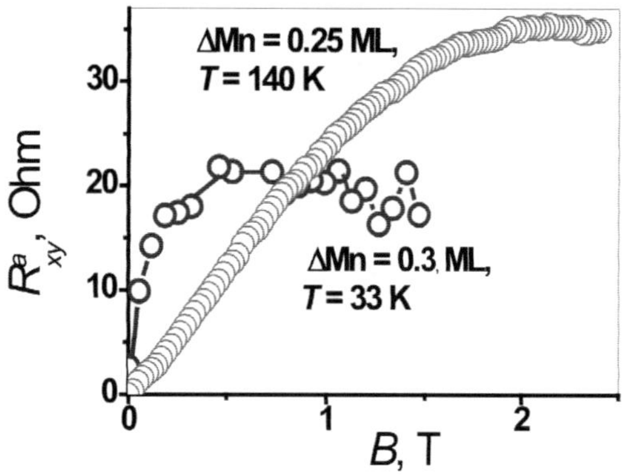

Fig. 2. Anomalous Hall effect resistance versus magnetic field for samples with Mn content ΔMn = 0.3 ML and 0.25 ML, measured at T = 33 K and T = 140 K, correspondently, d = 3 nm.

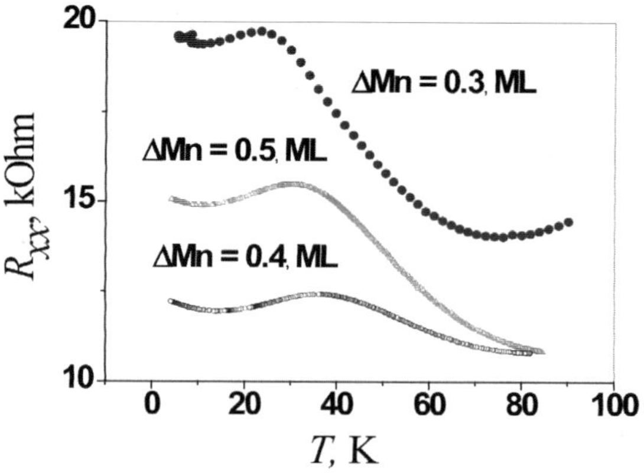

Fig. 3. Temperature dependence of resistance for samples with various Mn content ΔMn = 0.3, 0.4 and 0.5 ML, d = 3 nm. The hump on these dependences corresponds to T_c.

In this paper, the possible mechanisms of exchange interaction leading to the FM state in these systems will be discussed. To shed light on these mechanisms, we performed measurements at various Mn concentration, QW depth U_0 and spacer thickness d. These results, presented in Figs. 4 and 5, suggest the occurrence of two mechanisms contributing to the spin polarization of carriers in the QW. The first is related to the indirect exchange between carriers in the QW and Mn ions in δ-layer [11]. According to the second mechanism the magnetization of Mn δ-layer is due to holes inside it and the interaction between FM ordered Mn δ-layer and the carriers in the QW leads to carrier spin polarization [12].

The first model assumes that holes wave function has a tail which is extending out of the QW that could overlap with the Mn atoms in the Mn –layer. So exchange between Mn atoms could be mediated by free carriers (localized in the QW) as in the RKKY model, this mechanism provides in turns a way for the polarization of the mobile holes. However, the interaction between carriers and magnetic atoms is weaker than in the RKKY scenario, because the amplitude of the wave function tail decays exponentially outside the QW. On the other hand the mean free path of carriers in our structures is much longer than in bulk due to the high mobility value that helps for exchange.

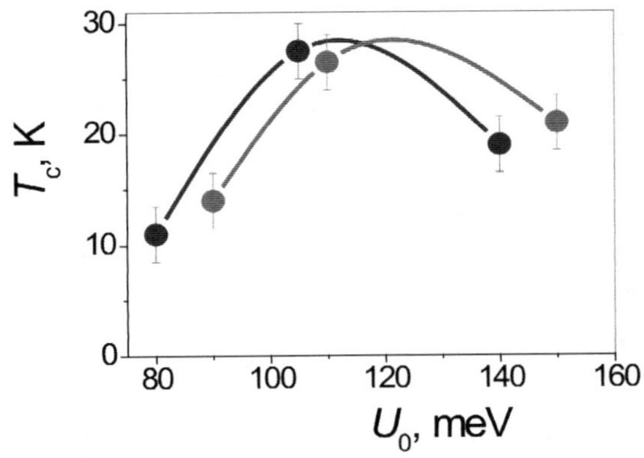

Fig. 4. Dependence of the Curie temperature on the depth of the QW U_0 controlled by In concentration for two set of samples with ΔMn = 0.3 ML (red) and 0.25 ML (blue), $d = 3$ nm.

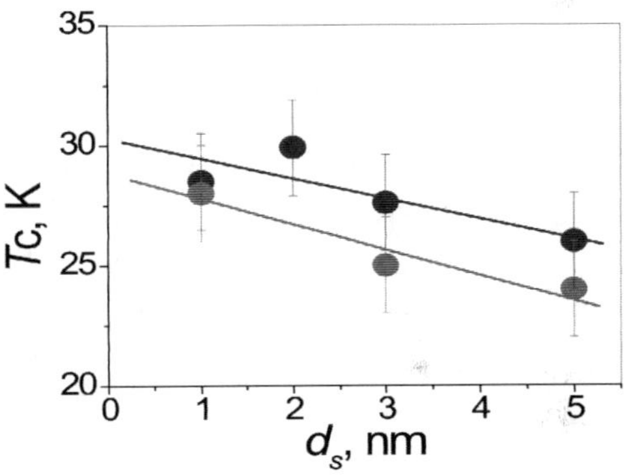

Fig. 5. Dependence of the Curie temperature on the spacer thickness d for two sets of samples with ΔMn = 0.3 ML (red) and 0.25 ML (blue), $U_0 \approx 110$ meV.

This mechanism could explain the nonmonotonous dependence of $T_C(U_0)$ determined as the temperature at which $R(T)$ has a hump, see Fig. 4. Holes in the QW are transferred from Mn acceptors. So if the QW is not deep enough the hole subband could be located below the acceptor level and no free holes could contribute to the magnetic exchange. With increasing U_0 free holes concentration in QW increases also and becomes large enough to give rise to the above depicted exchange interaction. Further increase of U_0 results in diminishing of the wave function tail amplitude outside the QW. That leads to decrease of the interaction between carriers and magnetic atoms and lowering of $T_C(U_0)$ at high enough U_0 values. This mechanism may be relevant for small thicknesses of spacer and should depend on its thickness. In support of this is the graph shown in Fig. 5, which clearly shows that the Curie temperature is a decreasing function of the spacer thickness.

However the dependence $T_C(d)$ presented at Fig. 5 is very weak, while the above mechanisms assumes the exponential behavior of Curie temperature versus spacer thickness. So another mechanism of the FM ordering and spin polarization of the carriers in the system should be suggested to explain this dependence. It was proposed in Ref. [12]. Due to diffusion, the Mn δ -layer is blurred and spread into

GaAs layers (spacer and cover) up to a thickness of 2-3 nm, but as demonstrated by X-ray measurements, does not overlap with the QW [3]. So the Mn δ -layer could be considered as a thin GaMnAs DMS layer. According to [12] it is sufficient for the occurrence of itinerant FM ordering in this layer. This mechanism of the ferromagnetic ordering does not need carriers from QW and T_C should not depend on *d*. As already stated, carriers in the QW come from Mn atoms located in the δ -layer and act as acceptors. The transition to FM state results in the lowering of the Fermi energy in the layer and so some carriers should be transferred from the QW back to acceptors, according to the new Fermi level position. That was observed and shown in Fig. 6 also proving the contribution of this mechanism of FM ordering. Tails of wave functions related to the FM state in the Mn layer could penetrate to QW causing the holes spin – polarization [12]. The large temperature range where hole concentration *p* is *T* dependant is related to the nonuniform spatial distribution of Mn. At high enough temperature the Mn layer is paramagnetic. When temperature decreases the phase separation occurs in this layer. It is divided in ferromagnetic and paramagnetic regions. The Curie temperature in each of FM regions (FM puddles) has its local value T_C^{loc} and these values are different for various puddles. As it was mentioned above the hole concentration in QW regions corresponding to the FM puddles drops down at the related T_C^{loc}. The averaged hole concentration diminishes also. As the temperature diminishes the interaction between FM droplets increases and their size increases so they overlap and the long range FM state is forming at $T = T_C$. So the hole concentration should go down in the temperature range between T_C^{loc} maximum and the T_C value. In accordance with this model $p(T)$ flattens at $T = T_c$ as it is seen in Fig. 6.

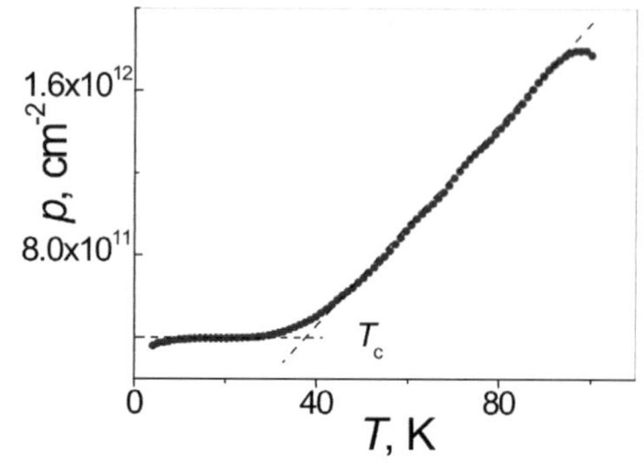

Fig. 6. The carrier concentration obtained from the normal Hall effect versus temperature for sample with Mn content 0.3 ML and $d = 3$ nm.

The amplitude of the T_C maximum at the nonmonotonous $T_C(U_0)$ dependence related to the first mechanism is about 15 K, while residual T_C values at the smallest U_0 and at the widest spacer are also in the range 15 – 20 K. So taking into account that the error of T_C measurements is about 2 K we conclude that both mechanisms provide a considerable contribution to the ferromagnetic ordering and to the holes spin polarization in our samples.

Our experimental results point out the intrinsic nature of the anomalous Hall effect (AHE) in 2D DMS, which is related to the renormalization of carrier group velocity under the influence of the spin-orbit interaction. The main reason for this statement is the observation of the change of the AHE sign with lowering temperature, as shown in Fig.7. The AHE magnitude is in a good agreement with the theoretical estimations for 2D structures, taking into account the intrinsic mechanism of AHE [13]. In the whole temperature range, we observed $\rho_{xy}^a \propto \rho_{xx}^2$ which is relevant for two AHE mechanisms: side-jump and intrinsic [14] (ρ_{xy}^a and ρ_{xx} are

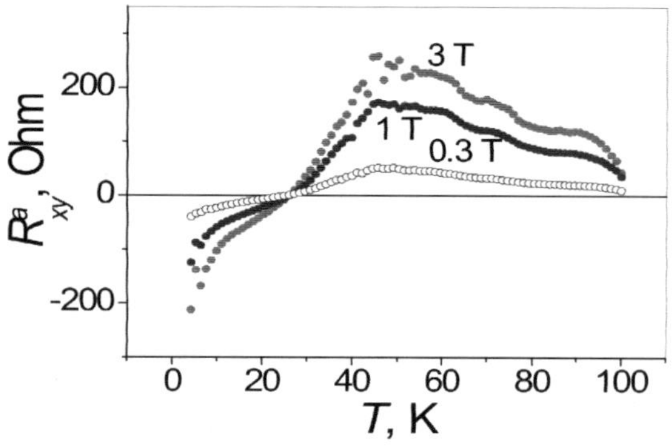

Fig. 7. Anomalous Hall effect resistance versus temperature measured for sample with ΔMn = 0.3 ML, $d = 3$ nm at fields 3 T, 1 T and 0.3 T.

the anomalous Hall and longitudinal resistivities). The change of the AHE sign may only be explained if one assumes two different mechanisms contributing to the AHE at low and high temperatures. So one of these mechanisms is intrinsic which plays the main role at low temperatures, when the scattering and so side-jump is less effective.

Conclusion

We have studied the electronic transport and magnetism in semiconductor heterostructures GaAs/InGaAs/GaAs quantum wells with a remote Mn δ-layer. In these structures, we observed the FM state and the quantum Hall effect. The FM state was detected by the presence of a broad maximum of the resistivity vs. temperature, anomalous Hall Effect and magnetization measurements. The measured T_C dependencies on the QW depth and the spacer thickness between QW and Mn δ-layer lead to conclusion that there are two mechanisms contributing to the FM state and holes spin polarization. According one of them the FM ordering in the Mn δ-layer is due to exchange between Mn ions through carriers in QW as a result of penetration of the "tails" of the carrier wave functions into Mn layer. Second mechanism is based on assumption that the actual Mn δ-layer could be represented as a thin GaMnAs layer due to diffusion of Mn atoms into surrounding GaAs blankets. So FM ordering could occur in this layer by itself without help of holes from QW, while the spin-polarization of carriers in QW is caused by a penetration into QW tails of wave functions related to the FM state in the Mn layer. It is logical to assume that in the studied structures FM transition occurs due to both mechanisms, which are complementary to each other and provide a considerable contribution.

2.1. Acknowledgement

The work was supported by the RFBR, project No. 11- 02-00363 and Russian – Israeli grant No. 11- 02 - 92478. BA is grateful to A. Gerber for fruitful discussions.

References

[1] Rupprecht B., Krenner W., Wurstbauer U., Heyn Ch., Windisch T., Wilde M. A., Wegscheider W.and Grundler D. 2010 *J. Appl. Phys.* **107** 093711

[2] Nishitani Y., Chiba D., Endo M., Sawicki M., Matsukara F., Dietl T. and Ohno H. 2010 *Phys. Rev. B* **81** 045208

[3] Aronzon B. A., Pankov M. A., Rylkov V. V., Meilikhov E. Z., Lagutin A. S., Pashaev E. M., Chuev M. A., Kvardakov V. V., Likhachev I. A., Vihrova O. V., Lashkul A. V., Lahdereanta E., Vedeneev A. S. and Kervalishvili P. 2010 *J. Appl.Phys.* **107** 023905

[4] Wojtowicz T., Lim W. L., Liu X., Dobrowolska M., Furdyna J. K., Yu K. M., Walukiewicz W., Vurgaftman I. and Meyer J. R. 2003 *Appl. Phys. Lett.* **83** 4220

[5] Nazmul A. M., Amemiya T., Shuto Y. and Tanaka M. 2005 *Phys. Rev. Lett.* **95** 017201

[6] Tripathi V., Dhochak K., Aronzon B. A., Rylkov V. V., Davydov A. B., Raquet Bertrand, Goiran Michel and Kugel K. I. 2011 *Phys. Rev. B* **84** 075305

[7] Jungwirth T., Sinova Jairo, Maˇsek J., Kuˇcera J., and MacDonald A. H. 2006 *Rev. Mod. Phys.* **78** 809

[8] Dorokhin M. V., Danilov Yu. A., Demina P. B., Kulakovskii V. D., Vikhrova O. V., Zaitsev S. V. and Zvonkov B. N. 2008 *J. Phys. D* **41** 245110

[9] Aronzon B.A., Kulbachinskii V.A., Gurin P.V., Davydov A.B., Rylkov V.V., Granovski A.B., Zvonkov B.N., Harikoshi Y. and Onomitsu K. 2007 *JETP letters* **85** 27

[10] Aronzon B., Lagutin A., Rylkov V., Pankov M., Lashkul A. and Laiho R. 2008 *Phys. Stat. Sol. (c)* **5** 814

[11] Meilikhov E. Z. and Farzetdinova R. M. 2010 *J. Appl. Phys.* **107** 114314

[12] Men'shov V. N., Tugushev V. V. and Caprara S. 2009 *Phys Rev. B.* **80** 035315

[13] Dugaev V. K., Bruno P., Taillefumier M., Canals B. and Lacroix C. 2005 *Phys. Rev. B* **71** 224423

[14] Nagaosa N., Sinova J., Onoda S., MacDonald A.H. and Ong N.P 2010 *Rev. Mod. Phys.* **82** 1539

Renormalization of effective mass in self-assembled quantum dots due to electron-electron interactions

A Babinski[1], M Korkusinski[2], P Hawrylak[2], M Potemski[3] and Z R Wasilewski[2]

[1] Faculty of Physics, University of Warsaw, ul. Hoza 69, 00-681 Warszawa, Poland
[2] Quantum Theory Group Security and Disruptive Technologies Emerging Technologies Division NRC, 1200 Montreal Rd, Ottawa, Ontario K1A0R6, Canada
[3] Laboratoire National des Champs Magnétiques Intenses, CNRS-UJF-UPS-INSA, 25, avenue des Martyrs, 38042 Grenoble, France

E-mail: babinski@fuw.edu.pl

Abstract. Magnetic-field dispersion of the multiexcitons related to the p shell of a single quantum dot (QD) is analysed in this work. The reduced cyclotron effective mass of carriers is determined from the energy splitting between the p_+- and p_-- related multiexcitonic emission lines. The reduced mass in the occupied QD was found to be larger than the mass related to the QD's single particle structure. The apparent increase of the reduced mass with increasing excitonic occupation of the dot is related to the mass renoralization due to electron-electron interactions within a multiexcitonic droplet.

1. Introduction

Semiconductor self-assembled quantum dots (QDs) have been intensely studied for almost 20 years. This interest is triggered by both scientific curiosity and possible applications [1]. The confinement of carriers within a space of the size comparable to their de Broglie wavelength allows to study electronic properties of strongly interacting carriers. Important for theoretical description of QDs properties is their single-particle (SP) energy structure. It has been shown that the structure can be well described using the harmonic lateral confinement potential [2], with characteristic confinement energy and effective mass of carriers. One of the approaches to determine the reduced effective mass $\mu^*_\pm$ of an electron-hole pair confined to a dot relies on the magnetic field-induced energy splitting between the resonances related to the p_+ and p_- subshells of a dot [3, 4, 5, 6]. In this communication we report on determination of the reduced mass using the single-dot optical magnetospectroscopy. We compare the result to the reduced mass μ^*, which characterizes the SP structure of the dot and we discuss the effect of electron-electron interactions on the reduced cyclotron mass in a highly occupied QD.

2. Experimental procedure

The structure investigated in this work comprised InAs/GaAs QDs grown using molecular beam epitaxy with application of the In-flush procedure [7]. The sample contains a single layer of dots, grown on a 600 nm undoped GaAs buffer layer, In-flushed at 5 nm and capped with a 100 nm GaAs layer [8]. Dots in the as-grown structure were found to be flat disks of 3.5 nm height and approx. 10 nm radius. This morphology results in vertical confinement significantly larger than the lateral confinement, which allows to assume the investigated dots to be two-dimensional. The structure was annealed after growth (30 s at 850 °C) in order to shift the emission energy into the sensitivity range of a CCD camera [9]. The sample was placed on top of piezo-driven attocube positioners and kept in gaseous helium at T=4.2 K [10]. A confocal microscope with a Y-shaped single mode fiber assembly was used in the measurements. Excitation was provided by laser light (λ=647 nm) coupled to one branch of the fiber and transmitted to the microscope located directly above the sample. The photoluminescence (PL) was collected by the microscope objective and extracted from the other branch of the fiber, dispersed by a 0.5 m monochromator, and detected by a liquid nitrogen-cooled CCD camera. The sample was placed inside a resistive magnet at National Laboratory of High Magnetic Fields in Grenoble, France. Magnetic field in Faraday configuration up to 19T was applied during measurements.

3. Experimental results

The PL spectra of a single QD excited with increasing laser-power density measured at zero magnetic field are shown in Fig.1. The weakly excited PL spectrum consists of two emission lines attributed to a neutral exciton (X) and a trion (X^*). Both: X and X^* excitonic emission lines are often observed in non-resonantly excited spectra of QDs, which charge state fluctuates during the spectrum collection. The double-line pattern is characteristic for the investigated structure [8] and the charge state attribution of the observed lines is based on results of polarization-sensitive measurements of the sample. Weak features accompanying main lines are due to respective biexcitons.

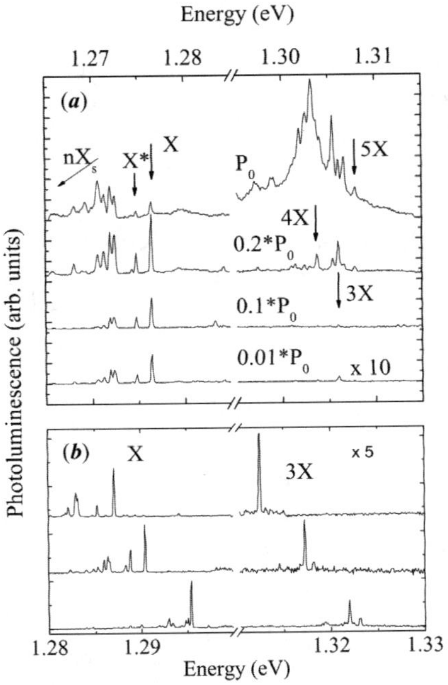

Figure 1. (a) Photoluminescence from a single InAs/GaAs QD at $B = 0T$ as a function of excitation power. The estimated excitation power density $P_0 \approx 70W/cm^2$.(b) Emission spectra from several weakly excited single dots in the investigated structure.

Excitation of the dot with higher power density results in the increase of the X and X^* lines intensities and in the appearance of additional emission lines. In particular an emission line $3X$

related to the higher lying SP level (p-shell) can be observed at the energy approximately 30 meV higher than the s-shell emission energy. Additional spectroscopic measurements confirmed the $3X$ attribution to the neutral charge state of a QD. This allows us to relate the $3X$ line to a complex of three excitons (triexciton) in the QD. Two carriers of each sign (electron and hole) occupy the lowest s-shell of a dot in such a complex. The third carrier of each sign occupies the p-shell because of the Pauli exclusion limiting the occupation of a particular SP level to two carriers of anti-parallel spins. The $3X$ emission line is related to the transition which leaves the dot in the ground state of a biexcitonic complex. The triexciton can also recombine to an excited state of a biexciton, which results in an additional emission line in the lowest energy region (s-shell) [11]. The occupation of the dot with more excitons further complicates the observed spectral pattern. The $4X$ ($5X$) emission line which emerges with increasing excitation power is attributed to recombination of four (five) excitonic complexes. The recombination leaves the dot in the ground state of a three (four) excitonic droplet. Both multiexcitonic configurations can also recombine leaving the $N-1$ excitonic complex in an excited state, therefore the number of possible transitions increases significantly with the number of confined excitons. This results in significant broadening of the observed emission spectra.

The magnetic field evolution of the PL from a highly excited dot can be analyzed in order to determine the reduced effective mass of an electron-hole pair $\mu^*_\pm$ [3, 4, 5, 6]. The single-dot-resolved optical experiments uncover also another feature of the evolution [12, 13], which is the zero-field splitting of the $3X$ and $5X$ (see Fig. 2(a)). The evolution of the PL emission from the investigated dot is shown in Fig.2(a). The $3X-$ and $5X-$ related emission lines are denoted with open symbols for more clarity. The $3X$ ($5X$) emission line follows the p_- (p_+) shell structure of SP energy levels of a charge carrier confined in a harmonic lateral potential (referred to as the Fock-Darwin (FD) spectrum[14, 15]) except for the lowest magnetic fields. The low-field deviation is due to the scattering of the electron-hole pair between the equivalent SP p-shell configrations of a three (five) exciton complex [16]. The energy separation between the $3X$ and $5X$ emission lines 2 (b) in the intermediate magnetic field was fitted with a linear dependence: $\Delta E_\pm = \alpha + (\hbar eB/\mu^*_\pm)$ with the result of the fitting shown with a continuous line in Fig.2 (b). The resulted reduced effective mass is equal to $\mu^*_\pm = 0.061 \pm 0.0005 m_0$.

The obtained cyclotron mass $\mu^*_\pm$ can be compared to the reduced effective mass μ^*, which characterises the SP energy level structure of the dot. The latter reduced mass can be found from the analysis of the magnetic-field dispersion of the X and $3X$ emission lines in a broad magnetic field range [17]. The magnetic-field (Faraday configuration) evolution of the PL spectrum with those lines is shown in Fig.3 in the form of a surface plot. The spin Zeeman splitting of emission lines can be identified at highest magnetic fields. The splitting is a nonlinear function of magnetic field, which is characteristic for the investigated type of dots [8, 18]. Both X and $3X$ field-dispersions are analysed in within the approximation of equal spatial extents of electron and hole wavefunctions. As a result of the fitting procedure the reduced effective mass μ^* was determined to be equal to $\mu^* = 0.053 m_0$, which is lower than the cyclotron reduced effective mass $\mu^*_\pm$. The reduced effective mass well corresponds to the mass obtained from **kp** calculations for a similar structure [5]. If the usually observed in III-V materials 2:1 conduction band-valence band offset is assumed, the effective electron (hole) mass equals then to $m_e = 0.080 m_0$ ($m_h = 0.16 m_0$). The X and $3X$ dispersions fitted with the reduced mass μ^* are shown in Fig. 3 with dotted lines.

The apparent difference between the cyclotron reduced effective mass $\mu^*_\pm$ and the "SP-structure" reduced effective mass of the QD μ^* results in our opinion from the effect of magnetic field on electron-hole interactions within a multiexcitonic droplet. The effect coresponds to the previously observed effective mass renormalization due to many-body interactions in two-dimensional electron-hole systems systems [19, 20]. The renormalization of the reduced effective mass makes the system different from a dot occupied with electrons. If only electrons are present in the dot, the optical spectra are insensitive to the electron number and to details of electron-

electron interactions [21]. This results from the coupling of electromagnetic radiation to the center-of mass degrees of freedom only [22]. Moreover if the QD confinement is well approximated by a parabolic potential, the center-of-mass and relative motion separate due to the generalized Kohn's theorem and the intraband absorpition spectra are those of a single particle [23].

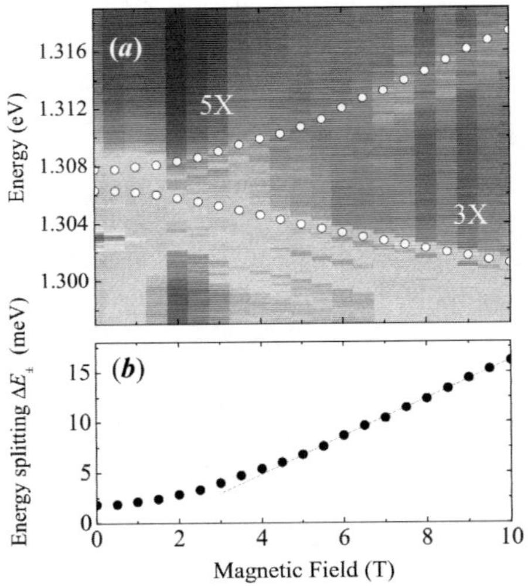

Figure 2. (color online) (a) Photoluminescence of a highly excited single dot with energies of the $3X$ and $5X$ emission lines denoted with open points. (b) The energy difference between the $3X$ and $5X$ emission (points) and the fit to the dependence (line).

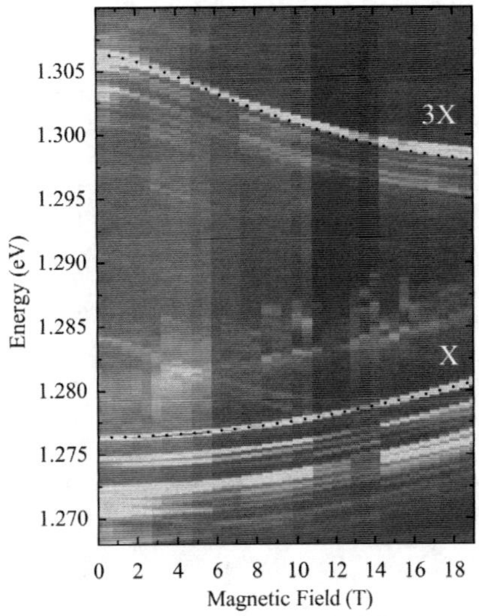

Figure 3. (color online) The evolution of the emission spectrum from a single InAs/GaAs quantum dot in perpendicular magnetic field. Simulations of the X and $3X$ emission lines obtained with parameters described in the text are denoted with dotted lines.

4. Summary

In conclusion, we have investigated the s- and p-shell related emission from a single InAs/GaAs QD in magnetic fields. We have shown that the cyclotron reduced effective mass $\mu^*_{\pm}$ obtained from the analysis of the dot occupied with at least five excitons is higher than the reduced mass μ^* resulting from the SP structure of the dot. We relate the difference to the effect of magnetic field on the Coulomb interactions within a dot.

Acknowledgments

The work has been supported in part by the EuroMagNET II consortium under the EU contract no. 228043 and by Polish Funds for Science 2009-2012.

References

[1] for a recent review on QDs properties see e.g. 2008 *Handbook of Self Assembled Nanostructures for Novel Devices in Photonics and Electronics* M.Hennni ed. (Amsterdam:Elsevier)
[2] Wojs A, Hawrylak P, Fafrad S and Jacak L 1996 *Phys. Rev.* B **54** 5604
[3] Rinaldi R, Guigno P V, Cingolani R, Lipsanen H, Sopanen M, Tulkki J and Ahopelto J 1996 *Phys. Rev. Lett.* **77** 342
[4] Bayer M, Schmidt A, Forchel A, Faller F, Reinecke T L, Knipp P A , Dremin A A and Kulakovskii D V 1995 *Phys. Rev. Lett.* **74** 3439
[5] Paskov P P, Holtz P O, Monemar B, Garcia J M, Schoenfeld W and Petroff P M 2000 *Phys. Rev.* B **62** 7344
[6] Awirothananon S, Sheng W D, Babinski A, Studenikin S, Raymond S, Sachrajda A, Potemski M, Fafard S, Ortner G and Bayer M 2004 *Jap. Journal of Appl. Phys.* **43** 2088
[7] Wasilewski Z R, Fafard S and McCaffrey J P 1999 *J. Cryst. Growth* **201** 1131
[8] Babinski A, Ortner G, Raymond S, Potemski M, Bayer M, Sheng W, Hawrylak P, Wasilewski Z, Fafard S and Forchel A 2006 *Phys. Rev.* B **74** 075310
[9] Fafard S and Allen C N 1999 *Appl. Phys. Lett.* **75** 2374
[10] Babinski A, Awirothananon S, Lapointe J and Wasilewski Z 2005 *Physica* E **26** 190
[11] Bayer M, Stern O, Hawrylak P, Fafard S and Forchel A 2000 *Nature (London)* **405** 923
[12] Babinski A, Potemski M, Raymond S, Lapointe J and Wasilewski Z R 2006 *Phys. Rev.* B **74** 155301
[13] Babinski A, Potemski M, Raymond S, Lapointe J and Wasilewski Z 2006 *phys. stat. sol.* (c) **3** 3748
[14] Fock V 1928 *Z. Phys.* **47** 446, Darwin C G 1930 *Proc. Cambridge Philos. Soc.* **27** 86
[15] Raymond S *et al* 2004 *Phys. Rev. Lett.* **92** 187402
[16] Hawrylak P 1999 *Phys. Rev.* B **60** 5597
[17] Molas M *et al* to be published
[18] Sheng W and Babinski A 2007 *Phys. Rev.* B **75** 033316
[19] Potemski M, Maan J C, Ploog K and Weimann G 1990 *Solid State Commun.* **75** 185
[20] Cingolani R, La Rocca G C, Kalt H, Potemski M and Maan J C 1991 *Phys. Rev.* B **43** 9662
[21] Drexler H, Leonard D, Hansen W, Kotthaus J P and Petroff P M 1994 *Phys. Rev. Lett.* **73** 2252
[22] Maksym P A and Chakraborty T 1990 *Phys. Rev. Lett.* **65** 108
[23] Wojs A and Hawrylak P 1996 *Phys. Rev.* B **53** 10841

Journal of Physics: Conference Series **456** (2013) 012003 doi:10.1088/1742-6596/456/1/012003

Magnetoresistance in a High Mobility Two-Dimensional Electron System as a Function of Sample Geometry

L. Bockhorn[1], A. Hodaei[1], D. Schuh[2], W. Wegscheider[3], R. J. Haug[1]

[1] Institut für Festkörperphysik, Leibniz Universität Hannover, D-30167 Hannover, Germany
[2] Institut für Experimentelle und Angewandte Physik, Universität Regensburg, D-93053 Regensburg, Germany
[3] Laboratory for Solid State Physics, ETH Zürich, CH-8093 Zürich, Switzerland

E-mail: bockhorn@nano.uni-hannover.de

Abstract. In a high mobility two-dimensional electron gas (2DEG) realized in a GaAs / $Al_{0.3}Ga_{0.7}As$ quantum well we observe changes in the Shubnikov-de Haas oscillations (SdHO) and in the Hall resistance for different sample geometries. We observe for each sample geometry a strong negative magnetoresistance around zero magnetic field which consists of a peak around zero magnetic field and of a huge magnetoresistance at larger fields. The peak around zero magnetic field is left unchanged for different geometries.

1. Introduction

The increased mobility of the two-dimensional electron gas (2DEG) has allowed not only the observation of the fractional quantum Hall effect (FQHE)[1],[2] at low magnetic fields but also many new effects. While the mobility and quality of the samples increases the electron-electron interaction becomes more important for the magnetotransport at low magnetic fields. One effect influenced by the electron-electron interaction is the observation of a strong negative magnetoresistance around zero magnetic field. This strong negative magnetoresistance consists of a peak around zero magnetic field and of a huge magnetoresistance at larger magnetic fields, both effects show a parabolic field dependence [3].

We study here the influence of the sample geometry on the strong negative magnetoresistance, whereas in previous measurements the temperature dependence and the electron density dependence of the strong negative magnetoresistance was analyzed [3, 4].

2. Sample Information

Our samples were cleaved from a wafer of a high-mobility GaAs/$Al_{0.3}Ga_{0.7}As$ two-dimensional electron system (2DES) grown by molecular-beam epitaxy. The quantum well has a width of 30 nm and is Si-doped from both sides. The 2DES is located 150 nm beneath the surface and has an electron density of $n_e = 3.2 \cdot 10^{11}$ cm^{-2} and a mobility of $\mu_e = 11.9 \cdot 10^6$ cm^2/Vs. The different geometries were defined by photolithography and by wet etching. Our measurements were performed in a dilution refrigerator with a base temperature of 20 mK. The magnetotransport measurements were carried out by using low-frequency (13 Hz) lock-in technique.

Journal of Physics: Conference Series **456** (2013) 012003 doi:10.1088/1742-6596/456/1/012003

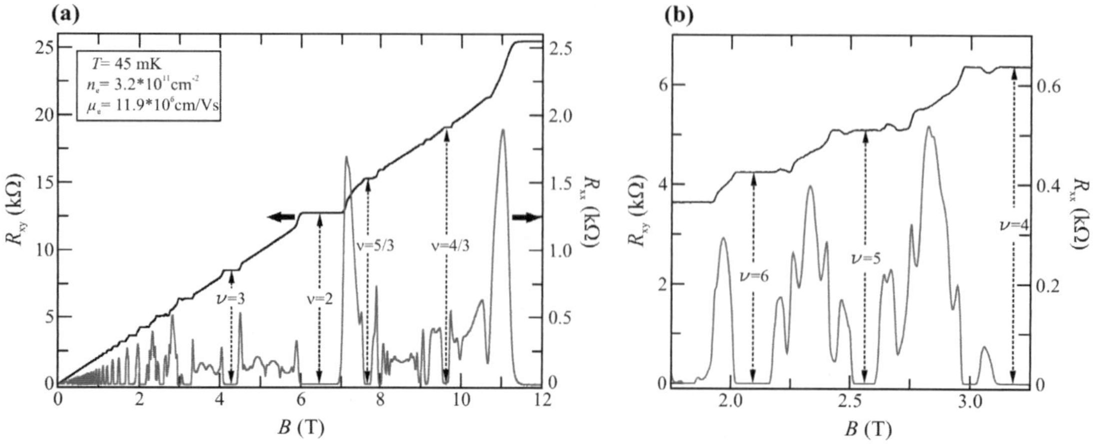

Figure 1. **(a)** The Hall resistance R_{xy} and the longitudinal resistance R_{xx} vs. the magnetic field B. **(b)** Observation of the reentrant integer quantum Hall effect (RIQHE).

The different geometries produced from the high-mobility 2DES will be discussed in the following section. In the next sections we present our magnetotransport measurements for a Hall geometry, especially for the magnetic field range of the strong negative magnetoresistance. The second geometry consists of four Hall bars with different length-to-width ratios.

3. Hall Geometry

The Hall bars have a total length of 1.2 mm, a width of $w = 200$ μm and a potential probe spacing of $l = 300$ μm. Different ungated and gated samples were used for the magnetotransport measurements. In case of the gated sample there is an additional layer of 600nm PMMA between the Hall bar and the metallic topgate to avoid leakage current.

Figure 1 (a) shows the Hall resistance R_{xy} and the longitudinal resistance R_{xx} vs. magnetic field B to demonstrate the quality of our samples. The longitudinal resistance decreases to zero between the integer filling factors $\nu = 4$ and $\nu = 6$ (see Figure 1 (b)), but the corresponding Hall plateaus are quantized at integer values, e.g. at $B = 3.0$ T. This phenomenon is called the reentrant integer quantum Hall effect (RIQHE) and is often observed in high mobility samples [5, 6]. We observe a clear minimum in the longitudinal resistance for the filling factor $\nu = 5/2$ at $B = 5.1$ T. The filling factor $\nu = 5/2$ is only observed in high mobility samples and low temperatures. Also a series of different fractional quantum Hall states is observed for filling factors $\nu < 2$. The fractional filling factors $\nu = 5/3$ and $\nu = 4/3$ are marked in Figure 1 (a).

We observe a strong negative magnetoresistance around zero magnetic field which can be divided into two regions (see Figure 2). The two regions consist of a peak around zero magnetic field and of a huge magnetoresistance at larger fields.

3.1. *Peak around zero magnetic field*

The longitudinal resistivity vs. the magnetic field is shown in Figure 2. The peak around zero magnetic field is fitted to a parabolic magnetic field dependence. In previous experiments we observed that the curvature of the peak is left unchanged by increasing the temperature to 0.8 K [3]. This temperature independence of the peak is a sign for the absence of weak localization. Also for different electron densities the curvature of the peak is left unchanged. We assume according to Mirlin *et al.* [7] that the peak is induced by an interplay of smooth disorder and rare strong scatteres.

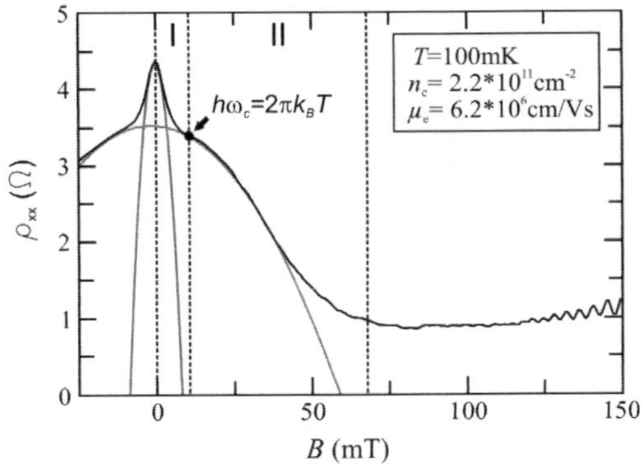

Figure 2. The longitudinal resistivity ρ_{xx} vs. the magnetic field B. The strong negative magnetoresistance is divided into two sections.

3.2. Huge Magnetoresistance

Previous publications reported on a strong negative magnetoresistance, which vanished by increasing the temperature [3, 4]. The huge magnetoresistance in Figure 2 depends not only strongly on the temperature, it depends also strongly on the electron density and it becomes more pronounced by decreasing the electron density. Also the huge magnetoresistance is fitted by a parabolic magnetic field dependence.

We examine the electron-electron interaction correction to the conductivity in the situation of a long-range fluctuation potential and in the regime of ballistic transport [8]-[11] to describe the huge magnetoresistance. In accordance with Gornyi *et al.* [11] we assume a model of mixed disorder to fit the huge magnetoresistance. The electron interaction induced correction to the conductivity considering the model of mixed disorder is expressed by

$$\rho_{xx} = \rho_0 - \rho_0 \frac{c_0}{n_e^2 \, \pi \, h} \sqrt{\frac{\hbar}{T\tau \, k_B}} \cdot \alpha \cdot B^2 \tag{1}$$

with $c_0 = 0.276$. ρ_0 is the longitudinal resistivity at zero magnetic field, n_e is the electron density, τ is the transport scattering time, T is the electron temperature and k_B is the Boltzmann constant. The factor α is in accordance with Gornyi *et al.* [11] expressed by $\alpha = 4(\tau_{sm}/\tau)^{-1/2}$ with $\tau_{sm} = (k_F d)^2 \tau_q$ depending on the quantum scattering time τ_q. The expected temperature dependence of $T^{-1/2}$ from the electron interaction correction to the conductivity is observed for temperatures below 400 mK. Above 400 mK the curvature is approximately proportional to T^{-1} as long as the huge magnetoresistance is observable.

We observed a discrepancy between our measurement and theory. The factor α determined from the curvature of the huge magnetoresistance is larger than the expected value. A possible origin of this discrepancy between theory and experiment is that the quantum scattering time τ_q is dominated by smooth disorder. The influence of the strong scatterers should also be considered in the quantum scattering time because of the description of the peak around zero magnetic field.

4. Four-in-a-row Hall bar

The second geometry we discuss here consists of four Hall bars with different length-to-width ratios (Figure 3). On the basis of this special geometry we could measure the influence of the geometry on the strong negative magnetoresistance in one cooling cycle. Several samples were produced and measured and all of them showed similar results.

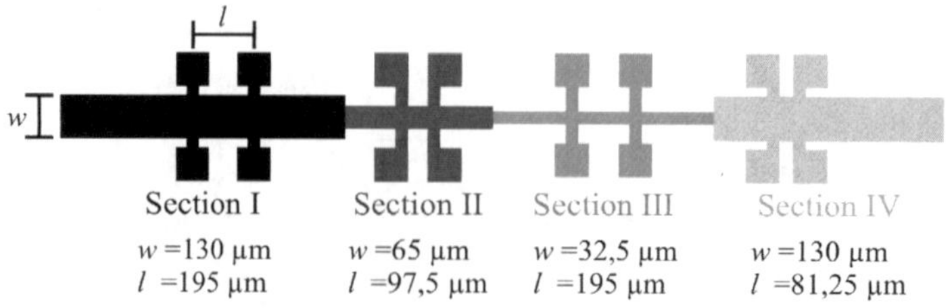

Figure 3. Schema of the four-in-a-row Hall bar. Each section of this geometry has a different width w and a different potential probe spacing l, which are note below the corresponding section.

4.1. ***Geometry Information***

The four-in-a-row Hall bar has a total length of 2.7 mm and is separated into four sections. Each section consists of a Hall bar with a different length-to-width ratio. The first section in Figure 3 has a width of $w = 130\ \mu$m and a potential probe spacing of $l = 195\ \mu$m. The second section has the same length-to-width ratio as section I. The width of the section II is $w = 65\ \mu$m and the potential probe spacing is $l = 97.5\ \mu$m. Section III has the smallest width ($w = 32.5\ \mu$m) and a potential probe spacing of $l = 195\ \mu$m. The section IV has a width of $w = 130\ \mu$m and a potential probe spacing of $l = 82.25\ \mu$m.

Two types of contact geometries are used for the four-in-a-row Hall bar. The potential probe spacing of section I and of section III is large enough to put the ohmic contacts in the middle of the side contact. The situation is changed for the other sections. The potential probe spacing is here too small. The size of the ohmic contacts can't be reduced, so the ohmic contacts are shifted to one side.

4.2. ***Magnetotransport Measurement***

Figure 4 (a) shows the longitudinal resistivity ρ_{xx} and the Hall resistance R_{xy} vs. the magnetic field B for section I of the four-in-a-row Hall bar (red). Also a corresponding measurement of a reference Hall bar (black) is shown. The length-to-width ratio of section I is the same as for the reference Hall bar, so one could expect a similar behavior of the longitudinal resistivity and the Hall resistance. We observe instead that the minima of the longitudinal resistivity of section I increase till filling factor $\nu = 4$, then the longitudinal resistivity decreases. Also the slope of the Hall resistance of section I changes at small magnetic fields and the values of the Hall plateaus are lower than usually. The Hall plateaus are the expected ones only at filling factor $\nu = 4/3$ for magnetic fields $B > 12$ T. This astonishing behavior is observed for each section of the four-in-a-row Hall bar for different samples. The minima of the longitudinal resistivities increase till filling factor $\nu = 4$ for each section of the geometry and then drop to zero. Also the Hall resistances deviate at small magnetic fields for each length-to-width ratio from the linear behavior with magnetic field. The increase of the longitudinal resistivity is stronger for the smallest length-to-width ratio, while the Hall plateaus decrease to lower values for the highest length-to-width ratio.

A possible origin of the unexpected behavior considers charge puddles in the doping layer. The influence of the charge puddles for wide samples is negligible as long the extension of the charge puddles is much smaller than the width of the sample. In the situation of the narrowest section of the four-in-a-row Hall bar the size of the charge puddles could be in the range of the width of the Hall bar and short-circuit the sample in this way. Hence all magnetotransport measurements of this geometry will be influenced by these charge puddles due to the short-circuit in the narrowest part and the interconnection of the whole sample in the quantum Hall

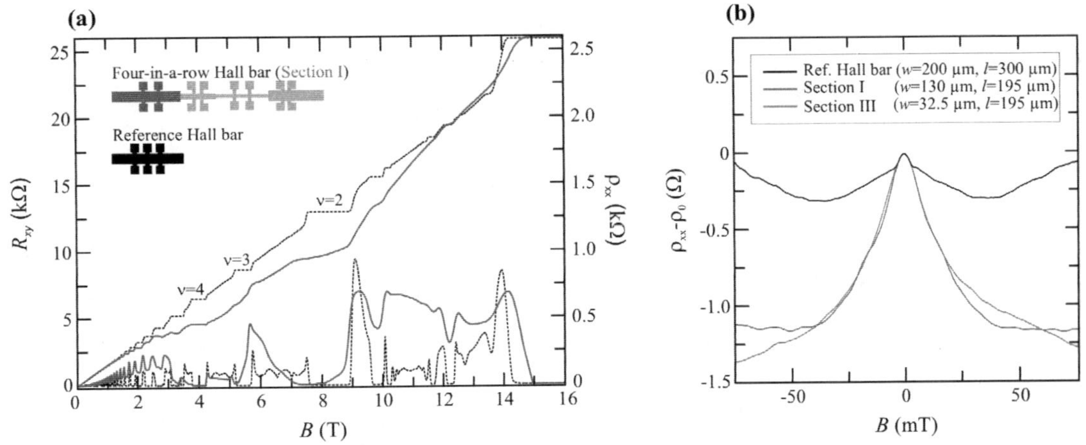

Figure 4. **(a)** The Hall resistance R_{xy} and the longitudinal resistivity ρ_{xx} of section I vs. the magnetic field B at 45 mK. The dotted line is the corresponding measurement of a reference Hall bar. **(b)** The longitudinal resistivity ρ_{xx} reduced by the corresponding ρ_0 around zero magnetic field for different geometries.

regime.

Figure 4 (b) shows the longitudinal resistivity ρ_{xx} reduced by the corresponding ρ_0 vs. the magnetic field B around zero magnetic field for different geometries. The strong negative magnetoresistance around zero magnetic field is observed for each section of the geometry. The curvature of the peak around zero magnetic field is left unchanged for different length-to-width ratios, while the height of the peak is changed. Differences in the curvature are observed for different sample materials. The situation of the huge magnetoresistance is here more complicated and a conclusion for a geometry dependence is not possible.

5. Conclusion

We observe for each sample geometry a strong negative magnetoresistance around zero magnetic field which consists of a peak around zero magnetic field and of a huge magnetoresistance at larger fields. The huge magnetoresistance depends strongly on temperature and electron density, while the peak around zero magnetic field is left unchanged. The peak around zero is left unchanged also for different geometries.

This work was financially supported by the Cluster of Excellence QUEST.

6. References

[1] D. C. Tsui, H. L. Stormer, and A. C. Gossard, *Phys. Rev. Lett.* **48**, 1559 (1982).
[2] R. B. Laughlin, *Phys. Rev. Lett.* **50**, 1395 (1983).
[3] L. Bockhorn, P. Barthold, D. Schuh, W. Wegscheider, and R. J. Haug, *Phys. Rev. B* **83**, 113301 (2011).
[4] A. T. Hatke, M. A. Zudov, J. L. Reno, L. N. Pfeiffer, and K. W. West, *Phys. Rev. B* **85**, 081304 (2012)
[5] J. P. Eisenstein, *Solid State Commun.* **117**, 123 (2001)
[6] J. P. Eisenstein, K. B. Cooper, L. N. Pfeiffer, and K. W. West *Phys. Rev. Lett.* **88**, 076801 (2002)
[7] A. D. Mirlin, D. G. Polyakov, F. Evers, and P. Wölfle, *Phys. Rev. Lett.* **87**, 126805 (2001).
[8] M. A. Paalanen, D. C. Tsui, and J. C. M. Hwang, *Phys. Rev. Lett.* **51**, 2226 (1983).
[9] I. V. Gornyi and A. D. Mirlin, *Phys. Rev. Lett.* **90**, 076801 (2003).
[10] L. Li, Y. Y. Proskuryakov, A. K. Savchenko, E. H. Linfield, and D. A. Ritchie, *Phys. Rev. Lett.* **90**, 076802 (2003).
[11] I. V. Gornyi and A. D. Mirlin, *Phys. Rev. B* **69**, 045313 (2004).

HMF-20 IOP Publishing
Journal of Physics: Conference Series **456** (2013) 012004 doi:10.1088/1742-6596/456/1/012004

Effect of an in-plane magnetic field on microwave photoresistance and Shubnikov-de Haas effect in high-mobility GaAs/AlGaAs quantum wells

A. Bogan[1,2], A. T. Hatke[3], S. A. Studenikin[1,2], A. Sachrajda[1], M. A. Zudov[3], L. N. Pfeiffer[4], K. W. West[4]

[1]National Research Council of Canada, Ottawa, ON K1A 0R6, Canada
[2]Department of Physics and Astronomy, University of Waterloo, Waterloo, ON N2L 3G1, Canada
[3]School of Physics and Astronomy, University of Minnesota, Minneapolis, MN 55455, USA
[4]Princeton University, Department of Electrical Engineering, Princeton, NJ 08544, USA

E-mail: `Alexander.Bogan@nrc-cnrc.gc.ca`

Abstract. A recent study of the decay of microwave-induced resistance oscillations (MIRO) in GaAs/AlGaAs quantum wells due to an in-plane magnetic field $B_{\|}$ has revealed the dominant role of a quadratic-in-$B_{\|}$ correction to the quantum scattering rate. In the present study, we examine the evolution of Shubnikov-de Haas oscillations (SdHO) with increasing tilt angle in the same sample. Even though we find that the SdHO also diminish at high tilt angles, this decay is qualitatively different from that of MIRO, possibly indicating a different physical origin.

1. Introduction

Among a variety of transport phenomena [1–14] recently discovered in very high Landau levels of a high mobility two-dimensional electron gas (2DEG), microwave-induced resistance oscillations (MIRO) [1,2] and zero-resistance states [5,6] have received the most attention, both theoretically [15–23] and experimentally [24–41]. It is now well established that at low microwave power and in overlapping Landau levels, MIRO are described by [42]

$$\delta\rho(\epsilon) \propto -\epsilon\lambda^2 \sin 2\pi\epsilon, \tag{1}$$

where $\epsilon = \omega/\omega_c$, $\omega = 2\pi f$ and $\omega_c = eB_\perp/m^\star$ are the microwave and cyclotron frequencies, respectively, $m^\star$ is the electron effective mass, $\lambda = \exp(-\pi/\omega_c\tau_q^0)$ is the Dingle factor, and τ_q^0 is the quantum lifetime. While τ_q^0 is often obtained from Shubnikov-de Haas oscillations (SdHO), this method often leads to underestimated τ_q^0, presumably, due to extra suppression of the SdHO amplitude by macroscopic density fluctuations [27]. MIRO, on the other hand, are not affected by such fluctuations and thus offer a preferred way to obtain the quantum lifetime. As a result, MIRO are an attractive tool to study the dependence of the quantum lifetime on $B_{\|}$, the understanding of which is important for other systems, such as lateral quantum dots [43].

The effect of $B_{\|}$ on MIRO has been investigated several years ago by two experimental groups [44,45] with conflicting outcomes; in Ref. [44] MIRO remained unchanged up to $B_{\|} \simeq 1$ T, but in Ref. [45] MIRO were strongly suppressed by $B_{\|} \simeq 0.5$ T. Stimulated by this controversy,

Figure 1. Photograph showing sample, 3 mm antenna, and an Allen Bradley carbon thermo-resistor with DPPH, used to calibrate B.

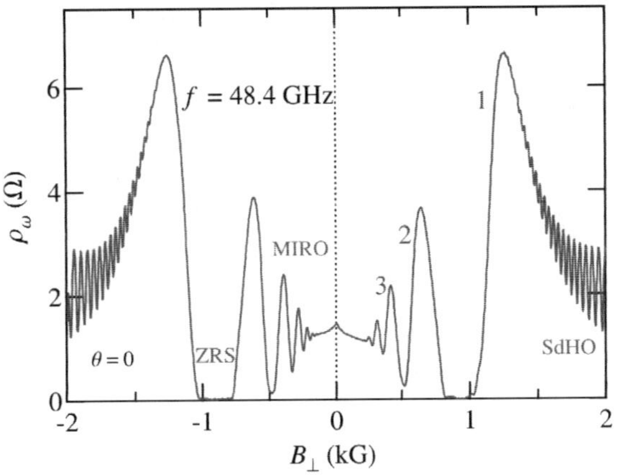

Figure 2. Magnetoresistivity $\rho_\omega(B_\perp)$ measured at $f = 48.4$ GHz, $T \simeq 0.3$ K, and $\theta = 0$, showing MIRO, a zero-resistance state, and SdHO.

we have revisited the issue [46] and found that the decay of the MIRO amplitude with increasing tilt angle can be understood in terms of a $B_\|$-induced increase of the single particle scattering rate, which acquires a quadratic-in-$B_\|$ correction. As a result of this correction, the lower order MIRO decay faster than the higher orders. Here, we extend our study, performed on the same sample, to the evolution of SdHO with increasing $B_\|$. As discussed below, SdHO also decay with increasing tilt angle, but, in contrast to MIRO, the decay appears to be roughly the same for all filling factors studied.

2. Experimental details

Our sample is a 200 μm-wide Hall bar etched from a symmetrically doped GaAs/$Al_{0.24}Ga_{0.76}As$ 30 nm-wide quantum well grown by molecular beam epitaxy. After illumination with a red light-emitting diode, the electron density and the mobility were $n_e \approx 3.6 \times 10^{11}$ cm^{-2} and $\mu \approx 1.3 \times 10^7$ cm^2/Vs, respectively. Microwave radiation of $f = 48.4$ GHz was delivered to the sample via a semirigid coaxial cable terminated with a 3 mm antenna (Fig. 1) [27, 32, 46]. A split-coil superconducting solenoid allowed us to change the magnetic field direction *in situ*, by rotating the ^{3}He insert, without disturbing the distribution of the microwave field. The resistivity was recorded using low-frequency (a few hertz) lock-in technique under continuous irradiation in sweeping magnetic field at a constant coolant temperature of $T \simeq 0.3$ K.

3. Experimental results and discussion

In Fig. 2 we present the magnetoresistivity $\rho_\omega(B_\perp)$ recorded at $\theta = 0°$, which shows pronounced MIRO and a well developed zero-resistance state, attesting to the high quality of our sample. MIRO orders are marked by integers. A Dingle plot analysis performed on these data reveals a quantum lifetime $\tau_q^0 = 21$ ps, which is typical of a high mobility 2DEG.

In Fig. 3 we present the resistivity ρ_ω as a function of the perpendicular field $B_\perp$ measured at different tilt angles: (a) $\theta = 57.5°$, (b) $\theta = 74.9°$, and (c) $\theta = 82.3°$. Each panel also includes the data recorded at $\theta = 0°$, demonstrating that the MIRO period depends only on the perpendicular component of the magnetic field, $B_\perp = B\cos\theta$, in agreement with earlier studies [44, 45]. At the same time, direct comparison with the $\theta = 0°$ data reveals that MIRO monotonically decay away with increasing tilt angle. As a result, the zero-resistance state is no longer observed at the highest angle.

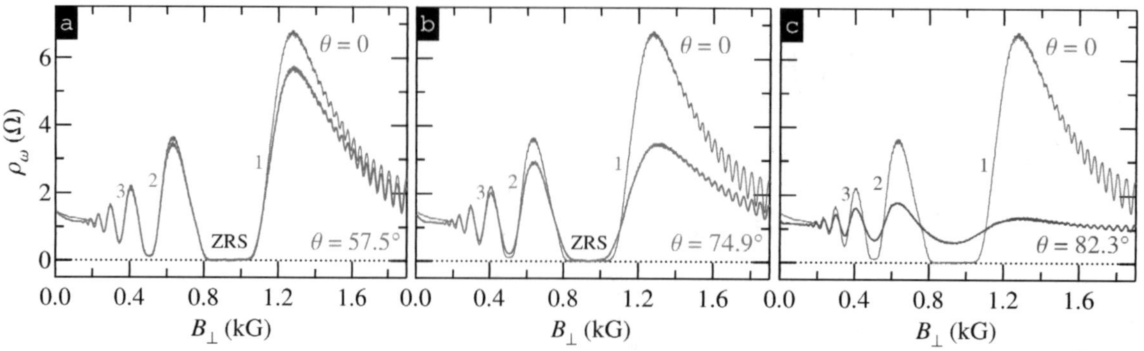

Figure 3. Magnetoresistivity $\rho_\omega(B_\perp)$ measured under microwave irradiation of $f = 48.4$ GHz at different tilt angles: (a) $\theta = 57.5°$, (b) $\theta = 74.9°$, and (c) $\theta = 82.3°$. For comparison each plot also contains $\rho_\omega(B_\perp)$ measured at $\theta = 0$. Adapted from Ref. [46].

The observed MIRO decay with increasing θ, however, is clearly not uniform and depends on the oscillation order (and, hence, on ϵ) [46]. For example, the data at $\theta = 57.5°$ and $\theta = 74.9°$ [Fig. 3(a) and Fig. 3(b), respectively], clearly reveal that the first (fundamental) oscillation decays faster then the second, and that there is virtually no suppression of higher order oscillations. As a result, in contrast to the data obtained at $\theta = 0°$, where the MIRO amplitude monotonically increases with $B_\perp$, the data obtained at higher tilt angles show more complicated behavior. Indeed, in the data obtained at $\theta = 82.3°$ [Fig. 3(c)], the first oscillation becomes considerably weaker than the second, while the second oscillation appears roughly the same as the third. At still higher tilt (not shown) the lower order oscillations virtually disappear, while the higher order (lower $B_\perp$) oscillations can still be observed. All these findings indicate that the degree of suppression is determined by an in-plane component of the magnetic field, $B_\parallel$. Indeed, the in-plane magnetic field is proportional to $B_\perp$, and, as a result, lower order (higher $B_\perp$) MIRO are subject to a larger $B_\parallel$ for a given θ.

As discussed in Ref. [46], the main mechanism of the MIRO decay in tilted magnetic fields appears to be the same as that responsible for the decay of Hall field-induced resistance oscillations [4, 47–49]. More specifically, the suppression of the oscillations can be interpreted in terms of a $B_\parallel$-induced correction to the quantum scattering rate, which is quadratic in $B_\parallel$ [49],

$$1/\tau_{\rm q} = 1/\tau_{\rm q}^0 + \delta(1/\tau_{\rm q}), \quad \delta(1/\tau_{\rm q}) = (1/{\tau_{\rm q}}^0) \cdot (B_\parallel/B_0)^2 . \qquad (2)$$

Here, $\delta(1/\tau_{\rm q})$ is the correction to the zero-tilt scattering rate $1/\tau_{\rm q}^0$, induced by $B_\parallel$, and B_0 is a characteristic in-plane magnetic field, which corresponds to doubling of the quantum scattering rate. As the obtained B_0 is a few kG [46, 49], the increase of $1/\tau_{\rm q}$ may be relevant to such areas as spin-based quantum computing [43, 50], where parallel fields of several Tesla are typical.

Further examination of the data in Fig. 3 reveals that SdHO also monotonically diminish with increasing tilt angle. To compare the decays of MIRO and SdHO, we construct Fig. 4, which shows (a) MIRO amplitude and (b) SdHO amplitude for different θ, as marked, normalized to their values at $\theta = 0$, as a function of ϵ and filling factor ν, respectively. It is clear that the decay of the MIRO amplitude is most pronounced at lower ϵ (*higher* $B_\parallel$) [cf. Eq. (2)] and at the highest tilt angle, in agreement with Eqs. (1),(2). The SdHO decay, on the other hand, shows a very different behavior. Indeed, at all tilt angles studied, the SdHO amplitude appears to be damped rather uniformly, i.e., the degree of suppression is roughly independent of the filling factor (and, hence, of $B_\parallel$).

We notice, however, that the SdHO data are confined to a rather narrow range of perpendicular magnetic fields (1.7 kG $\le B_\perp \le$ 2 kG), in contrast to MIRO which are studied

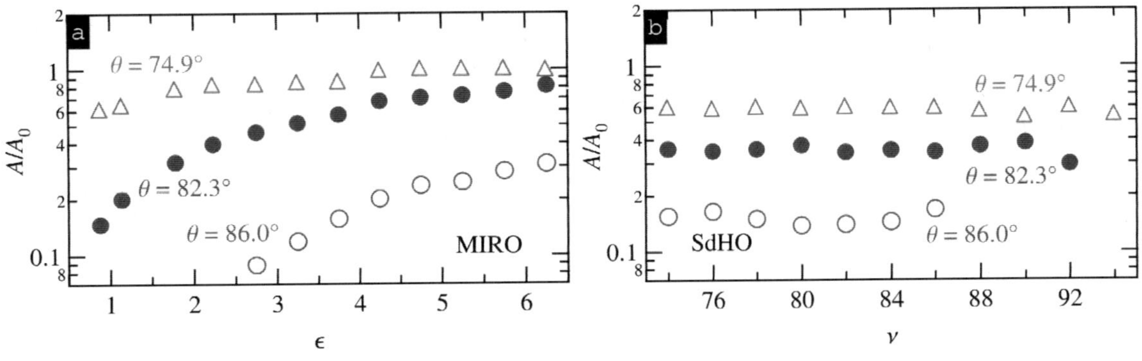

Figure 4. (a) MIRO amplitude and (b) SdHO amplitude for different θ, as marked, normalized to its value at $\theta = 0$, as a function of ϵ and filling factor ν, respectively.

over a much wider range (0.2 kG $\leq B_\perp \leq$ 1 kG). This fact clearly limits our ability to detect possible dependence on ν from the data presented in Fig. 4(b). In addition, the SdHO range of $B_\perp$ definitely falls into the regime of well separated Landau levels, where the density of states can no longer be described by a single harmonic with the amplitude given by the Dingle factor. As a result, in order to see if the modification of the quantum lifetime can account for the observed SdHO suppression, the theoretical description of the SdHO should first be extended to the regime of separated Landau levels.

Finally, we notice that in a recent study [51], the reduced SdHO amplitude in tilted magnetic fields was attributed to a $B_\parallel$-induced increase of the spin splitting. In another report [52], however, the SdHO were not suppressed until much higher tilt angles (and higher $B_\parallel$) than those used in the present study, suggesting that the effect might be sample dependent.

4. Summary

In summary, we have compared the effects of an in-plane magnetic field on MIRO and on SdHO in a high-mobility 2DEG. The decay of the MIRO amplitude with increasing tilt angle can be understood in terms of a $B_\parallel$-induced increase of the single particle scattering rate which acquires a quadratic-in-$B_\parallel$ correction. The decay of SdHO, on the other hand, shows a uniform suppression at all filling factors studied. However, because of the limited range of $B_\perp$ and the fact that SdHO appears in the regime of separated Landau levels, further investigations are needed to identify the origin of the SdHO suppression.

5. Acknowledgments

The work at Minnesota was supported by the DOE Grant DE-SC0002567. The work at Princeton was funded by the Gordon and Betty Moore Foundation and by the NSF MRSEC Program through the PCCM (DMR-0819860). A. B. and S. S. acknowledge financial support from NSERC. A. H. acknowledges support from the NSF Grant No. DMR-0548014.

References

[1] Zudov M A, Du R R, Simmons J A and Reno J L 2001 *Phys. Rev. B* **64** 201311(R)
[2] Ye P D, Engel L W, Tsui D C, Simmons J A, Wendt J R, Vawter G A and Reno J L 2001 *Appl. Phys. Lett.* **79** 2193
[3] Zudov M A, Ponomarev I V, Efros A L, Du R R, Simmons J A and Reno J L 2001 *Phys. Rev. Lett.* **86** 3614
[4] Yang C L, Zhang J, Du R R, Simmons J A and Reno J L 2002 *Phys. Rev. Lett.* **89** 076801
[5] Mani R G, Smet J H, von Klitzing K, Narayanamurti V, Johnson W B and Umansky V 2002 *Nature (London)* **420** 646
[6] Zudov M A, Du R R, Pfeiffer L N and West K W 2003 *Phys. Rev. Lett.* **90** 046807

[7] Kukushkin I V, Akimov M Y, Smet J H, Mikhailov S A, von Klitzing K, Aleiner I L and Falko V I 2004 *Phys. Rev. Lett.* **92** 236803
[8] Zhang W, Zudov M A, Pfeiffer L N and West K W 2007 *Phys. Rev. Lett.* **98** 106804
[9] Zhang W, Zudov M A, Pfeiffer L N and West K W 2008 *Phys. Rev. Lett.* **100** 036805
[10] Khodas M, Chiang H S, Hatke A T, Zudov M A, Vavilov M G, Pfeiffer L N and West K W 2010 *Phys. Rev. Lett.* **104** 206801
[11] Wiedmann S, Gusev G M, Raichev O E, Bakarov A K and Portal J C 2010 *Phys. Rev. Lett.* **105** 026804
[12] Hatke A T, Chiang H S, Zudov M A, Pfeiffer L N and West K W 2010 *Phys. Rev. B* **82** 041304(R)
[13] Dai Y, Du R R, Pfeiffer L N and West K W 2010 *Phys. Rev. Lett.* **105** 246802
[14] Hatke A T, Zudov M A, Pfeiffer L N and West K W 2011 *Phys. Rev. B* **83** 121301(R)
[15] Andreev A V, Aleiner I L and Millis A J 2003 *Phys. Rev. Lett.* **91** 056803
[16] Durst A C, Sachdev S, Read N and Girvin S M 2003 *Phys. Rev. Lett.* **91** 086803
[17] Lei X L and Liu S Y 2003 *Phys. Rev. Lett.* **91** 226805
[18] Vavilov M G and Aleiner I L 2004 *Phys. Rev. B* **69** 035303
[19] Dmitriev I A, Mirlin A D and Polyakov D G 2003 *Phys. Rev. Lett.* **91** 226802
[20] Dmitriev I A, Vavilov M G, Aleiner I L, Mirlin A D and Polyakov D G 2005 *Phys. Rev. B* **71** 115316
[21] Khodas M and Vavilov M G 2008 *Phys. Rev. B* **78** 245319
[22] Dmitriev I A, Khodas M, Mirlin A D, Polyakov D G and Vavilov M G 2009 *Phys. Rev. B* **80** 165327
[23] Hatke A T, Khodas M, Zudov M A, Pfeiffer L N and West K W 2011 *Phys. Rev. B* **84** 241302(R)
[24] Zudov M A 2004 *Phys. Rev. B* **69** 041304(R)
[25] Willett R L, Pfeiffer L N and West K W 2004 *Phys. Rev. Lett.* **93** 026804
[26] Studenikin S A, Potemski M, Coleridge P T, Sachrajda A S and Wasilewski Z R 2004 *Solid State Commun.* **129** 341
[27] Studenikin S A, Potemski M, Sachrajda A, Hilke M, Pfeiffer L N and West K W 2005 *Phys. Rev. B* **71** 245313
[28] Smet J H, Gorshunov B, Jiang C, Pfeiffer L, West K, Umansky V, Dressel M, Meisels R, Kuchar F and von Klitzing K 2005 *Phys. Rev. Lett.* **95** 116804
[29] Zudov M A, Du R R, Pfeiffer L N and West K W 2006 *Phys. Rev. B* **73** 041303(R)
[30] Zudov M A, Du R R, Pfeiffer L N and West K W 2006 *Phys. Rev. Lett.* **96** 236804
[31] Bykov A A, Bakarov A K, Islamov D R and Toropov A I 2006 *JETP Lett.* **84** 391
[32] Studenikin S A, Sachrajda A S, Gupta J A, Wasilewski Z R, Fedorych O M, Byszewski M, Maude D K, Potemski M, Hilke M, West K W and Pfeiffer L N 2007 *Phys. Rev. B* **76** 165321
[33] Hatke A T, Chiang H S, Zudov M A, Pfeiffer L N and West K W 2008 *Phys. Rev. B* **77** 201304(R)
[34] Hatke A T, Chiang H S, Zudov M A, Pfeiffer L N and West K W 2008 *Phys. Rev. Lett.* **101** 246811
[35] Hatke A T, Zudov M A, Pfeiffer L N and West K W 2009 *Phys. Rev. Lett.* **102** 066804
[36] Bykov A A, Marchishin I V, Goran A V and Dmitriev D V 2010 *Appl. Phys. Lett.* **97** 082107
[37] Dorozhkin S I, Pfeiffer L, West K, von Klitzing K and Smet J H 2011 *Nature Phys.* **7** 336
[38] Hatke A T, Zudov M A, Pfeiffer L N and West K W 2011 *Phys. Rev. B* **83** 201301(R)
[39] Hatke A T, Zudov M A, Pfeiffer L N and West K W 2011 *Phys. Rev. B* **84** 241304(R)
[40] Hatke A T, Zudov M A, Watson J D and Manfra M J 2012 *Phys. Rev. B* **85** 121306(R)
[41] Hatke A T, Zudov M A, Watson J D, Manfra M J, Pfeiffer L N and West K W 2012 *unpublished*
[42] Dmitriev I A, Mirlin A D, Polyakov D G and Zudov M A 2012 *Rev. Mod. Phys.* **84** 1709
[43] Pioro-Ladrière M, Abolfath M R, Zawadzki P, Lapointe J, Studenikin S A, Sachrajda A S and Hawrylak P 2005 *Phys. Rev. B* **72** 125307
[44] Mani R G 2005 *Phys. Rev. B* **72** 075327
[45] Yang C L, Du R R, Pfeiffer L N and West K W 2006 *Phys. Rev. B* **74** 045315
[46] Bogan A, Hatke A T, Studenikin S A, Sachrajda A, Zudov M A, Pfeiffer L N and West K W 2012 *Phys. Rev. B* **86** 235305
[47] Zhang W, Chiang H S, Zudov M A, Pfeiffer L N and West K W 2007 *Phys. Rev. B* **75** 041304(R)
[48] Hatke A T, Zudov M A, Pfeiffer L N and West K W 2009 *Phys. Rev. B* **79** 161308(R)
[49] Hatke A T, Zudov M A, Pfeiffer L N and West K W 2011 *Phys. Rev. B* **83** 081301(R)
[50] Elzerman J M, Hanson R, Beveren L H W v, Witkamp B, Vandersypen L M K and Kouwenhoven L P 2004 *Nature* **430** 431
[51] Romero N, McHugh S, Sarachik M P, Vitkalov S A and Bykov A A 2008 *Phys. Rev. B* **78** 153311
[52] Hatke A T, Zudov M A, Pfeiffer L N and West K W 2012 *Phys. Rev. B* **85** 241305(R)

Nonequilibrium currents in the quantum Hall effect regime spatially resolved by transport experiment

M V Budantsev[1,2], A G Pogosov[1,2], D A Pokhabov[1,2], E Yu Zhdanov[1,2], A K Bakarov[1] and A I Toropov[1]

[1] Institute of Semiconductor Physics, SB RAS, Novosibirsk, Russia
[2] Novosibirsk State University, Novosibirsk, Russia

E-mail: budants@isp.nsc.ru

Abstract. Eddy currents in a constriction placed in a two-dimensional electron gas (2DEG) have been experimentally studied. The constrictions fabricated by the electron-beam lithography demonstrate quantum Hall effect in high magnetic field. Pronounced hysteresis of magnetoresistance is observed at Hall plateaus corresponding to integer filling factors ν = 1, 2 and 4 of a macroscopic 2DEG. The dependence of magnetoresistance hysteresis on the constriction width is obtained. The dependence contains the threshold value of width that points to the edge nature of eddy currents and enables us to determine the eddy currents area width, which is about 0.5 μm.

1. Introduction

A two-dimensional electron gas (2DEG) in the quantum Hall effect (QHE) regime demonstrates nonequilibrium state, manifested itself in hysteretic behavior of different 2DEG characteristics, such as magnetization [1], charge [2, 3], local electrostatic potential [4, 5] as a function of magnetic field in condition of σ_{xx}=0. For a long time this nonequilibrium state remained unnoticed in transport experiments. Recently this state has been revealed and investigated in conventional magnetoresistance measurements in the simple system — 2DEG with a constriction [6, 7, 8, 9, 10].

Such unusual behavior is explained by long-lived nonequilibrium currents of magnetization, induced in a 2DEG, that are often referred to as eddy currents. The term "eddy currents" designates currents induced in a conductor by sweep of the magnetic field. However it has been shown in Refs. [11, 12] that the same currents, responsible for the nonequilibrium magnetization of a 2DEG, can be induced by the sweep of the gate voltage as well as the sweep of the magnetic field. Therefore in Refs. [11, 12] they are referred to as "nonequilibrium currents". This term is more general and emphasizes their nonequilibrium nature.

One of the most important unsolved problems is the problem of eddy currents spatial distribution. In the QHE regime the 2DEG bulk is occupied by incompressible electron liquid and the magnetic field sweep induces charge transfer between the bulk and the edge [13] that in its turn establishes radial electric field which generates large azimuthal eddy currents in disk-shaped 2DEG. In Ref. [14] these eddy currents are assumed to be distributed over the whole plane of a 2DEG, while Ref. [5] where the distribution of local electrostatic potential has been experimentally studied points rather to the edge nature of eddy currents.

Journal of Physics: Conference Series **456** (2013) 012005 doi:10.1088/1742-6596/456/1/012005

It has been shown that the magnetoresistance of a constriction placed in the 2DEG bath demonstrates hysteretic behavior in magnetic field. The dependence of hysteresis of the magnetoresistance on the constriction width has been experimentally studied. It has been found that the hysteresis vanishes at certain critical width. This points to the edge nature of eddy currents. The obtained experimental results allow us to define the width of the area, where eddy current is located. This area width has been found to be about 0.5 μm.

2. Experimental details

Experimental samples have been fabricated on the basis of GaAs/AlGaAs heterojunction, containing a 2DEG grown by means of molecular-beam epitaxy. The heterostructure with 2DEG is presented in Figure 1. The electron mobility is 0.8×10^6 $cm^2/(V{\cdot}s)$ and the 2DEG density is $1.8\div2.2\times10^{11}$ cm^{-2} at 4.2 K.

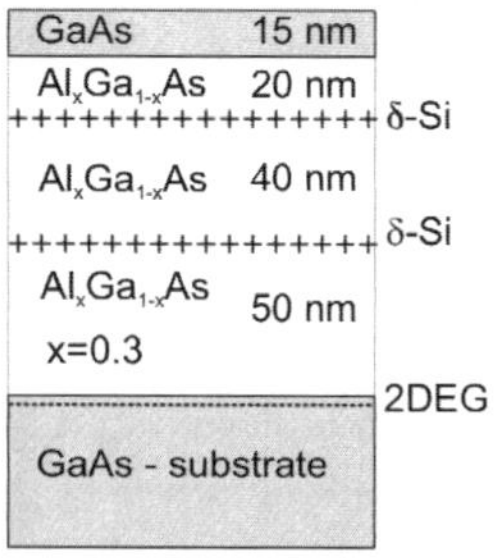

Figure 1. Heterostructure GaAs/AlGaAs with a 2DEG.

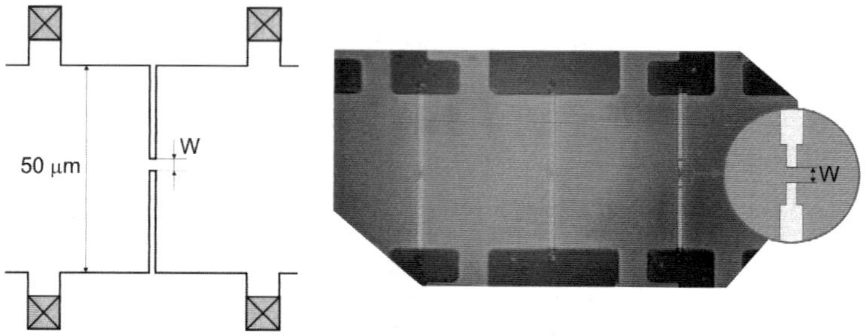

Figure 2. Schematic representation and microphotography of the Hall bar with the constrictions.

Constrictions of different widths have been fabricated by means of electron-beam lithography with following plasma-chemical etching on the Hall bar of 50 μm width. Six constrictions having lithographic widths in a range $W = 0.8\div1.3$ μm have been fabricated. The microphotography of the sample is represented in Figure 2. Such samples allow measuring the Hall resistance of the macroscopic 2DEG as well as longitudinal magnetoresistance of constrictions. All the constrictions have a length 0.6 μm which is much less than the electron mean free path at given 2DEG parameters.

The measurements were carried out by means of lock-in technique in the linear response regime on the alternating current of the magnitude 10 nA and the frequency 7 Hz at temperature of 0.48 K. The magnetic field was directed perpendicularly to the 2DEG plane and covered a range $0\div11$ T. The magnetic field sweep rate in the experiment was 0.01 T/s.

3. Experimental results and discussion

All the constrictions have low resistance ranged from 0.2 kΩ to 0.8 kΩ at zero magnetic field and demonstrate QHE regime in high magnetic field.

The minima of the constriction magnetoresistance is shifted to the lowest field relative to the centers of the Hall plateau of a macroscopic 2DEG (Fig. 3). It can be attributed to the difference between the electron density in the constrictions and that of the macroscopic 2DEG. This difference reached 10%.

The hysteresis of magnetoresistance of all the constrictions have been observed at the filling factors $\nu = 1$, 2 and 4 of the macroscopic 2DEG (Fig. 3). The hysteresis is practically independent on sweep rate. The increase of sweep rate in 30 times does not cause any significant change in the hysteresis [7].

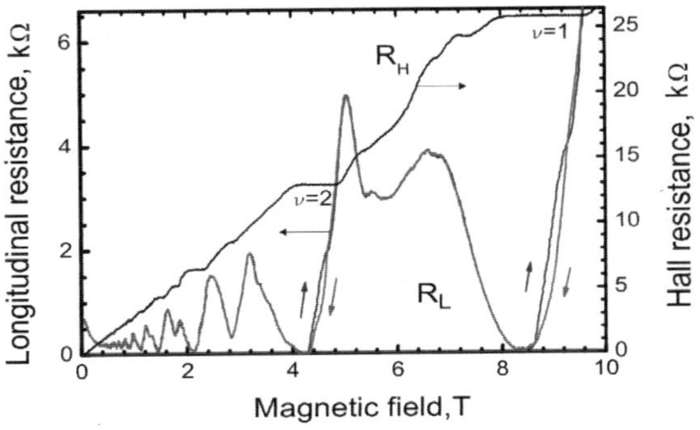

Figure 3. The longitudinal and Hall resistance in the constriction of width 1 μm as a function of the magnetic field. Arrows indicate the magnetic field sweep direction.

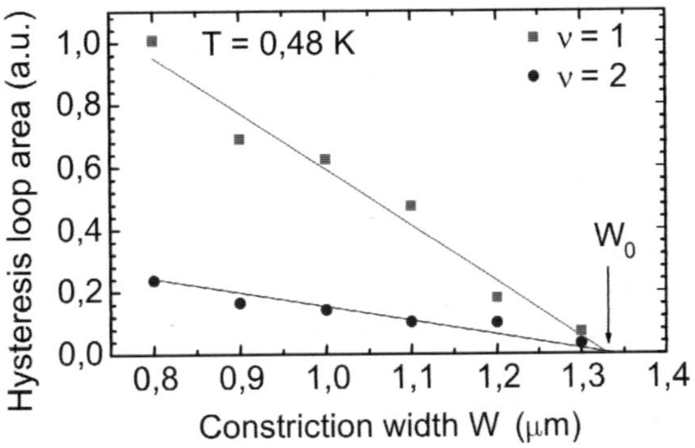

Figure 4. The hysteresis loop area as a function of the constriction lithographic width at the filling factors ν = 1 and 2.

Figure 4 shows the hysteresis loop area as a function of the lithographic constriction width at filling factors ν =1 and 2. The experimental points are well fitted by straight lines. One can see that the loop area decreases with increasing width and vanishes at certain critical width. Moreover these critical widths coincide for both filling factors and is about 1.35 μm. Note that the slope of the dependence for $\nu = 1$ is 4 times steeper than that for $\nu = 2$ with a good accuracy.

The nonequilibrium state is not observable in transport measurements in a wide Hall bar. A constriction brings together the opposite 2DEG edges and makes eddy currents propagating in opposite directions to interact. If the eddy currents are distributed over the whole plane of a 2DEG, we would not observe such the clear cut-off in the dependence. The found threshold width points to edge nature of eddy currents.

To estimate lateral width of eddy currents the depletion width W_{depl} should be determined. The effective width of the conductive channel is $W_{eff} = W - 2 \times W_{depl}$. Due to the fact that the electron mean free path ($\sim$ 5 μm) is much larger than the constriction length (0.6 μm) the constrictions could be considered as quantum point contacts and their conductance could be found from the equation:

$$G = \frac{2e^2}{h} \cdot \frac{k_F \times W_{eff}}{\pi}. \tag{1}$$

Taking into account the constrictions resistance at zero magnetic field it has been found that the effective width W_{eff} is about 0.4 μm less than the lithographic width W for all the constrictions (Fig. 5). Thus the depletion layer width W_{depl} is about 0.2 μm.

The constriction of critical width W_0 contains two counter propagating eddy currents each of them has width W_{EC} and two depletion layers of width W_{depl}:

$$W_0 = 2 \times W_{depl} + 2 \times W_{EC}. \tag{2}$$

This gives $W_{EC} \approx 0.5$ μm for the studied samples.

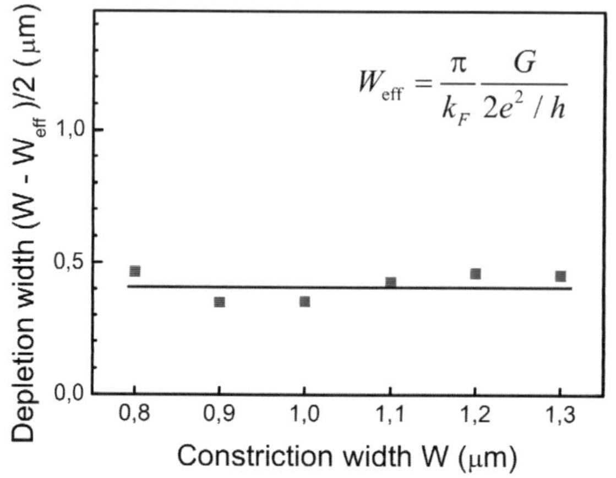

Figure 5. The depletion width calculated from zero-field conductance by the given formula as a function of the constriction lithographic width W.

It is well known that magnetic field sweep generates azimuthal electric field, which in its turn in the condition of $\sigma_{xx} = 0$ induces the charge transfer between the bulk and the edge [13, 3]. Thereby, the area of incompressible liquid is changed: it increases with decreasing the magnetic field and decreases with increasing the magnetic field. A constriction can be sensitive to such changes. Two different cases can be realized in the constriction at the same magnetic field:

(i) The area of incompressible liquid increases at down magnetic field sweep direction (Fig. 5 (a)). Edge currents in a constriction move away from each other, the backscattering is suppressed resulting in the resistivity decrease.

(ii) The area of incompressible liquid decreases at up magnetic field sweep direction (Fig. 5 (b)). Edge currents in a constriction move toward each other, the backscattering increases resulting in the resistivity increase.

This can be considered as a topological transition. At $\partial B/\partial t < 0$ the edge current transmits through the constriction forming the single loop and incompressible liquid forms simply connected space, while at $\partial B/\partial t > 0$ edge current scatters in the constriction and incompressible liquid splits into two spaces divided by compressible strip in the constriction.

It should be noted that the measurements of hysteresis of conductance of quantum point contact [10] have been carried out in a tunnel regime. The QHE is not realized in the mentioned case. The hysteresis is observed on the background of large magnetoresistance. In this case there are no topological transitions in the constriction. Such measurements are not able to provide the information about spatial distribution of eddy current since eddy current does not transmit through the constriction.

The range of widths of the constrictions studied in present paper allows us to establish the edge nature of NECs and determine eddy currents area width.

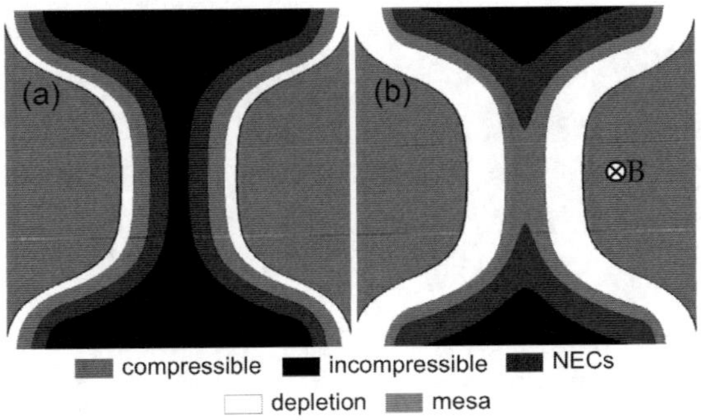

Figure 6. The schematic representation of Hall liquid structure in the constriction at down (a) and up (b) magnetic field sweep direction.

4. Conclusions

The original and simple method of study eddy currents induced in a 2DEG in the QHE regime has been suggested. This method is based on conventional measurements of magnetoresistance of constriction placed in a 2DEG bath. It has been shown that the constriction width is a critical parameter responsible for the observation of the hysteresis. The hysteresis of magnetoresistance as a function of the constriction width has been studied. The found critical width points to edge nature of eddy currents. The results obtained in present paper as well as in Ref. [5] are the experimental evidence of the fact that eddy currents are induced in a narrow area along the edge rather than distributed over the whole plane of the 2DEG as it proposed in Ref. [14]. The found width of area in which the eddy currents are induced is about 0.5 μm.

Acknowledgments

The work is supported by RFBR (grant 12-02-00532), Program of DNIT RAS (project 3.2), Program of Presidium of RAS (grant 24.19).

References

[1] Usher A, and Elliott M 2009 *J. Phys.: Condens. Matter* **21** 103202
[2] Pudalov V M, Semenchinsky S G and Edelman V S 1984 *Sol. St. Commun.* **51** 713
[3] Dolgopolov V T, Shashkin A A, Zhitenev N B, Dorozhkin S I and von Klitzing K 1992 *Phys. Rev. B* **46** 12560
[4] Huels J, Weis J, Smet J, von Klitzing K and Wasilewski Z R 2004 *Phys. Rev. B* **69** 085319
[5] Klaffs T, Krupenin V A, Weis J and Ahlers F J 2004 *Physica E* **22** 737
[6] Budantsev M V, Pogosov A G, Plotnikov A E, Bakarov A K, Toropov A I and Portal J C 2007 *JETP Lett.* **86** 264
[7] Budantsev M V, Pogosov A G, Plotnikov A E, Bakarov A K, Toropov A I and Portal J C 2009 *JETP Lett.* **89** 46
[8] Budantsev M V, Pogosov A G, Bakarov A K, Toropov A I and Portal J C 2009 *JETP Lett.* **89** 92
[9] Budantsev M V, Pogosov A G, Pokhabov D A, Zhdanov E Yu, Bakarov A K, Toropov A I and Portal J C 2011 *AIP Conf. Proc.* **1399** 601
[10] Pioro-Ladriere M, Usher A, Sachrajda A S, Lapointe J, Gupta J, Wasilewski Z, Studenikin S and Elliott M 2006 *Phys. Rev. B* **73** 075309
[11] Ruhe N, Stracke G, Heyn Ch and Heitmann D 2009 *Phys. Rev. B* **80**, 115336
[12] Ho L H, Taskinen L J, Micolich A P, Hamilton A R, Atkinson P and Ritchie D A 2010 *Phys. Rev. B* **82**, 153305
[13] Laughlin R B 1981 *Phys. Rev. B* **23** 5632
[14] Matthews A J, Kavokin K V, Usher A, Portnoi M E, Zhu M, Gething J D, Elliott M, Herrenden-Harker W G, Phillips K, Ritchie D A, Simmons M Y, Sorensen C B, Hansen O P, Mironov O A, Myronov M, Leadley D R and Henini M 2004 *Phys. Rev. B* **70** 075317

Journal of Physics: Conference Series **456** (2013) 012006 doi:10.1088/1742-6596/456/1/012006

Quantum Hall effect in monolayer, bilayer and trilayer graphene

C. Cobaleda[1], E. Diez[1], M. Amado[1,†], S. Pezzini[1,2], F. Rossella[2], V. Bellani[2], D.López-Romero[3], D. K. Maude[4,‡]

[1]Laboratorio de Bajas Temperaturas, Universidad de Salamanca, E-37008 Salamanca, Spain
[2]Dipartimento di Fisica and CNISM, Università degli studi di Pavia, I-27100 Pavia, Italy
[3]CT-ISOM, Universidad Politécnica de Madrid, E-28040 Madrid, Spain
[4]Laboratoire National des Champs Magnétiques Intenses, CNRS-UJF-INSA-UPS,F-38042 Grenoble, France
[†]Present address: SNS-NEST and CNR, Piazza San Silvestro 12, 56127 Pisa, Italy
[‡]Present address: LNCMI, CNRS-UJF-INSA-UPS, F-31400 Toulouse, France

E-mail: ccobaleda@usal.es

Abstract. We have performed magneto-transport experiments in monolayer, bilayer, trilayer and four layered graphene, at temperatures between 2 and 190 K and magnetic fields up to 28 T. In particular, in monolayer graphene we studied the quantum Hall effect and the metal-insulator transition. On the other hand, in bilayer graphene we observed quantum Hall plateaus at filling factor $\nu = 4, 8, 12, 16, 20, ...$ and the $\nu = 6$ plateau in trilayer graphene, studying their temperature dependence. We have also studied the symmetry properties which are related with different contact configurations describing the method used to study inhomogeneus samples. Finally, four layered graphene we did not found quantum Hall plateaus, but we observed and investigated an ambipolar conduction effect.

Graphene is a material which has been studied in several investigations because of its unique electronic properties, which are interesting due to the theoretical concepts involved and its potential applications [1, 2, 3, 4, 5]. The electronic properties of graphene induced by its electronic band structure leads to unusual effects, such as the massless Dirac fermions and the anomalous quantum Hall effect (QHE) [1, 2, 3, 6, 4, 5, 7, 8], which results in the quantum Hall plateaus series $\nu = \pm 2, \pm 6, \pm 10, \pm 14, ...$ for monolayer graphene and $\nu = \pm 4, \pm 8, \pm 12, \pm 16, ...$ for bilayer graphene. Because of the experimental observation of these series for the QHE in graphene [6, 7, 8] the electronic properties of graphene and multilayer graphene has been extensively investigated. In particular, QHE in trilayer graphene (TLG) has also been investigated since its band structure depends on the interlayer stacking sequence [9, 10, 11, 12].

Here we study the QHE in monolayer, bilayer and trilayer graphene. In particular, we investigate the metal-insulator transition in monolayer graphene, the serie of the quantum Hall plateaus for bilayer graphene and the presence of the $\nu = \pm 6$ quantum Hall plateau in trilayer graphene. Furthermore, we study an asymmetry of the conductivity arising from the presence of a density gradient, which has been previously observed in 2DEG [13] and more recently in graphene [14, 15]. Finally, we have also investigated the magnetotransport properties of four layered graphene.

The graphene flakes have been fabricated by means of mechanical exfoliation of natural

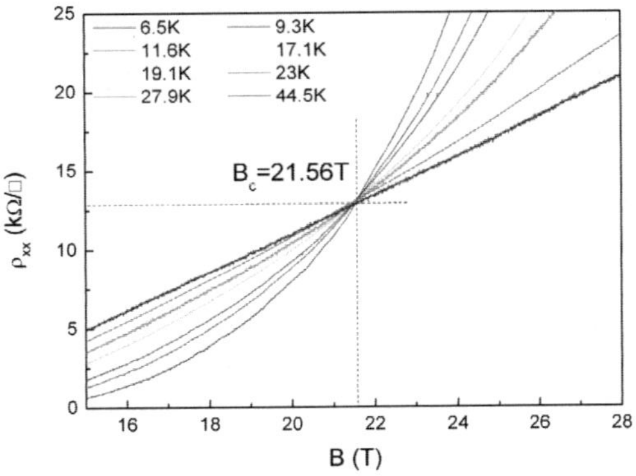

Figure 1. Longitudinal resistivities measured at gate voltage 0 V using the magnetic field as driving parameter at temperatures from 6.5 K to 44.5 K.

graphite and were placed on top of SiO_2. Their number of layer was determined by Raman spectroscopy. The processing of the samples was done as follows: the Hall bar geometry was defined by means of electron beam lithography and the contacts (50 Å of Ti followed by 500 Å of Au) were deposited by e-beam evaporation. The undesired parts of the flakes were removed using O_2/Ar plasma etching. The magneto-transport measurements have been carried out at temperatures between 1.9 K and 190 K, with sample placed in a ^{4}He cryostat. All the measurements were performed using standard ac lock-in techniques and four probe measurements, using an excitation current of $I = 50$ nA which was obtained by applying 5 V with frequency below 15 Hz to a 100 MΩ series resistance. This current flowed from contact S to D (see inset in Fig. 3(**b**)), pre-amplified with a gain of $\simeq 100$. We measured simultaneously the voltage drop V_{1-3}, V_{1-2} and V_{3-4} between the contact pairs $1-3$, $1-2$ and $3-4$ (see inset of Fig. 3(**b**)). A 10 MW resistive magnet was used to apply a magnetic field up to 28 T perpendicularly to the samples.

We measured the longitudinal resistivity using the back gate voltage as driving parameter of the charge density in a monolayer graphene sample on top of SiO_2, finding a charge neutrality point (CNP) at $V_{CNP} = 4$ V. We placed the gate voltage enough close to the Dirac point in order to have only the last Landau Level $\nu = -2$ below the Fermi energy, since $\nu = nh/(eB)$. We then studied, at different temperatures, the metal-insulator transition from the state $\nu = -2 \rightarrow \nu = 0$ which is characterized by a crossing point at a critical magnetic field B_c (see Fig. 1).

The longitudinal resistance exhibits a temperature independent crossing point at $B_c = 21.56$ T. Using the standard scaling theory analysis $R_{xx} \propto \exp(\Delta\nu/\nu_0(T))$ with $\Delta\nu = 1/B - 1/B_c$ (where $\Delta\nu$ should not be confused with the filling factor) we have obtained the critical exponent of the transition $\kappa = 0.58 \pm 0.01$ [16].

For the bilayer sample, the Dirac point was at 25 V (see figure 2(**a**)). We define the longitudinal resistance as $R_{xx} = (V_{1-2}/I)$ whereas the Hall resistance is defined as $R_{xy} = (V_{1-3}/I)$ (the labels of the contacts are shown in the inset in figure 3(**b**)). In figure 2(**b**) we show the QHE measured in the bilayer sample. This QHE differs from the QHE observed previously and reveals a new serie of plateaus $\nu = \pm 4, \pm 8, \pm 12, \pm 16, ...$ which have not been observed in monolayer graphene.
We characterized the bilayer sample by fitting the slope of the linear part of Hall resistances

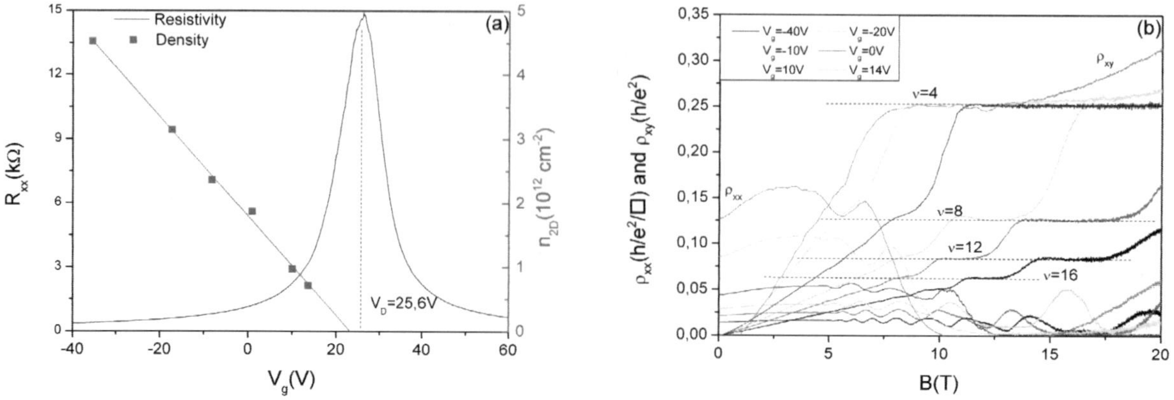

Figure 2. a) Longitudinal resistance measured at zero magnetic field using the gate voltage as driving parameter. The Dirac point was found at $V_G = 25$ V. The carrier density in the hole-like regime is shown in red.
b) Measurements of the quantum Hall effect performed in a bilayer graphene flake at 2 K using the magnetic field as driving parameter at different values of the gate voltage, from -40 V up to 14 V. The series of quantum Hall plateaus $\nu = 4, 8, 12, 16, 20, ...$ is observed.

measured at different gate voltages, since, at low magnetic field, it can be assumed that $R_{xy} \sim B/(ne)$ (figure 2**(a)**). The values of both the mobility (given by $\mu = 1/(neR_0)$) and the density at the different gate voltages are reported in the following table:

V$_{gate}$(V)	-40	-20	-10	0	10	14
n$_{2D}$(10^{12}cm^{-2})	4.5	3.2	2.4	1.6	0.9	0.7
μ(cm^2/Vs)	3800	3600	3557	3400	3217	2800

For the trilayer sample, we obtain the Hall resistivities (denoted as "left" and "right") using $\rho^l_{xy} = V_{1-3}/I$ and $\rho^r_{xy} = V_{2-4}/I$; whereas the longitudinal resistivites (denoted as "top" and "bottom") are obtained using $\rho^t_{xx} = (V_{1-2}/L)\,/\,(I/W)$ and $\rho^b_{xx} = (V_{3-4}/L)\,/\,(I/W)$. In Fig.3 we show the Hall and longitudinal resistivities as a function of the magnetic field.
We obtain the mobility μ and 2D carrier density from the Hall and the longitudinal resistances at low magnetic field, finding that the carrier density increases with increasing temperature, as expected for the semi-metallic TLG [9]. At the base temperature of 4.5 K the density and the mobility of the sample were $n_{2D} = 2.04 \times 10^{12}\text{cm}^{-2}$ and $\mu = 3750$ cm^2/Vs. Unfortunately a leak in the SiO_2 layer did not allow the use of the back gate in order to change the carrier density. In Fig. 3**(b)** we show that the "right" Hall resistivity shows the QHE with the presence of the $\nu = \pm 6$ plateau. It is also observed that the slope of ρ^r_{xy} (s^r) is higher than the slope of ρ^l_{xy} (s^l), such that $s^r \sim 3s^l$. This fact indicates a large inhomogeneity along the sample, since the density follows $n_{2D} \propto 1/s$. This inhomogeneity results in the asymmetry of the longitudinal resistivity upon the reversal of the magnetic field (namely, $\rho_{xx}(-B) = \rho_{xx}(B)$), since, as it is shown in figure 3**(a)**, $\rho^{t,b}_{xx}(B) \neq \rho^{t,b}_{xx}(-B)$, but $\rho^{t,b}_{xx} = \rho^{b,t}_{xx}(-B)$ [17]. That is, when the polarity of the magnetic field is inversed, the "top" longitudinal resistivity and the "bottom" longitudinal resistivity are swapped.

Finally, we have measured the magnetotransport properties of a four layered graphene sample at a base temperature of 4.5 K. We measured the longitudinal resistivity ρ_{xx} using the back gate voltage as a driving parameter at zero magnetic field. We found that the longitudinal resistance

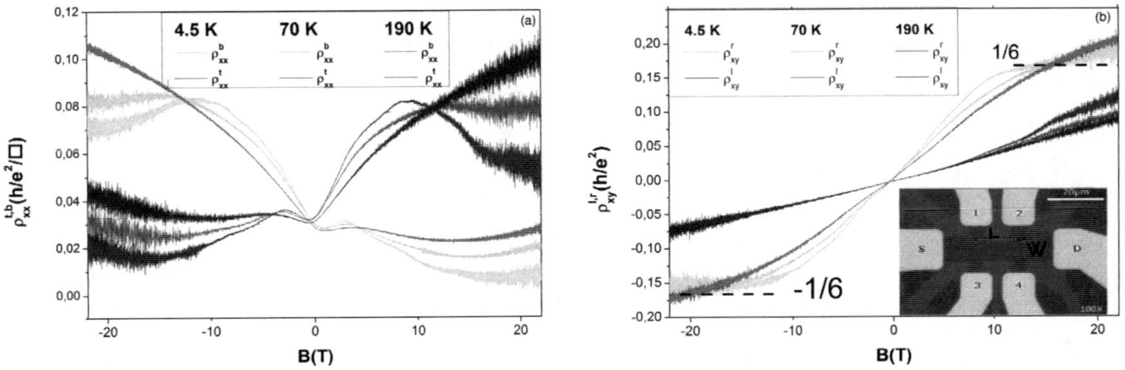

Figure 3. **a)** Longitudinal resistivities ρ_{xx} versus magnetic field at temperatures 4.5, 70 and 190 K.
b) Hall resistivity (ρ_{xy}) versus B at T = 4.5, 70 and 190 K. The quantum Hall plateau at $\nu = \pm 6$ corresponding to $\rho_{xy} = h/6e^2$ is visible up to 70 K.

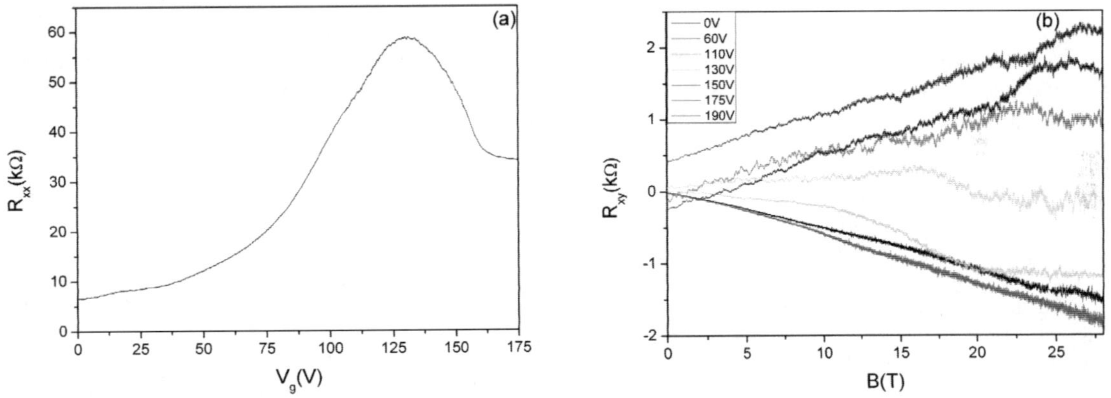

Figure 4. **a)**Longitudinal resistivity versus the back gate voltage at a temperature of 4.5 K. It can be seen a maximum, the so called charge neutrality point, at 130 V.
b) The Hall resistivities versus the magnetic field at different back gate voltage and at a temperature of 4.5 K. We note the linear behavior with the magnetic, with different positive and negative slopes, indicating that the density and the sign is tuned by the gate voltage.

shows a maximum at V_G = 130 V, followed by a saturation regime, as shown in figure 4(**a**). We believe that this feature indicates the presence of the CNP and that the saturation is caused by the inhomogeneities of the sample.

We think that the maximum located at 130 V is indeed the CNP point since, as we observe in figure 4(**b**), for back gate voltages below the CNP the Hall resistivity shows a linear dependence with the magnetic field with a negative slope. Nonetheless, we observe that for gate voltages closer to the CNP the Hall resistivity deviates slightly from this non linear behaviour. When the back voltage is exactly at the CNP, the Hall resistivity is almost constant with the magnetic field; which is an indication of the presence of the electron-hole puddle regime. As the gate voltage is further increased, the resistance shows again a linear dependence with the magnetic field, but with a positive slope, which is the expected behaviour found in graphene when the back gate voltage passes through the Dirac peak and the transition from the hole-like regime to the electron-like regime.

In conclusion, we have measured the quantum Hall effect in monolayer, bilayer and trilayer graphene samples observing the presence of the corresponding Hall plateaus.
We have observed the quantum Hall plateaus resulting for monolayer graphene $\nu = \pm 2, \pm 6, \pm 10...$ and the metal-insulator transition. We have measured the critical exponent of this transition, resulting in $\kappa = 0.58 \pm 0.01$ [16].
The resulting series of the quantum Hall plateaus for bilayer graphene is $\nu = \pm 4, \pm 8, \pm 12, \pm 16, \pm 20, ...$
We have observed the $\nu = \pm 6$ plateaus for trilayer graphene and studied the asymmetry of the longitudinal resistances, which arises from an inhomogeneity of carrier density along the sample [17].
Finally, we have studied the magnetotransport properties of a four layered graphene flake, observing an ambipolar effect.
Thus, we remark that the quantum Hall effect is a notable tool to distinguish monolayer, bilayer and trilayer graphene, since their plateaus series are different. As far as four layerered graphene is concerned, more experiments with higher quality samples are required.

Acknowledment

This work was supported by the projects Cariplo Foundation QUANTDEV, MEC FIS2009-07880, PPT310000-2009-3, JCYL SA049A10-2, the FPU program by MEC and by EuroMagNET II under the EU contract n. 228043.

References

[1] A. K. Geim and K. Novoselov, *Nature Mater.* **6** 183 (2007)
[2] A. H. Castro Neto, F. Guinea, N. M. R. Peres, K. S. Novoselov and A. K. Geim, *Rev. Mod. Phys.* **81**, 109 (2009)
[3] M. Vozmediano, M. Katsnelson and F. Guinea, *Phys. Rep.* **496** 109 (2010)
[4] A. Cresti, G. Grosso and G. P. Parravicini, *Phys. Rev. B* **77** 115408 (2008)
[5] N. M. R. Peres, *Rev. Mod. Phys.* **82** 2673 (2010)
[6] K. S. Novoselov, A. K. Geim, S. V. Morozov, D. Jiang , M. I. Katsnelson, I. V Grigorieva, S. V. Dubonos and A. A. Firsov, *Nature* **438** 197 (2005)
[7] Y. Zhang, Y.-W. Tan, H. L. Stormer and P. Kim, *Nature* **438** 201 (2005)
[8] K. S. Novoselov, E. McCann, S. V. Morozov, V. I. Fal'ko, M. I. Katsnelson, U. Zeitler, D. Jiang, F. Schedin, A. K. Geim, *Nat. Phys.* **2** 177 (2006)
[9] M. F. Craciun, S. Russo, M. Yamamoto, J. B. Oostinga, A. F. Morpurgo and S. Tarucha, *Nature Nanotech.* **4** 383 (2009)
[10] W. Zhu, V. Perebeinos, M. Freitag and P. Avouris, *Phys. Rev. B* **80** 235402 (2009)
[11] Y. Liu , S. Goolaup, C. Murapaka, W. S. Lew and S. K. Wong, *ACS Nano* **4** (2010)
[12] W. Bao, Z. Zhao, H. Zhang, G. Liu , P. Kratz, L. Jing, J. Velasco, D. Smirnov and C. N. Lau, *Phys. Rev. Lett.* **105** 246601 (2010)
[13] E. Peled, Y. Chen, E. Diez, D. C. Tsui, D. Shahar, D. L. Sivco and A. Y. Cho, *Phys. Rev. B* **69** 241305 (2004)
[14] D.-K. Ki and H.-J. Lee, *Phys. Rev. B* **79** 195327 (2009)
[15] D.-K. Ki, S. Jo and H.-J. Lee, *Appl. Phys. Lett.* **94** 162113 (2009)
[16] M. Amado, E. Diez, F. Rossella, V. Bellani, D. López-Romero and D. K. Maude, *J. of Phys. Cond. Mat.* **24** 305302 (2012)
[17] C. Cobaleda, F. Rossella, S. Pezzini, E. Diez, V. Bellani, D. K. Maude and P. Blake, *Physica E* **44** 530 (2011)

The quantum percolation model of the scaling theory of the quantum Hall effect: a unifying model for plateau-to-plateau transitions.

N. A. Dodoo-Amoo, K. Saeed, L. H. Li, E. H. Linfield, A. G. Davies, J. E. Cunningham

School of Electronic and Electrical Engineering, University of Leeds, Leeds, LS2 9JT, UK

E-mail: el06nada@leeds.ac.uk

Abstract. We present a unifying model of plateau-to-plateau transitions in the quantum Hall effect based on results from high resolution frequency scaling experiments. We show that as the frequency or quantum coherence length of the two-dimensional electron system is varied, one observes a crossover between classical percolation and quantum percolation in the measured values of the critical scaling exponents of the plateau-to-plateau transitions. This crossover is dependent on the relationship between certain relevant length scales of a two-dimensional system and can be explained using a quantum percolation model. The model explains why quantum criticality can be observed in some systems, but is absent from others.

1. Introduction

The scaling theory of plateau-to-plateau transitions (PPTs) in the quantum Hall effect (QHE) has been of great interest ever since its discovery [1]. The scaling theory has its origins in the early work of Abrahams *et al.* [2], and is based on the observation that, as the centre of a Landau level (LL) is approached, the localization length ξ diverges to infinity, forming a singularity at the centre of the LL. This singularity which can be described as

$$\xi \propto |E - E_c|^{-\gamma}, \tag{1}$$

indicates the existence of a second order quantum phase transition between the two insulating phases of a plateau-to-plateau transition (PPT) [3]. As expected with all phase transitions, this singularity is associated with certain critical universal scaling exponents, such as the critical exponent of the localization length γ in Eq. (1). If this exponent is a quantum critical universal constant it should be independent of heterostructure, LL or microscopic sample details, such as the mobility and carrier concentration.

The value of γ in two-dimensional electron systems (2DESs) has been the subject of sustained theoretical interest [4, 5]. The most numerically accurate value so far as pointed out in a recent review by Kramer *et al.* [4] is $\gamma = 2.33 \pm 0.03$ from a Monte Carlo calculation for chiral fermions [5].

Though the value of γ has been theoretically established and refined over many decades of analytical and numerical investigations, the experimentally determined value of γ has been in dispute [5-7] ever since the first reported experimental studies by Wei *et al.* [8].

Journal of Physics: Conference Series **456** (2013) 012007 doi:10.1088/1742-6596/456/1/012007

The aim of the work presented here is to clarify and illuminate the reasons for the apparent contradiction between the experimentally determined value of γ and its theoretically value which is expected to be $\gamma \approx 2.3$. We achieve this by considering electron transport within the bulk of the 2DES as being through networks of percolating clusters of electrons. We will show that using this percolation framework, one can describe the crossover between regions of universal quantum criticality, and regions where this quantum criticality is absent.

2. Classical and quantum critical percolation of plateau-to-plateau transitions

In 2DESs formed in heterostructures, there are two main sources of disorder which are both connected with impurities. The first is generated by the ionized dopants, which in some heterostructures are separated from the 2DES by a spacer layer. The second is caused by the residual ions located in the vicinity of the 2DES. These two sources of disorder both form a background random potential fluctuation, which in a linear screening regime can be effectively screened by small redistributions of the electron density [9]. Close to integer filling factors, however, screening becomes non-linear and the electron density exhibits strong inhomogeneities across the surface of the 2DES [10]. The background potential fluctuations tear apart the Fermi liquid into isolated clusters of electrons and, as a result, at sufficiently low temperatures the dissipative conductivity tends to zero.

According to the percolation model of the quantum Hall effect, in a 2DES with a slow varying background random potential, these isolated clusters consist of contours of closed equipotential lines. In a sufficiently strong magnetic field, electrons perform small oscillatory orbits while the centre of these orbits propagate along these equipotential lines [11]. Since the electrons are effectively localized within these clusters, the localization length can be defined as the correlation length (the width) of the electron clusters ξ_p. As the critical energy E_c found at the centre of a LL band is approached, percolation theory states that ξ_p increases according to a power law:

$$\xi_p(E) \propto |E - E_c|^{-\gamma_p}, \tag{2}$$

with critical universal exponent $\gamma_p = 4/3$ [11]. In other words, as the critical point or percolation threshold is approached the clusters grow in size, merging together, and then begin to form extended clusters of larger sizes, until at the percolation threshold enough clusters merge to form a percolating equipotential which extends across the entire system. This point determines the onset of dissipative conductivity.

Chalker and Codington modified this quasiclassical view of the QHE to describe universal quantum criticality in 2DESs, by considering the effects of quantum tunnelling and interference [12]. They showed that just before two neighbouring clusters coalesced, points of minimum potential between the two clusters are formed (termed saddle points). These points which occur between the outmost equipotential of the clusters, allow electrons to quantum tunnel between neighbouring clusters as shown in Figure 1. Quantum tunnelling therefore introduces a delocalization mechanism that, when considered within the quasiclassical model, modifies the critical exponent to the value expected in a quantum critical system of $\gamma = 7/3$ or ≈ 2.3 [12]. It is argued [12-14] that it is the inclusion of saddle point tunnelling that distinguishes the two universality classes of quantum critical and classical percolation. Tunnelling is thus crucial to the experimental observation of quantum criticality within a 2DES. A detailed treatment of the model can be found in a recent report by Kramer *et al.* [4]. We argue further, that by taking this model into account, one can explain many of the disagreements found between the quantum critical exponents obtained using the scaling theory of plateau-to-plateau transitions, and experimental observations.

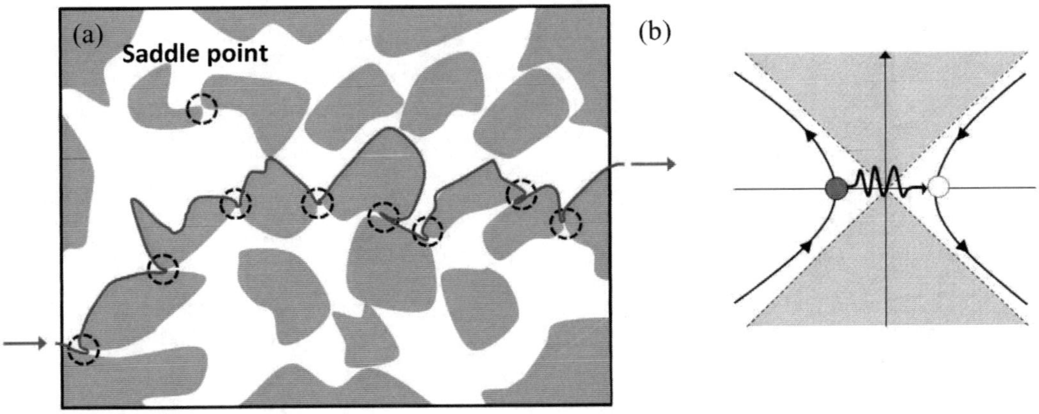

Figure 1. (a) A percolation view of dissipative transport in the bulk of a 2DES, electrons percolate through interlinked networks of saddle points. (b) Saddle point tunnelling of an electron between the outermost trajectories (equipotentials) of two isolated clusters, the shaded area denotes regions of high potential.

3. Quantum verses classical crossover

In this section we present experimental results that determine γ based on the change of the width of the conductivity peak of a LL band as a function of frequency. It was shown by Pruisken [15] that the width of the dissipative conductivity peak vanishes according to a temperature dependent power law of the form

$$\Delta B \propto T^{1/\gamma}. \quad (3)$$

The critical exponent γ can therefore be determined by measuring the temperature dependence of the dissipative conductivity according to Eq. (3).

Using a high resolution frequency technique [16] to investigate the quantum criticality of PPTs, we have been able to find a clear crossover from the expected quantum critical behaviour, to a classical behaviour in the critical exponent. Just as described above for temperature, the LL conductivity peak also has a similar frequency dependence [3],

$$\Delta B \propto f^{1/\gamma}. \quad (4)$$

The overwhelming majority of experimental investigations on the quantum criticality of PPTs have been conducted using temperature dependent experiments [6, 8, 17, 18]. Here, we use a frequency dependent experiment employing a coplanar waveguide which is formed on top of the 2DES and attached to a microwave network analyzer. Details of the experimental technique used in this work is reported in Ref. [16]. This experimental setup provides an extremely high degree of precision in varying the coherence length through the frequency parameter – a degree of precision that cannot be achieved in temperature dependent experiments owing to the limitations of cryogenic temperature controllers. The high resolution frequency dependent data we obtain provides a highly accurate description of the evolution of the width of a LL conductivity peak.

Figure 2 shows frequency dependent results obtained from a sample with carrier density $n_e \approx 2.89 \times 10^{11}/\text{cm}^2$ and mobility $\mu \approx 38 \times 10^4 \text{cm/Vs}$. All results reported here are obtained at a temperature of 35 mK where it was experimental verified that the temperature was sufficiently low for temperature dependent effects to be negligible and for frequency to be the dominant scaling parameter

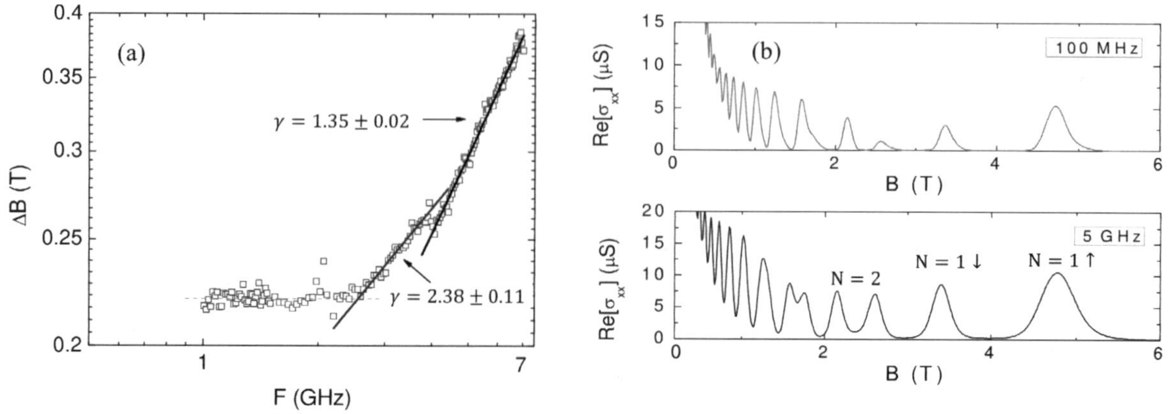

Figure 2. Plot (a) shows a high resolution frequency plot showing the quantum criticality observed in a 2DES. A crossover between a quantum and classical behaviour is observed as a function of frequency on the $N = 1 \downarrow$ LL. Plot (b) shows the magnetic field dependence of the real part of conductivity obtained from the high frequency response of the 2DES.

of PPTs. The plot in Figure 2(a) shows the dependence of the width of the conductivity peak of a LL band according Eq. (4). The plot consists of 3 parts: a saturated region, followed by a slope with a quantum critical gradient of $\gamma = 2.38$, then a crossover to a gradient in good agreement with the classically expected value of $\gamma \approx 1.33$. The result shows a clear crossover from a quantum critical behaviour at lower frequencies to a classical behaviour at higher frequencies. The significance of this result to the contradictions reported in the literature is discussed below.

4. Impact of disorder

It has long been argued that quantum criticality within a 2DES is affected by the correlation length of disorder found within the 2DES [5]. For example, it has been observed that quantum criticality is generally found within InGaAs/InP heterostructures [8], where decoherence is predominantly caused by short-range alloy-disorder scattering within the vicinity of the 2DES, whereas the critical exponent in GaAs/AlGaAs heterostructures in general deviates from quantum criticality [5, 6]. Scattering in GaAs/AlGaAs heterostructures are dominated by long-range disorder potentials originating from remote ions in the spacer layer. This relation between criticality and the correlation length of disorder, though apparently generally accepted [5, 8] and empirically observed [18], has yet to be properly explained.

Using the same frequency dependent experiment described above [16], we measure quantum criticality in two different types of 2DESs [16]. We used the device discussed in Figure 2(a) as a reference, while in a second device a modulation doped GaAs/AlGaAs heterostructure was grown with Al impurities introduced into the GaAs layer. This results in a $Al_xGa_{1-x}As/Al_{0.33}Ga_{0.67}As$ heterostructure, where x refers to the concentration of the introduced Al impurities, which in our sample was $x = 0.15\%$. It has been shown by Li *et al.* [19] that this level of impurity doping is sufficient to change what would otherwise have been a sample exhibiting long-range scattering into a sample with short-range alloy scattering. The resulting heterostructure had carrier density $n_e \approx 2.57 \times 10^{11}/\text{cm}^2$ and mobility $\mu \approx 30 \times 10^4 \text{cm/Vs}$.

The results are shown in Figure 3. Modifying the scattering mechanism in the 2DES changes the gradient of the slope as determined by Eq. (4) and results in a value of $\gamma = 2.38 \pm 0.34$ measured over two decades of frequency, in agreement with the expected quantum critical value. The value of γ determined in the long-range system deviates from the expected value of $\gamma \approx 2.3$, and tends towards a value of $\gamma = 1.67 \pm 0.14$, which is closer to the classically expected value of the critical exponent which is $\gamma \approx 1.3$.

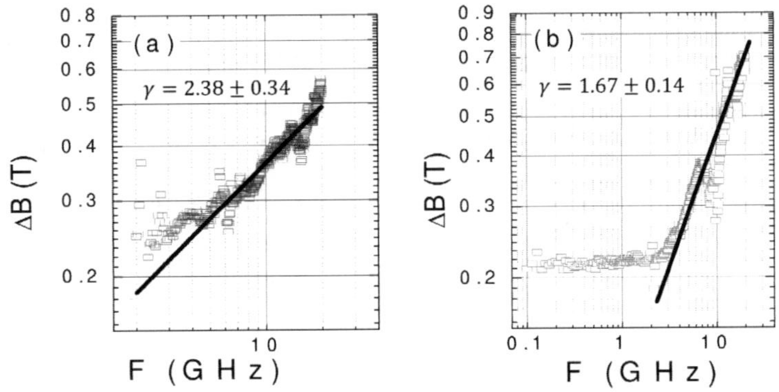

Figure 3. The two plots show results from the $N = 1 \downarrow$ LL from a short-range disordered and a long-range disordered sample. In plot (a), γ is found to be quantum critical when disorder was increased by introducing Al impurities in the vicinity of the 2DES. The plot in (b) shows the result from the reference long-range disordered GaAs/AlGaAs sample (with no added impurities), where the critical exponent tends towards a classical value. The result from this plot is obtained from the extended frequency range (up to 20 GHz) of the sample measured in Figure 2(a).

5. Discussion

We explain the results presented above using a percolation model which allows us to form a unifying description of PPTs within 2DESs. It is important to first define the three key length scales that form the basis of this model. These are the correlation length of the system ξ_p, which describes the typical width of electron clusters within the system, the localization length ξ, which is the length that defines the quantum coherent displacement of electrons within the system, and the phase coherent length l_φ, which represents the dephasing length of the system. ξ and ξ_p are Fermi energy or magnetic field dependent, as described by Eqs. (1) and (2) respectively. As the magnetic field approaches the critical point at the centre of a LL band ξ_p grows, describing larger cluster sizes close to the critical point. Similarly, ξ diverges, signifying longer extensions of quantum coherent displacements of electrons through interlinked networks of saddle points within the system. It is intuitive that ξ_p depends on the random background potential, or disorder within system. The greater the disorder, the more fragmented the electron clusters, and thus the smaller the cluster widths. l_φ, on the other hand, is a measure of inelastic scattering within the system, and is therefore dependent on the temperature or frequency of the system. An increase in frequency causes a decrease in l_φ and an increase in the inelastic scattering rate. We will henceforth describe increases in frequency of the 2DES simply as a decrease in $l_\varphi(f)$.

As discussed above, according to the percolation theory of the quantum Hall effect, saddle point tunnelling between adjacent electron clusters is the factor which distinguishes quantum critical behaviour from classical percolation. Without quantum tunnelling, ξ is limited to ξ_p, and thus such a system behaves like any classical fluid percolating through a random landscape. Quantum critical percolation can then be simply described as $\xi > \xi_p$, which is enabled by saddle point tunnelling.

Let us consider a system close to the percolation threshold (i.e. near the centre of the LL band) and at a temperature approaching absolute zero. Dissipative conductivity is possible through tunnelling of electrons between clusters since $\xi > \xi_p$. For any practical size of 2DES, l_φ will be very large compared to the sample size L_{2D}, and will theoretically extend beyond the boundaries of the sample. As l_φ is decreased (i.e. as the frequency is increased) dissipative conductivity within the system will remain unchanged as long as $l_\varphi > L_{2D}$ since on average an electron is able to propagate across the entire sample before it is scattered. This region can be seen as the saturated section observed in Figure 2(a).

Once $l_\varphi < L_{2D}$, dissipative conductivity will be influenced by scattering. Though tunnelling is still dominant, dissipative conduction no longer requires electrons to propagate through lengthy interlinked networks of saddle points, but rather electrons can be displace to remote clusters through phonon induced scattering in-between saddle point tunnelling events. The coexistence of tunnelling and scattering enhances dissipative conductivity which increases as l_φ is decreased and allows changes in conductivity to be experimentally observed as a function of l_φ (frequency). Since tunnelling still occurs, crucially $\xi > \xi_p$ and quantum critically is still persistent. This region is depicted in the low-frequency region in Figure 2(a), which shows a slope consistent with a quantum critical exponent of $\gamma \approx 2.38$. The relationship between the onset of inelastic scattering within the system and the sample size means that the saturation region can be eliminated or minimized with larger size samples, as shown experimentally by Koch *et al.* [20]. It is also noted that in this region ξ is now limited by l_φ as electrons will lose coherence beyond l_φ. The effect of a finite sample size is also evident in results reported by Li *et al.* [21], where the occurrence of the saturation region due the finite-size effect is discussed in some detail.

As l_φ (and as a result ξ) is further decreased the condition $l_\varphi, \xi < \xi_p$ will be reached. In other words, ξ will be smaller than the width of a cluster and an electron will on average be scattered before it is able to escape its cluster through a saddle point. The absence of tunnelling reduces the system to a purely classical fluid in a random potential exhibiting no quantum mechanical behaviour. Quantum critically is lost and thus the critical exponent in this region is determined to be classical. This can be seen in Figure 2(a) as a slope with $\gamma \approx 1.35$ in accordance with expectations of classical percolation [22].

The discussion above highlights the difficulties in the experimental observation of quantum criticality. When using relatively low resolution temperature dependent techniques, it is possible for the quantum critical region to be hidden, especially if the region is very narrow. The possibility of observing quantum critically can however be significantly increased by using samples of larger sizes, which reduces the range in frequency of the saturation region and brings about an earlier onset (in frequency) of the quantum critical region. On the other hand, the observation of quantum criticality can also be enhanced by delaying the quantum-classical crossover point, i.e. by pushing it to higher frequencies. As described above this crossover point occurs when the ξ is equal to the cluster size or $l_\varphi, \xi = \xi_p$. In a practical sense, this can be enhanced by having clusters of smaller sizes (reducing ξ_p), which as we have shown can be achieved by increasing the disorder within the system (as shown in Figure 3, where changing the disorder mechanism from long-range correlation to short-range alloy scattering increased the disorder within the system, and as a result the exponent changed to the expected quantum critical value). This may explain why quantum critically is easily observed systems with short-range scattering, such as those formed in InGaAs/InP heterostructures, but rarely observed in long-range GaAs/AlGaAs heterostructures. A convincing investigation of the dependence of the crossover point on disorder has been report by Li *et al.* [23] in further support of this theory.

Experimental investigations that do not observe this crossover run the risk of applying a fit that covers both the quantum and the classical region when determining γ, and will thus measure a value which is between the quantum and classical values, ie. $2.3 \geq \gamma \geq 1.3$.

Acknowledgements
We are grateful for the support of the EPSRC, the Royal Society, the Wolfson Foundation and the ERC programme, *NOTES*.

References

[1] K. v. Klitzing, G. Dorda, and M. Pepper, Physical Review Letters **45**, 494 (1980).
[2] E. Abrahams *et al.*, Physical Review Letters **42**, 673 (1979).
[3] S. L. Sondhi *et al.*, Reviews of Modern Physics **69**, 315 (1997).
[4] B. Kramer, T. Ohtsuki, and S. Kettemann, Physics Reports **417**, 211 (2005).
[5] B. Huckestein, Reviews of Modern Physics **67**, 357 (1995).
[6] S. Koch *et al.*, Physical Review B **43**, 6828 (1991).
[7] N. Q. Balaban, U. Meirav, and I. Bar-Joseph, Physical Review Letters **81**, 4967 (1998).
[8] H. P. Wei *et al.*, Physical Review Letters **61**, 1294 (1988).
[9] A. L. Efros, Solid State Communications **65**, 1281 (1988).
[10] A. L. Efros, Solid State Communications **67**, 1019 (1988).
[11] S. A. Trugman, Physical Review B **27**, 7539 (1983).
[12] J. Chalker, and P. Coddington, Journal of Physics C: Solid State Physics **21**, 2665 (1988).
[13] Y. Dubi, Y. Meir, and Y. Avishai, Physical Review Letters **94**, 156406 (2005).
[14] Y. Dubi, Y. Meir, and Y. Avishai, Physical Review B **71**, 125311 (2005).
[15] A. M. M. Pruisken, Physical Review Letters **61**, 1297 (1988).
[16] K. Saeed *et al.*, Physical Review B **84**, 155324 (2011).
[17] H. P. Wei *et al.*, Physical Review B **45**, 3926 (1992).
[18] W. Li *et al.*, Physical Review Letters **94**, 206807 (2005).
[19] W. Li *et al.*, Applied Physics Letters **83**, 2832 (2003).
[20] S. Koch *et al.*, Physical Review Letters **67**, 883 (1991).
[21] W. Li *et al.*, Physical Review Letters **102**, 216801 (2009).
[22] D. Stauffer, Physics Reports **54**, 1 (1979).
[23] W. Li *et al.*, Physical Review B **81**, 033305 (2010).

Journal of Physics: Conference Series **456** (2013) 012008 doi:10.1088/1742-6596/456/1/012008

One-dimensional model for the fractional quantum Hall effect

M I Dyakonov
Laboratoire Charles Coulomb, Université Montpellier II, CNRS, France

E-mail: michel.dyakonov@gmail.com

Abstract. A simple one-dimensional model is proposed, in which N spinless repulsively interacting fermions occupy $M>N$ degenerate states. It is argued that the energy spectrum and the wavefunctions of this system strongly resemble the spectrum and wavefunctions of 2D electrons in the lowest Landau level (the problem of the Fractional Quantum Hall Effect). In particular, Laughlin-type wavefunctions describe ground states at filling factors $\nu = N/M = 1/q$, q odd.. Within this model the complimentary wavefunction for $\nu=1-1/q$ is found explicitly, and extremely simple ground state wavefunctions for arbitrary odd-denominator filling factors are proposed.

1. Introduction

Thirty years after the discovery of the Fractional Quantum Hall Effect [1] the understanding of this extraordinary phenomenon is still only partial and a consistent theory is absent. After Laughlin's famous article [2], the intriguing concept of composite fermions advanced and developed by Jain [3, 4] became the generally accepted physical description: the fractional QHE is the integer QHE of new particles, moving in a reduced magnetic field. While what exactly a composite fermion is, remains unclear ("electron + $2p$ zeros of the many-electron wavefunction" is not really comprehensible), it was proved by numerical calculations that the proposed ground states for fractional fillings are virtually exact [4], and this justifies the rather bizarre construction of these states involving higher Landau levels which physically are irrelevant.

Currently, we are still in an awkward position: on the one hand many experimental facts support Jain's idea [3] of composite fermions moving in a reduced effective magnetic field, and this is the only physical description available. On the other hand [5], nobody has really shown theoretically, apart from what may be described as wishful thinking, the existence of composite fermions, as (quasi) free particles. So, the nature of the object whose properties are measured experimentally remains a mystery.

While Laughlin's idea [2] certainly gave a clue to understanding the FQHE, some important questions remain unanswered. One of them concerns the $\nu=2/3$ state (and, generally, all the $\nu=1-1/q$ states with odd q). Because of the electron-hole symmetry, this state, $\Psi_{2/3}$, can be regarded as the $\nu=1/3$ *hole* state described by the Laughlin wavefunction, $\Psi_{1/3}$, depending on the coordinates of N holes in a completely filled Landau level. The physical properties should be (and, in fact, are) quite similar to those at $\nu=1/3$.

Suppose, however, that one wants to have a look at the $\Psi_{2/3}$ function written in terms of $2N$ *electron* coordinates. To do this, one must (i) write down the Laughlin function $\Psi_{1/3}$ as a superposition of $N\times N$ determinants involving one-particle hole wavefunctions, and (ii) leaving the coefficients in the superposition unchanged, replace each determinant by its complimentary $2N\times 2N$ electron determinant. The resulting unwieldy expression, which nobody knows how to write down explicitly, will represent the ν=2/3 ground state, $\Psi_{2/3}$. It will go to zero at $z_i \to z_j$ as (z_i-z_j), just like any antisymmetric function, and we will hardly be able to understand why this function should minimize the interaction energy!

This shows the existence of wavefunctions that are as good as the Laughlin function, but which do not have higher order zeros when the electron coordinates coincide. In this sense the $\Psi_{2/3}$ function resembles the wavefunctions for other rational fillings, such as $\Psi_{2/5}$. It remains an open question, what are the relevant properties of these ground state wavefunctions, and this is a clear signal that our understanding is not complete.

2. One-dimensional model for FQHE

It was argued [5, 6] that the energy spectrum responsible for the FQHE arises whenever N spin polarized (or "spinless") fermions with appropriate repulsive interaction occupy $M>N$ initially degenerate states. Thus it makes sense to look for other, more simple, problems of this kind. In Ref. 6 a one-dimensional model with M degenerate states on a circle was proposed with $\psi_k(\varphi) = (2\pi)^{-1/2}\exp(ik\varphi)$, k=0..$M-1$. Alternatively, another basis of one-particle states $\Phi_n(\varphi)$ localized at $\varphi = 2\pi n/M$ can be used [6].

The Laughlin-like N-particle wavefunction $\Psi_{1/q}$ (q odd) has the familiar form:

$$\Psi_{1/q}(\varphi_1...\varphi_N) = A\prod_{1\le i<j\le N}\left[\exp(i\varphi_i)-\exp(i\varphi_j)\right]^q, \tag{1}$$

This can be rewritten in the $\Phi_n(\varphi)$ basis as

$$\Psi_{1/q}(\varphi_1...\varphi_N) = \sum_{(n)} C(n_1...n_N)\Phi_{n_1}(\varphi_1)...\Phi_{n_N}(\varphi_N), \tag{2}$$

where the coefficients C (providing just another representation of the same function) are given by:

$$C(n_1...n_N) = A\prod_{1\le i<j\le N}(\omega^{n_i} - \omega^{n_j})^q;\ \omega = \exp\left(\frac{2\pi i}{M}\right), \tag{3}$$

with a *known* normalization constant A [6] (it was calculated by Dyson [7, 8] a long time ago).

3. The wavefunctions for ν=1−1/q

Rather surprisingly, within my model it can be *proved* that the wavefunctions $\Psi_{1-1/q}$, corresponding to fillings 2/3, 4/5, 6/7… have coefficients C given by ***exactly the same*** expression (3), the powers of ω^n higher than $M-1$ being automatically reduced to the interval [0, $M-1$].

In summary, starting from the Laughlin-like wavefunction (1) for $\nu=1/q$ we find the following *exact result*. For both cases when either $qN=M$, $\nu=1/q$ (q odd), or $qN=(q-1)M$, $\nu=1-1/q$, the wavefunction in the Φ_n representation (which from now on we denote as Ψ) has the <u>same</u> form:

$$\Psi(n_1...n_N) = \prod_{1\le i<j\le N}(\omega^{n_i} - \omega^{n_j})^q;\ \omega = \exp(2\pi i/M), \tag{4}$$

the only difference being in the number of variables (n_i).

3. Conjecture

On the basis of this striking result, it is tempting to make the following *conjecture*: in our model *all* the ground states for *arbitrary* rational fillings $\nu=p/q$, or $qN=pM$ (q odd, p and q do not have common divisors) are described by the extremely simple and universal formula (4). As we have seen, this is true for $p=1$ and $p=q-1$. We now *suggest that this is true for all p.*

This conjecture is further supported by the fact that is self-consistent. Indeed, it can be proved that *if* it is true for some filling ν, it is also true for filling $1-\nu$.

Of course, only numerical calculations with small numbers of particles within the proposed model will show whether this conjecture is correct or not.

4. Relation between the model and the true FQHE problem

Such a relation is provided by some fascinating properties of the matrix $\omega^{nm}=\exp(2\pi inm/M)$. Consider a set of one-particle wavefunctions $\psi_n(m)$ of a discrete variable m, labelled by the index n:

$$\psi_n(m)=\frac{\omega^{nm}}{\sqrt{M}},\qquad n,m=0\ldots M-1. \tag{5}$$

This is analogous to plane waves, where both the momentum and the coordinate are discrete. Now introduce the operators X and Y, which shift by 1 the numbers m and n respectively, so that

$$\begin{aligned} X\psi_n(m)&=\omega^n\psi_n(m),\\ Y\psi_n(m)&=\psi_{n+1}(m). \end{aligned} \tag{6}$$

Obviously, $X^M=1$, $Y^M=1$, and $XY=\omega YX$. These are exactly the properties of the elementary *magnetic translations* in the x and y directions for an electron in magnetic field in the (twisted) torus geometry, if M is the number of magnetic fluxes through the surface of the torus (number of degenerate states in a given Landau level).

While Eqs. (5, 6) provide a direct link to the true FQHE problem on a torus, it still remains to be seen whether the true 2D problem can be in some sense reduced to the 1D model considered here.

5.References

[1] Tsui D C, Stormer H L, and Gossard A C 1982, *Phys. Rev. Lett.* **48**, 1559
[2] Laughlin R 1983, *Phys. Rev. Lett.* **50** 1395
[3] Jain J K 1989, *Phys. Rev. Lett.* **63** 199
[4] Jain J K 2007, *Composite Fermions* (New York, Cambridge University Press)
[5] Dyakonov M I 2002, Twenty years since the discovery of the fractional Hall effect, *Recent Trends in Theory of Physical Phenomena in High Magnetic Fields, ed E Schöll Wiley* (*Preprint* arXiv:cond-mat/0209206)
[6] Dyakonov M I 2002, *J. Phys. IV France* **12**, Pr9-373; (*Preprint* arXiv:1203.4838)
[7] Dyson F J 1902, *J. Math. Phys.* **3** 140
[8] Mehta M L 1991, *Random Matrices* (London, Academic press)

HMF-20 IOP Publishing
Journal of Physics: Conference Series **456** (2013) 012009 doi:10.1088/1742-6596/456/1/012009

Exciton Transport in a Bilayer Quantum Hall Superfluid

J.P. Eisenstein[1], A.D.K. Finck[1], D. Nandi[1], L.N. Pfeiffer[2], and K.W. West[2]

[1]Department of Physics, California Institute of Technology, Pasadena, CA 91125
[2]Department of Electrical Engineering, Princeton University, Princeton, NJ 08544

Abstract. Bilayer quantum Hall systems at $\nu_T = 1$ support an excitonic ground state. In addition to the usual charged quasiparticles, this system possesses a condensate degree of freedom: exciton transport. Detection of this neutral transport mode is facilitated by the use of the Corbino multiply-connected geometry in which charge transport is suppressed. We here summarize our recent experiments on Corbino devices which directly demonstrate exciton transport across the bulk of the incompressible $\nu_T = 1$ quantum Hall state.

Quantum Hall systems are topological insulators. The bulk of the 2D electron system is an electrical insulator with an energy gap due to either single particle or many-body effects. At the boundary of the 2D system there are chiral conducting edge states. The number of such edge states is a topological invariant which determines the Hall resistance of the system. The quantized Hall state which appears in closely-spaced bilayer 2D systems at total Landau level filling $\nu_T = 1$ is no exception to this rule. In this case Coulomb interactions between the layers are comparable to interactions within each layer. As a result, at $\nu_T = 1$, there is spontaneous interlayer phase coherence; electrons are coherently spread between both layers even in the limit of zero single particle tunneling. A single edge state circulates at the boundary and the Hall resistance is quantized at $\rho_{xy} = h/e^2$. Charge transport via this edge state consists of parallel electrical currents in the individual layers.

In addition to ordinary parallel charge transport, the $\nu_T = 1$ bilayer quantum Hall system is unique in possessing an additional, entirely different mode of transport. This new mode occurs within the system condensate rather than its charged excitations above the energy gap. The condensate of the $\nu_T = 1$ quantum Hall system may be viewed as a coherent ensemble of interlayer excitons [1]. The excitons are analogous to the Cooper pairs in a superconductor, with the important distinction that they are charge neutral. Being neutral the excitons are not confined to the sample boundaries and should be able to move throughout the bulk of the bilayer system. Exciton transport is expected to be dissipationless [2], although this has not yet been completely confirmed. Unlike transport of the charged excitations of the system, which amounts to parallel current flow in the two layers, exciton transport is equivalent to oppositely directed, or *counterflowing* electrical currents in the layers.

Detecting exciton transport at $\nu_T = 1$ is complicated by several factors. First, there is the obvious problem of how to control and detect the motion of neutral objects using only electrical means. Fortunately, this difficulty can be overcome if separate electrical contacts to the two layers are available [3]. With such contacts it is possible to force oppositely directed currents through the two 2D electron layers and thereby potentially gain access to the condensate degree of freedom. Second, depending on circumstances, condensate transport may have to compete with ordinary charged quasiparticle transport. Indeed, early counterflow experiments were performed using Hall bar geometries in which all electrical contacts were placed along the single outside edge of the device [4, 5, 6]. In this case the highly conducting $\nu_T = 1$ edge state is available for charge transport. Furthermore, with contacts only on a single edge it was clearly not possible to assert that exciton transport was taking place throughout the bulk of the system.

As first recognized by Tiemann *et al.* [7, 8] and by Su and MacDonald [9], the multiply-connected Corbino device geometry (roughly an annulus) offers a significant advantage over simply-connected Hall bars for the direct detection of bulk exciton transport. In this case the edge states on the two boundaries of the annulus are separated by the bulk of the 2D system. At low temperatures (and sufficiently small bias voltage) charge transport between these edges is heavily suppressed by the quantum Hall energy gap. At $\nu_T = 1$ however, the bulk of the 2D system remains essentially transparent to exciton transport. We thus arrive at the remarkable conclusion that even though charge transport via parallel currents in the two layers across the bulk is suppressed, oppositely directed layer currents should be able to move freely by making use of the condensate degree of freedom.

In a recent experiment, Finck *et al.* [10] presented convincing evidence for exciton transport across the bulk of a Corbino ring at $\nu_T = 1$. With four ohmic contacts on the outside rim of the annulus (two on each 2D layer) and two more on the inner rim (one on each layer) a variety of transport measurements could be performed. At low temperatures and sufficiently small effective interlayer separation ($d/\ell \sim 1.5$, with d the physical separation of the layers and ℓ the magnetic length) the $\nu_T = 1$ bilayer quantum Hall effect was robust in their sample. This was clearly evident from, among other signatures, the exponentially suppressed parallel charge conductivity $\sigma_{xx}^{||}$ across the annulus.

In order to search for bulk exciton or counterflow conductivity a bias voltage V was applied between top and bottom layer contacts (the source and the drain) on the outer rim of the annulus and the resulting current I_1 was recorded. (The two additional contacts on the outer rim were used to provide a 4-terminal measurement of the interlayer voltage difference, V_i.) On the inner rim the remaining top and bottom layer contacts could be connected together via a shunt resistor R_s or left disconnected. With the shunt disconnected this is a simple interlayer tunneling configuration. With the shunt in place additional current might flow from source to drain if counterflowing layer currents cross the otherwise insulating bulk of the 2D system and pass through the shunt. The current I_2 passing through the shunt is easily determined from the measured voltage drop across R_s. Figure 1a schematically illustrates the configuration with the shunt connected.

With the magnetic field applied perpendicular to the 2D planes and the shunt disconnected, Finck *et al.* observed the well-known Josephson-like tunneling response of the bilayer system [11]. As the bias voltage V was increased from zero the current I_1 tunneling between the layers steadily increased up to a critical current of about $I_c \sim 1.5$ nA. Simultaneous measurements of the interlayer voltage V_i showed that it remained very close to zero for currents below the critical current. When the bias voltage was increased further, the interlayer voltage abruptly jumped to a finite value and the current dropped below the critical value. As observed previously [11, 8], these two states strongly resemble the superconducting and resistive states of a conventional Josephson junction.

Connecting the shunt on the inner rim had no effect on the above results until the system was driven into the resistive state. For small bias voltages, where $I_1 < I_c$ and $V_i \sim 0$, no current was observed to flow through the shunt. In this situation it would appear that all of the bias current was simply tunneling between the layers and not counterflowing across the bulk to the shunt. At higher bias, in the resistive state, current does begin to flow through the shunt.

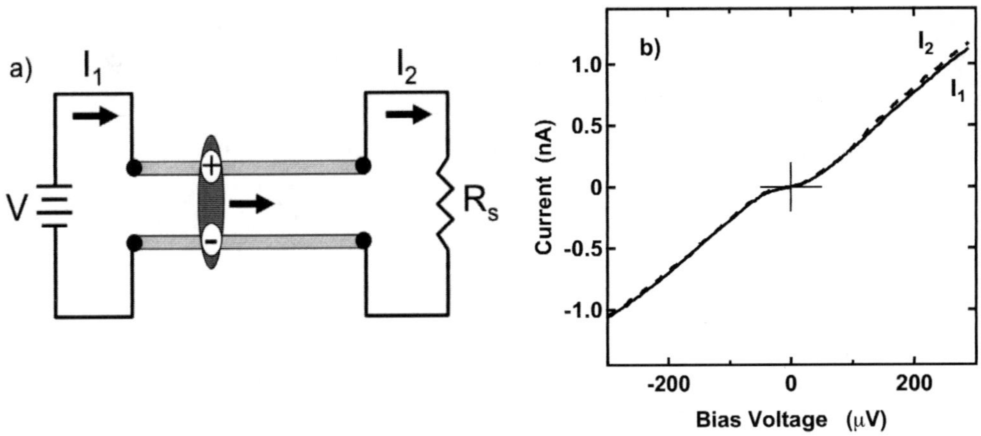

Figure 1. a) Schematic illustration of the circuit geometry used by Finck *et al.* [10] to detect exciton transport at $\nu_T = 1$. The bilayer electron system is shown in cross-section, with the solid dots on the left being on the outer rim of the Corbino annulus and the solid dots on the right being on the inner rim. In the ideal situation shown, only exciton transport is occurring; there is no interlayer tunneling and no charged quasiparticle transport. In these ideal conditions, the currents I_1 and I_2 are precisely equal. Additional resistances (due to contacts, etc.) have been omitted for clarity. b) Measured bias and shunt currents at $\nu_T = 1$, with $d/\ell = 1.5$ and $T \sim 20$ mK and the sample tilted to suppress coherent tunneling. The near perfect equality of I_1 and I_2 demonstrates the dominance of exciton transport across the insulating bulk of the 2D system.

Expecting these results, Finck *et al.* then rotated the sample relative to the magnetic field so that a significant in-plane component $B_{||} \sim 1$ T was present in addition to the perpendicular component $B_\perp \sim 2$ T used to establish $\nu_T = 1$. This drastically reduces the critical current for coherent Josephson-like tunneling [12] but otherwise has little effect on the $\nu_T = 1$ QHE. Indeed, with the shunt disconnected very little current was observed to flow in response to the bias voltage V_b. In contrast, when the interlayer shunt was connected large currents flowed and careful measurements revealed that the shunt current I_s almost perfectly matched the source-drain current I. In this case essentially all of the current is counterflowing across the bulk of the Corbino ring. This result constitutes direct proof of exciton transport across the insulating bulk of the $\nu_T = 1$ bilayer quantum Hall system. Further measurements showed that the magnitude of this excitonic current was overwhelmingly dominated by the various extrinsic resistances in the circuit; any dissipation in the exciton transport itself was undetectable.

In the above measurements the counterflowing currents are taking place at finite interlayer voltage, V_i. This voltage is inevitable in the circuit arrangement Finck *et al.* employed, if only because of the non-zero resistance of the shunt connection. Hence, the exciton transport was accompanied by a time-dependent condensate phase. We stress that this does not imply that dissipation is occurring in the exciton channel. So long as the interlayer voltage is uniform across the device (and the measurements suggested that this was the case), the order parameter winds at a constant rate with no phase slips being generated.

An alternative method for inducing exciton transport across the Corbino ring is to use a Coulomb drag configuration [13, 9]. In this case the source and drain contacts for the bias circuit are on the same 2D layer (the drive layer) but on opposite rims of the annulus. Meanwhile, inner and outer rim contacts on the opposite layer (the drag layer) are connected together through an external resistor, R_D. This arrangement is schematically illustrated in Fig. 2a. In the absence of strong interlayer correlations, driving a current I_1 through the drive layer will induce only a very tiny current I_2 in the drag layer. In contrast, at $\nu_T = 1$ interlayer correlations are very strong and exciton transport within the condensate offers an alternative: the current I_2 in the drag layer will precisely match the current I_1 in the drive layer, but be oppositely directed, as excitons flow uniformly across the Corbino ring. Furthermore, in the absence of dissipation in the exciton channel, the magnitude of these currents will determined only by the bias voltage V_b and the sum R_{tot} of all extrinsic resistances in the drive and drag loops. (We include in R_{tot} quantum Hall contact resistances of order h/e^2; see Su and MacDonald [9].) Unlike the experiment of Finck *et al.*, exciton transport in this case is generated without any explicit electrical connection between the two layers.

Recent experiments by Nandi *et al.* [14] have verified this scenario of exciton-mediated Coulomb drag. Figure 2b illustrates their basic findings, with the drive and drag currents, I_1 and I_2, plotted versus bias voltage V. These data were obtained at $\nu_T = 1$ and $T \sim 20$ mK and $d/\ell = 1.5$. At this effective layer separation the excitonic $\nu_T = 1$ QHE state is well-developed and measurements of the parallel charge transport conductivity, $\sigma_{xx}^{||}$, show it to be heavily suppressed. As before, the sample has been tilted relative to the applied magnetic field in order to suppress coherent interlayer tunneling. As the figure shows, for bias voltages below about 150 μV, equal currents flow across the bulk of the Corbino ring in the drive and drag loops; i.e. $I_1 \approx I_2$. In this situation, Coulomb drag is "perfect" and reflects a uniform flow of excitons across the bulk of the 2D sample.

We emphasize that when the drag layer is not closed upon itself via the resistor R_D virtually no current flows in *either* layer in response to small bias voltages. Indeed, were significant current to flow in the drive layer without a compensating opposite current in the drag layer, it

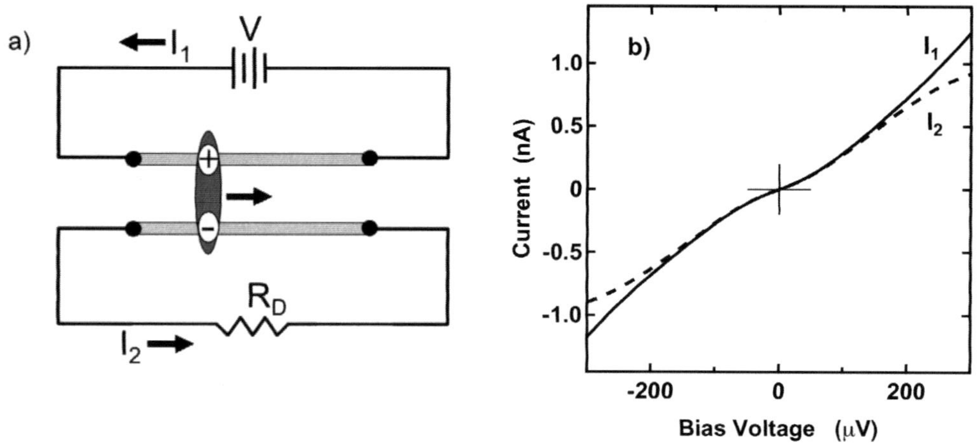

Figure 2. a) Schematic illustration of the circuit geometry for Corbino Coulomb drag measurements used by Nandi *et al.* [14]. b) Drive and drag currents, I_1 and I_2, at $\nu_T = 1$ and $d/\ell = 1.5$ and $T \sim 20$ mK. Under these conditions the excitonic phase is robust. At small bias voltages equal, but oppositely directed, currents flow in the two layers: $I_1 \approx I_2$. At higher bias voltages the two currents separate. This is due to enhanced charge transport across the bulk of the Corbino annulus. The sample has again been tilted relative to the magnetic field in order to suppress interlayer tunneling in the excitonic phase.

would conflict with the observation that $\sigma_{xx}^{||}$ is extremely small.

Above about $V \approx 150$ μV the drive current begins to exceed the drag current. In this regime net charge is being transported across the Corbino ring. Detailed measurements [14] show that $\sigma_{xx}^{||}$ grows significantly as the $\nu_T = 1$ QHE begins to "breakdown" with increasing bias voltage V. In fact, we find it possible to convincingly simulate the behavior of I_1 and I_2 with increasing V via independent measurements of $\sigma_{xx}^{||}$ in the non-linear regime. A similar analysis also explains the deviations from perfect Coulomb drag which arise from the temperature dependence of $\sigma_{xx}^{||}$.

Finally, we note that as expected, when the effective layer separation is increased to $d/\ell = 2.3$ and interlayer coherence is lost, essentially no current flows in the drag layer even though current is readily transported across the drive layer. In this case the bilayer system consists of essentially two independent 2D layers each at $\nu = 1/2$.

In summary, exciton transport across the insulating bulk of the $\nu_T = 1$ bilayer QHE state has been observed in two related, but distinct, transport experiments in Corbino geometry. While these experiments are so far consistent with dissipationless transport of the exciton ensemble, they do not place stringent upper bounds on it. Future multi-terminal measurements will address this important question.

This work was supported by the NSF under grant DMR-1003080.

References

[1] Eisenstein J P and MacDonald A H 2004 Bose-Einstein condensation of excitons in bilayer electron systems *Nature* **432** 691-694 and the references cited therein

[2] MacDonald A H and Girvin S M 1997 *Perspectives in Quantum Hall Effects* ed S Das Sarma and A Pinczuk (Wiley, New York) chapter 5 pp 161-224

[3] Eisenstein J P, Pfeiffer L N and West K W 1990 Independently contacted two-dimensional electron systems in double quantum wells *Appl. Phys. Lett.* **57** 2324-2326

[4] Kellogg M, Eisenstein J P, Pfeiffer L N and West K W 2004 Vanishing Hall resistance at high magnetic field in a double-layer two-dimensional electron system *Phys. Rev. Lett.* **93** 036801

[5] Tutuc E, Shayegan M and Huse D A 2004 Counterflow measurements in strongly correlated GaAs hole bilayers: evidence for electron-hole pairing. *Phys. Rev. Lett.* **93** 036802

[6] Wiersma R *et al.* 2004 Activated transport in the separate layers that form the $\nu_T = 1$ exciton condensate *Phys. Rev. Lett.* **93** 266805

[7] Tiemann L *et al.* 2008 Exciton condensate at a total filling factor of one in Corbino two-dimensional electron bilayers *Phys. Rev. B* **77** 033306

[8] Tiemann L, Dietsche W, Hauser M and von Klitzing K 2008 Critical tunneling currents in the regime of bilayer excitons *New J. of Phys.* **10** 045018

[9] Su Jung-Jung and MacDonald A H 2008 How to make a bilayer exciton condensate flow *Nature Physics* **4** 799-802

[10] Finck A D K, Eisenstein J P, Pfeiffer L N and West K W 2011 Exciton transport and Andreev reflection in a bilayer quantum Hall system *Phys. Rev. Lett.* **106** 236807

[11] Spielman I B, Eisenstein J P, Pfeiffer L N and West K W 2000 Resonantly enhanced tunneling in a double layer quantum Hall ferromagnet Phys. Rev. Lett. **84** 5808-5811

[12] Spielman I B, Eisenstein J P, Pfeiffer L N and West K W 2001 Observation of a linearly dispersing Goldstone mode in a quantum Hall ferromagnet *Phys. Rev. Lett.* **87** 036803

[13] Kogan V G and Travger B A 1971 *Physics of n-p Junctions and Semiconducting Devices* ed S M Ryvkin and Yu V Shmartsev (Consultants Bureau, New York) chapter 10 pp 39-45

[14] Nandi D, Finck A D K, Eisenstein J P, Pfeiffer L N and West K W 2012 Exciton condensation and perfect Coulomb drag *Nature* **488** 481-484

Spin Phase Transition Studies to Probe Spin Dynamics in Quantum Hall System

M. H. Fauzi[1], **S. Watanabe**[2], **and Y. Hirayama**[1,2]
[1] Department of Physics, Tohoku University, Sendai 980-8578, Japan
[2] ERATO Nuclear Spin Electronic Project, Sendai 980-8578, Japan
E-mail: fauzi@m.tohoku.ac.jp and hirayama@m.tohoku.ac.jp

Abstract. We investigate nuclear spin relaxation process in a canted antiferromagnetic state of the bilayer $\nu_{\text{tot}} = 2$ by measuring evolution of the spin phase transition peak of the filling $\nu = 2/3$. We observe a sudden change in the nuclear spin polarization when the nuclear spin is exposed to the canted antiferromagnetic state.

1. Introduction

The bilayer two-dimensional electron systems at total filling factor $\nu_{\text{tot}} = 2$ exhibits low-frequency electron spin fluctuation dynamics from which produce a gapless excitation mode or Goldstone mode. The produced Goldstone mode is due to the interplay between Coulomb energy, Zeeman energy, and inter-tunneling energy gap[1, 2]. An experimental signature of rapid nuclear spin relaxation rate T_1^{-1} probed by electrically-detected NMR has confirmed this notion[3, 4]. This rapid relaxation rate is proportional to the transverse components of electron-spin fluctuations at the Larmor frequency of nuclei

In spite of these interesting experimental proofs, the detection signal was obtained by measuring a change amplitude in resistance at a transition point of the filling $\nu = 2/3$[3, 4, 5, 6, 7]. Since the nuclear spin polarization is not only manifested in the resistance change, thus it only reveals partial information of the true nature of the physical system in questions. In contrast to that approach, we are able to map a minute change in the nuclear spin polarization distribution in a GaAs quantum well by measuring the profile of the transition peak. In this paper, we probe spin dynamics of bilayer GaAs quantum wells by using full record of the spin phase transition peak.

2. Sample and Experimental Methods

Experiments were carried out on two identical GaAs quantum wells sample with 20-nm in width and separated by a 2.2-nm insulation barrier. The sample was patterned into 30×100 μm Hall bar geometry. Ohmic contact pads were made with AuGeNi alloys annealed at a temperature of 420^0 C. The low temperature mobility is $\mu \sim 1.3 \times 10^6$ cm^{-2}/Vs at a carrier density of $n \sim 1.0 \times 10^{11}$ cm^{-2}. The electron density in the front- and the back-layer could be tuned from depletion to 3.5×10^{11} cm^{-2} by means of applying front and back gate bias voltages, respectively.

We used the spin phase transition peak of the filling $\nu = 2/3$ to induce and detect nuclear polarization. At the transition, two different electronic domains coexist namely spin-unpolarized

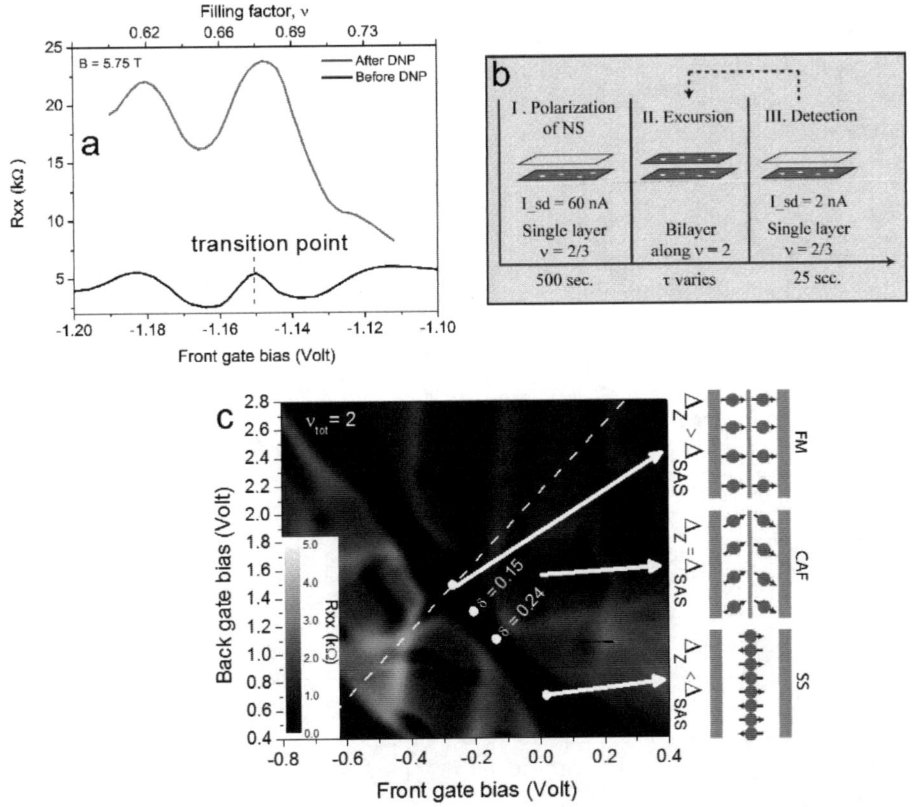

Figure 1. (a) Dynamic nuclear polarization at the transition of the filling factor $\nu = 2/3$. The nuclear spin polarization is manifested in the enhancement of the transition peak after applying a large current $\sim$ 60 nA for 500 seconds (red curve). (b) Schematic illustration of the timing sequence of our experiment. (c) Two dimensional plot of R_{xx} highlighted at the total filling factor $\nu_{tot} = 2$ as a function of a back- and front-gate bias voltage measured at 5.75 T and 50 mK. The corresponding electron spin states are shown at the right panel.

and spin-polarized domains[8]. Flowing a relatively high current across two domains results in electron spin flips at the domain wall boundaries which in turn flops the nuclear spin on the site to preserve total angular momentum. Figure 1a displays experimental realization of current induced nuclear polarization at the transition point. The amplitude of the transition peak increased dramatically and the width became very broad after applying a relatively high source-drain current $\sim$ 60 nA for 500 seconds (see red curve). The broadening of the transition peak reflects the spatial variation of the hyperfine field B_N.

The timing sequence of our experiment is shown in figure 1b. First we set the filling factor to a single layer $\nu_f = 0$ and $\nu_b = 2/3$ and applied a high current $I \sim 60$ nA for 500 seconds to dynamically polarize the nuclear spin ensemble. The transition developed into a huge peak and was brought out of equilibrium as depicted in figure 1a. Second, we set the electronic system to the bilayer $\nu_{tot} = 2$ with $\delta = 0.15,\ 0.24$ (see figure 1c) by changing the gate bias voltage. The polarized nuclear spin interacts with the canted antiferromagnetic state in this stage. We interrupted the process by temporary restoring the filling factor to a single layer $\nu_b = 2/3$ for a given interval of time "keeping time" and the remaining nuclear polarization was recorded by scanning the transition peak. The remaining polarization is a measure of nuclear spin relaxation

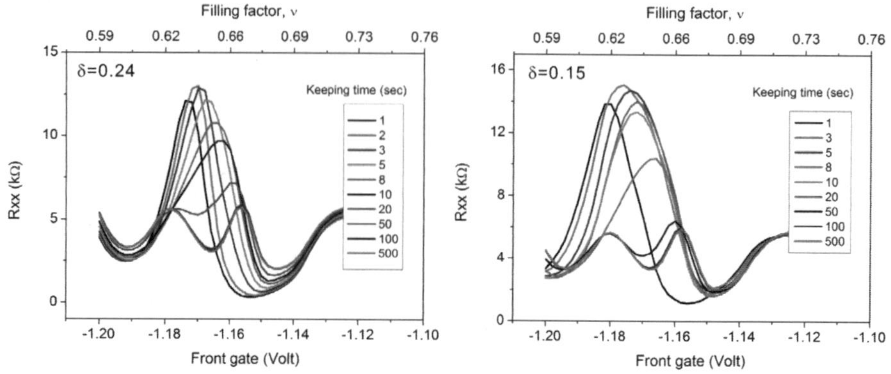

Figure 2. Evolution of the transition curve exposed to the CAF state with $\delta = 0.24$ (right) and $\delta = 0.15$ (left) at a temperature of 50 mK.

rate due to interaction with the bilayer electrons.

3. Experimental Results and Discussions

A two-dimensional map of the longitudinal resistance R_{xx} highlighted along $\nu_{tot} = 2$ as a function of the front- and the back-gate voltages is depicted in Fig. 1c. The magnetic field of 5.75 T was applied perpendicularly to the 2DES. The phase transition between different magnetic phases was driven by altering the charge density imbalance $\delta = n_f - n_b$ while the total density $n_{tot} = n_f + n_b$ was kept constant. The quantum Hall effect (QHE) was smoothly developed and preserved from the point of no charge imbalance $\delta = 0$, where the system possessed the ferromagnetic (FM) state, to the very large charge imbalance $\delta >> 0$, where the ground state altered to the spin single (SS) state as the tunneling gap Δ_{SAS} overwhelmed the Zeeman energy Δ_Z. In contrast, a single particle energy picture predicts no quantum Hall would be observed due to the expected level-crossing taken place between the crossover from FM to SS states at $\Delta_{SAS} \approx \Delta_Z$. Such apparent continuously connected QHE is a hallmark of a new class of magnetization namely the canted antiferromagnetic (CAF) state.

Unexpected nuclear spin relaxation behaviour was observed in the CAF state since the initial characteristics of DNP completely disappear as depicted in figure 2 for two different δ values. The very narrow transition curve shifted to a smaller filling factor $\nu_{tr} \rightarrow 0.62$ after the first second of exposure. As the time passed, the transition curve progressively shifted back to a higher filling factor (i.e equilibrium position at $\nu_{tr} \approx 0.66$) with the curve width remaining almost constantly narrow and extremely small nuclear spin relaxation time. The observed shift to a smaller filling factor with the narrow curve width clearly indicated a sudden change in nuclear spin polarization distribution. The underlying mechanism of this sudden change is not clear yet. This change could signal a sudden burst of the hyperfine field to the electron spins with the Goldstone mode. Alternatively, this change could possibly be due to rapid in-plane nuclear spin diffusion mediated by the Goldstone mode. Future studies will be necessary for both experiment and theory.

The long-ranged nature of the CAF state is expected to break down at high temperature due to vortex-pairs unbinding in the XY plane. At this stage the system undergoes a Kosterlitz–Thouless (KT) transition. U(1) symmetry is restored and the system becomes paramagnetic with no Goldstone mode present. We observed evidence that upon changing the temperature from 50 to 200 mK, the characteristic associated with the CAF state at $\delta = 0.24$ disappeared even though the relaxation time was quite fast(see Fig. 3). It is clearly apparent, particularly from

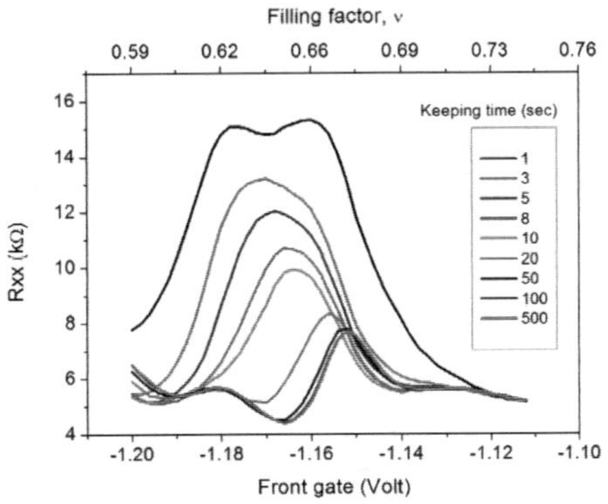

Figure 3. The evolution of the transition curve exposed to the CAF state with $\delta = 0.24$ at a temperature of 200 mK.

the first second of exposure, The transion curve became very broad making a strong contrast with those obtained in Fig. 2. This could be interpreted as a possibly straightforward signature of a transition between CAF and SS states. The electron spin ordering was destroyed, resulting in incoherent coupling between the nuclear and electron spin. We shall note that the shift of the transition peak equilibrium position to a higher filling factor as the temperature increased was due to the decrease of thermal nuclear polarisation.

4. Conclusions

In conclusion, by measuring the evolution of the transition peak of the filling factor $\nu = 2/3$ we found the sudden change in the nuclear spin polarization when the polarized nuclear spins were exposed to the bilayer CAF state. By ramping up the temperature up to 200 mK, we possibly observed KT transition at $\delta = 0.24$.

5. Acknowledgement

We thank K. Muraki of NTT Basic Research Laboratory for providing us high quality bilayer wafer.

References

[1] Das Sarma, S., Sachdev, S., & Zheng, L. *Phys. Rev. Lett.* **79**, 917 (1997).
[2] Zheng, L., Radtke, R. J. & Das Sarma, S. *Phys. Rev. Lett.* **78**, 2453 (1997).
[3] Kumada, N., Muraki, K., & Hirayama. Y. *Science* **313**, 329 (2006).
[4] Kumada, N., Muraki, K., & Hirayama. Y. *Phys. Rev. Lett.* **99**, 076805 (2007).
[5] K. Hashimoto, K. Muraki, T. Saku, and Y. Hirayama. *Phys. Rev. Lett.* **88**, 176601 (2002).
[6] Kobayashi, T., Kumada, N., Ota, T. Sasaki, S., & Hirayama, Y. *Phys. Rev. Lett.* **107**, 126807 (2011).
[7] Tiemann. L., Gamez. G., Kumada. N, & Muraki. K. *Science* **335**, 828 (2012).
[8] Stern, O. *et al. Phys. Rev. B* **70**, 0753187 (2004).

Terahertz Detectors based on graphene

Fathi Gouider[1], Majdi Salman[1,2], Markus Göthlich[1], Hennrik Schmidt[2,3], Franz-J Ahlers[4], Rolf Haug[3] and Georg Nachtwei[1]

[1] Institut für Angewandte Physik, Technische Universität Braunschweig, D-38106, Germany
[2] NTH Nano School for Contacts in Nanosystems, Germany
[3] Institut für Festkörperphysik, Universität Hannover, Hannover, D-30167, Germany
[4] Physikalisch-Technische Bundesanstalt, Braunschweig, D-38116, Germany

E-mail: `f.gouider@tu-bs.de`

Abstract. In this study we present magnetotransport an magnetooptical data obtained in the magnetic field range $0\mathrm{T} < B < 7\mathrm{T}$ at detectors patterned in Corbino geometry on epitaxial graphene wafer using a Ge detector. We observed the cyclotron resonance of charge carriers in these wafers by measurement of the transmission of THz wafes through the unpatterned squares (about $4\times4\mathrm{mm}^2$) of the wafers as a function of the magnetic field B applied perpendicular to the wafer. Further, we performed measurements of the photocunductivity of graphene-based devices shaped in Corbino geometry, induced by terahertz (THz) radiation generated by a p-Ge laser (emitting in the energy range $7.5\mathrm{meV} \leq E_{ph} \leq 11\mathrm{meV}$). Our photoconductivity measurement imply that graphene devices are suitable for the detection of terahertz radiation.

1. Introduction

THz radiation is very promising to open new opportunities in many fields such as for example the material research, medical technology and basic astro- and atmosphere physics. Despite these broad application possibilities very few of these opportunities have been exploited up to date. This is also due to the existence of the so called "terahertz gap" that describes the lack of both THz sources and detectors which are both easily and economicaly to operate. One possibility for the realization of a THz detector is by exploiting the Landau quantization in quantum Hall detectors. To achieve a quantization of 10 meV (corresponding to a wavelength of 124 μm a magnetic field strength of $B \approx 5\mathrm{T}$ is needed in GaAs [1] and of $B \approx 2\mathrm{T}$ in HgTe [2]. For graphene on the other hand—fabricated for the first time in 2004 by mechanical exfoliation[3]—calculations predict that a Landau-Level (LL) split of $\Delta E_L \approx 10\mathrm{meV}$ can be achieved with magnetic fields smaller than 0.5 T [4].

2. Experimental

We investigated two graphene samples (one exfoliated and one epitaxial). The samples were mounted together with a laser system for excitation in a Helium bath cryostat.

The laser system used is based on a p-Ge laser that can be tuned to emit in the energy range between 7.5–11meV (corresponding to a range in wafelength of 120–180μm). The laser is operated in pulsed mode with a pulse duration of 0.3–50μs and a repetition rate of 1Hz with a peak power of 1 W. Further details of the laser system can be obtained for example from Ref. [5].

One sample investigated was prepared by mechanical exfoliation from graphite. This sample was then contacted with four gold contacts and used for magnetotransport measurements (without laser excitation). These measurements are done at a temperature of $T \approx 2.8$K.

The other sample was epitaxially grown on SiC. This sample was then cleaned, etched with Ar/O-Plasma and contacted with a Corbino structure made of Ti/Au. It was then used for photoresponse measurements at a temperature of $T = 4$K.

3. Results and Discussion

3.1. Exfoliated graphene

In figure 1 we have plotted the resistence of the graphene sample as a function of the backgate voltage for 6 different magnetic fields. We can see the Dirac point at the position of the maximal resistance at around $V_{bg} = 5$V in the following called the 0th maximum. In the inset we plotted the difference of the backgate voltages of the 0th and 2nd maximum to the right. According to [6, 7] the energy of a Landau-Level M_n in graphene is given by

$$M_n = \text{sgn}(n)\sqrt{\Delta^2 + 2|n|\hbar v_F^2 eB}\ , \tag{1}$$

with Δ an energy gap, v_F the Fermi velocity, B the magnetic field and n the quantum number of the LL. In the inset we see that the measured voltage differences fit quite well to the square root dependence in (1) assuming $\Delta = 0$.

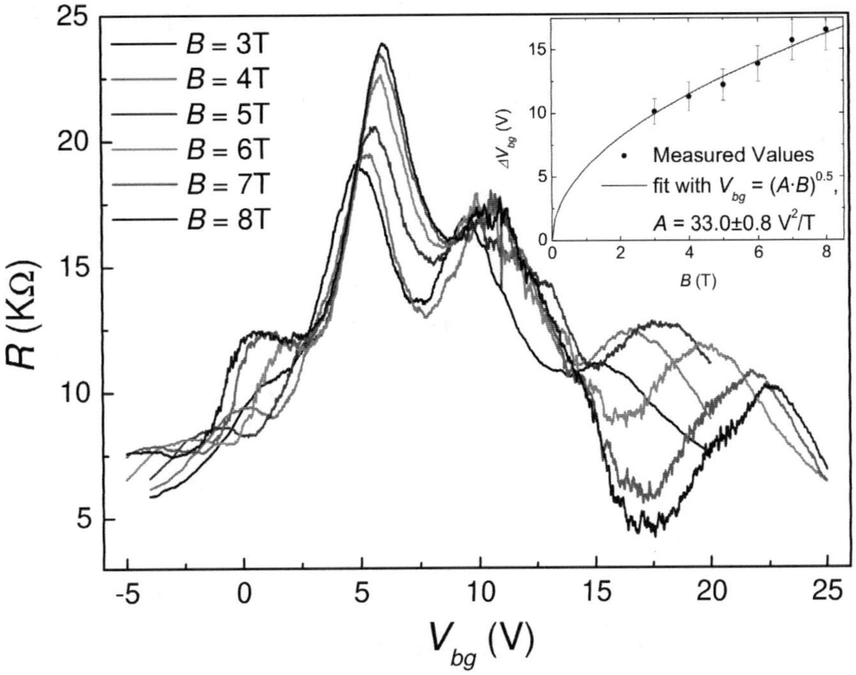

Figure 1: Transport measurements of exfoliated monolayer graphene on a SiO_2 substrate. The Dirac points are located at 5 V, the temperature is 2.8 K and the current is 95 nA. In the inset the difference of the backgate voltages at the 0th and the 2nd maximum is plottet versus the magnetic field.

3.2. Epitaxial graphene

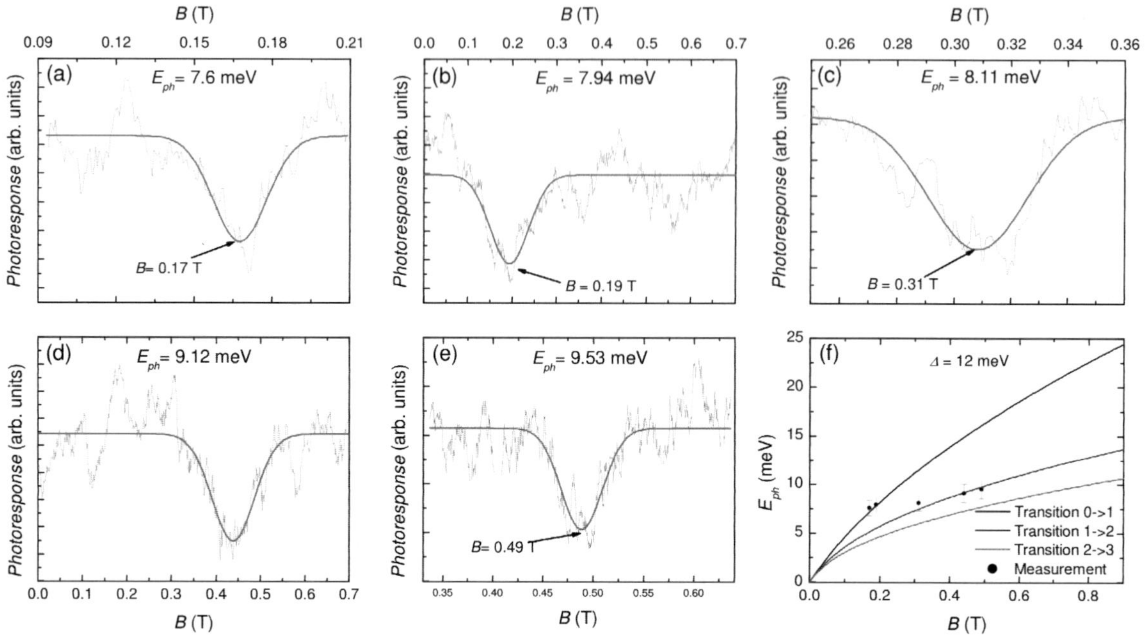

Figure 2: Photoresponse on epitaxial graphene. Figures (a–e) show the photoresponse at different photon energies E_{ph} as a function of the magnetic field B. Figure (f) shows the energy of the main photoresponse peak as a function of the applied magnetic field. Additionally the energies of the intraband Landau level transitions according to (2) assuming $\Delta = 12$meV are shown.

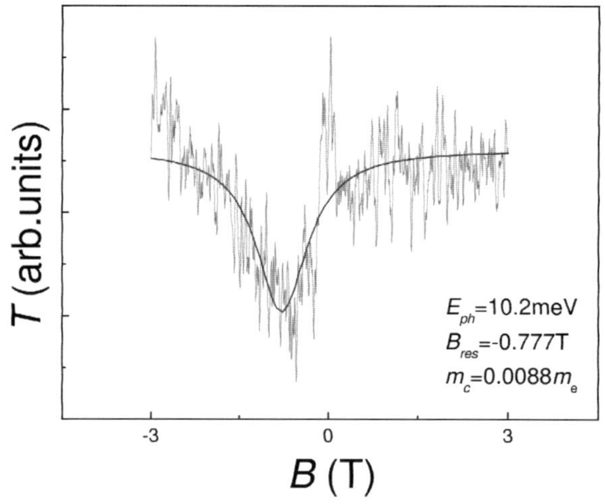

Figure 3: Phototransmission of an epitaxy sample as a function of the magnetic field.

In figure 2(a–e) the photoresponse of graphene for a set of photon energies is given. The photoserponse is strongest at the position of the minima of the photoresponse curves (see fig.2).

Using (1) the energy of the intraband transition between two LLs in the conduction band ist given by (only transitions with $\delta n = \pm 1$ are allowed):

$$E_{n+1,n} = \sqrt{\Delta^2 + 2(n+1)\hbar v_F^2 eB} - \sqrt{\Delta^2 + 2n\hbar v_F^2 eB} \quad (2)$$

In figure 2(f) the energies of the transition between two LLs in graphene are plotted as a function of the magnetic field. The measured values for the photoresponse fit quite well to the transitions according to eq.(2) assuming $\Delta = 12$meV. We attribute this to the opening of a dynamical gap by the intense electric field of the laser[8]. Another hint to the possibility of the opening of a dynamical gap can be found in the transmission curve presented in figure 3. From this transmission measurement we derive a cyclotron mass for the sample of $m_c = 0.0088 m_e$ (m_e the electron mass).

References

[1] Stellmach C, Vasile G, Hirsch A, Bonk R, Vasilyev Y B, Hein G, Becker C R and Nachtwei G 2007 *Phys. Rev. B* **76**(3) 035341
[2] Gouider F, Nachtwei G, Brüne C, Buhmann H, Vasilyev Y B, Salman M, Könemann J and Buckle P D 2011 *Journal of Applied Physics* **109** 013106 (pages 5)
[3] Novoselov K S, Geim A K, Morozov S V, Jiang D, Zhang Y, Dubonos S V, Grigorieva I V and Firsov A A 2004 *Science* **306** 666–669
[4] Salman M, Gouider F, Schmidt H, Vasilyev Y B, Haug R J and Nachtwei G 2011 *physica status solidi (c)* **8** 1208–1210
[5] Gouider F, Bugár M, Könemann J, Vasilyev Y B, Brüne C, Buhmann H and Nachtwei G 2009 *Journal of Physics: Conference Series* **193** 012066
[6] Gusynin V P, Sharapov S G and Carbotte J P 2007 *Phys. Rev. Lett.* **98** 157402
[7] Goerbig M O 2011 *Reviews of Modern Physics* **83** 1193–1243
[8] Zhou Y and Wu M W 2011 *Phys. Rev. B* **83** 245436

Journal of Physics: Conference Series **456** (2013) 012012 doi:10.1088/1742-6596/456/1/012012

Goldstone Modes in Bilayer Quantum Hall Systems at $\nu = 2$

Y. Hama[1,2], Y. Hidaka[2], G. Tsitsishvili[3], and Z. F. Ezawa[4]

[1] Department of Physics, The University of Tokyo, 113-0033, Tokyo, Japan
[2] Theoretical Research Division, Nishina Center, RIKEN, Wako 351-0198, Japan
[3] Department of Physics, Tbilisi State University, Tbilisi 0128, Georgia
[4] Advanced Meson Science Laborary, Nishina Center, RIKEN, Wako 351-0198, Japan

E-mail: hama@nt.phys.s.u-tokyo.ac.jp

Abstract. At the filling factor ν=2, the bilayer quantum Hall system has three phases, the spin-ferromagnet phase, the spin singlet phase and the canted antiferromagnet (CAF) phase, depending on the relative strength between the Zeeman energy and interlayer tunneling energy. We present a systematic method to derive the effective Hamiltonian for the Goldstone modes in these three phases. We then investigate the dispersion relations and the coherence lengths of the Goldstone modes. To explore a possible emergence of the interlayer phase coherence, we analyze the dispersion relations in the limit of zero tunneling energy. We find one gapless mode with the linear dispersion relation in the CAF phase.

1. Introduction

In the bilayer quantum Hall (QH) system, at total Landau filling factor $\nu = 2$, a rich phase structure emerges by the interplay between the spin and the layer (pseudospin) degrees of freedom [1, 2]. According to the one-body picture we expect to have two phases depending on the relative strength between the Zeeman gap Δ_Z and the tunneling gap Δ_{SAS}. One is the spin-ferromagnet and pseudospin-singlet phase (abridged as the spin phase) for $\Delta_Z > \Delta_{SAS}$; the other is the spin-singlet and pseudospin ferromagnet phase (abridged as the pseudospin phase) for $\Delta_{SAS} > \Delta_Z$. Instead, an intermediate phase, a canted antiferromagnetic phase (abridged as the CAF phase) emerges. This is a novel phase where the spin direction is canted and make antiferromagnetic correlations between the two layers[2].

The ground state structure of the $\nu = 2$ bilayer QH system has been investigated based on the SU(4) formalism [3, 4]. However, the effective Hamiltonian for the Goldstone modes has not been derived, especially in the CAF phase. On the other hand, experimentally, a role of a Goldstone mode has been suggested by nuclear magnetic resonance[5] in the CAF phase.

In this paper we develop a generic formalism to determine the symmetry breaking pattern and to derive the effective Hamiltonian for the Goldstone modes in the three phases of the $\nu = 2$ bilayer QH system. The symmetry breaking pattern reads, $SU(4) \to U(1) \otimes SU(2) \otimes SU(2)$, and there appear eight Goldstone modes in each phase. The corresponding Goldstone modes in the two phases match smoothly at the phase boundary.

This paper is organized as follows. In Sec. 2, we review the Coulomb interaction of the bilayer QH system projected to the lowest Landau level (LLL) and the SU(4) effective Hamiltonian after making the derivative expansion. We also review the ground state structure in the three phases. In Sec. 3, which is the main part of this paper, we develop a unified formalism to derive the effective Hamiltonian for the Goldstone modes. Then we discuss the SU(4) symmetry breaking pattern, the dispersion relations, and the coherence length in each phase. We study the dispersions and the coherence length in the limit $\Delta_{\text{SAS}} \to 0$, to explore a possible emergence of the interlayer phase coherence in the CAF phase. Remarkably, we find one gapless mode with the linear dispersion relation. Section 4 is devoted to discussion.

2. The SU(4) Effective Hamiltonian and the Ground State Structure

In the bilayer system an electron has two types of indices, the spin index $(\uparrow, \downarrow)$ and the layer index (f, b). They can be incorporated in 4 types of isospin index $\alpha =$ f↑,f↓,b↑,b↓. The electron field $\psi_\alpha(\boldsymbol{x})$ has four components, and the bilayer system possesses the underlying algebra SU(4) with having the subalgebra $\text{SU}_{\text{spin}}(2) \times \text{SU}_{\text{ppin}}(2)$. We denote the three generators of the $\text{SU}_{\text{spin}}(2)$ by τ_a^{spin}, and those of $\text{SU}_{\text{ppin}}(2)$ by τ_a^{ppin}. There are remaining nine generators $\tau_a^{\text{spin}}\tau_b^{\text{ppin}}$. Their explicit form is given in Appendix D in Ref.[1].

All the physical operators required for the description of the system are constructed as the bilinear combinations of $\psi(\boldsymbol{x})$ and $\psi^\dagger(\boldsymbol{x})$. There are 16 density operators

$$\rho(\boldsymbol{x}) = \psi^\dagger(\boldsymbol{x})\psi(\boldsymbol{x}), \quad S_a(\boldsymbol{x}) = \frac{1}{2}\psi^\dagger(\boldsymbol{x})\tau_a^{\text{spin}}\psi(\boldsymbol{x}),$$
$$P_a(\boldsymbol{x}) = \frac{1}{2}\psi^\dagger(\boldsymbol{x})\tau_a^{\text{ppin}}\psi(\boldsymbol{x}), \quad R_{ab}(\boldsymbol{x}) = \frac{1}{2}\psi^\dagger(\boldsymbol{x})\tau_a^{\text{spin}}\tau_b^{\text{ppin}}\psi(\boldsymbol{x}), \tag{1}$$

where S_a describes the total spin, $2P_z$ measures the electron-density difference between the two layers. The operator R_{ab} transforms as a spin under $\text{SU}_{\text{spin}}(2)$ and as a pseudospin under $\text{SU}_{\text{ppin}}(2)$.

The kinetic Hamiltonian is quenched, since the kinetic energy is common to all states in the LLL. The Coulomb interaction is decomposed into the SU(4)-invariant H_{C}^+ and SU(4)-noninvariant terms H_{C}^-. The tunneling and bias terms are summarized into the pseudo-Zeeman term. Combining the Zeeman and pseudo-Zeeman terms we have

$$H_{\text{ZpZ}} = -\int d^2x(\Delta_{\text{Z}}S_z + \Delta_{\text{SAS}}P_x + \Delta_{\text{bias}}P_z), \tag{2}$$

with the Zeeman gap Δ_{Z}, the tunneling gap Δ_{SAS}, and the bias voltage $\Delta_{\text{bias}} = eV_{\text{bias}}$. The total Hamiltonian is $H = H_{\text{C}}^+ + H_{\text{C}}^- + H_{\text{ZpZ}}$.

We project the density operators (1) to the LLL. What are observed experimentally are the classical densities, which are expectation values such as $\rho^{\text{cl}}(\boldsymbol{p}) = \langle\mathfrak{S}|\rho(\boldsymbol{p})|\mathfrak{S}\rangle$, where $|\mathfrak{S}\rangle$ represents a generic state in the LLL. The SU(4) effective Hamiltonian density are given by[3]

$$\begin{aligned}\mathcal{H}^{\text{eff}} = {} & J_s^d\left(\sum(\partial_k\mathcal{S}_a)^2 + (\partial_k\mathcal{P}_a)^2 + (\partial_k\mathcal{R}_{ab})^2\right) + 2J_s^-\left(\sum(\partial_k\mathcal{S}_a)^2 + (\partial_k\mathcal{P}_z)^2 + (\partial_k\mathcal{R}_{az})^2\right) \\ & + \rho_\phi[\epsilon_{\text{cap}}(\mathcal{P}_z)^2 - 2\epsilon_X^-\left(\sum(\mathcal{S}_a)^2 + (\mathcal{R}_{az})^2\right) - (\epsilon_X^+ - \epsilon_X^-)(\sum(\mathcal{S}_a)^2 + (\mathcal{P}_a)^2 + (\mathcal{R}_{ab})^2) \\ & - (\Delta_{\text{Z}}\mathcal{S}_z + \Delta_{\text{SAS}}\mathcal{P}_x + \Delta_{\text{bias}}\mathcal{P}_z) - (\epsilon_X^+ + \epsilon_X^-)],\end{aligned} \tag{3}$$

where J_s, J_s^d, ϵ_{cap} and $\epsilon_X^\pm$, are the intralyer stiffness, the interlayer stiffness, the capacitance energy and the exchange Coulomb energy, respectively, with the explicit formula given in Appendix A in Ref.[1]. $\rho_\Phi = \rho_0/\nu$ is the density of states, and we set $\rho^{\text{cl}}(\boldsymbol{p}) = \rho_0$, $S_a^{\text{cl}}(\boldsymbol{p}) = \rho_\Phi\mathcal{S}_a(\boldsymbol{p})$, $P_a^{\text{cl}}(\boldsymbol{p}) = \rho_\Phi\mathcal{P}_a(\boldsymbol{p})$, and $R_{ab}^{\text{cl}}(\boldsymbol{p}) = \rho_\Phi\mathcal{R}_{ab}(\boldsymbol{p})$ for the study of Goldstone modes.

It is to be remarked that all potential terms vanish in the SU(4) invariant limit, where perturbative excitations are gapless. They are the Goldstone modes associated with spontaneous breaking of the SU(4) symmetry. They get gapped in the actual system, since the SU(4) symmetry is explicitly broken. Nevertheless we call them the Goldstone modes.

The ground state is obtained by minimizing the effective Hamiltonian (3) for homogeneous configurations of the classical densities. The order parameters are the classical densities for the ground state. It has been shown[4] at $\nu = 2$ that they are given as

$$\mathcal{S}_z^0 = \frac{\Delta_Z}{\Delta_0}(1-\alpha^2)\sqrt{1-\beta^2}, \quad \mathcal{P}_x^0 = \frac{\Delta_{SAS}}{\Delta_0}\alpha^2\sqrt{1-\beta^2}, \quad \mathcal{P}_z^0 = \frac{\Delta_{SAS}}{\Delta_0}\alpha^2\beta,$$
$$\mathcal{R}_{xx}^0 = -\frac{\Delta_{SAS}}{\Delta_0}\alpha\sqrt{1-\alpha^2}\beta, \quad \mathcal{R}_{yy}^0 = -\frac{\Delta_Z}{\Delta_0}\alpha\sqrt{1-\alpha^2}\sqrt{1-\beta^2}, \quad \mathcal{R}_{xz}^0 = \frac{\Delta_{SAS}}{\Delta_0}\alpha\sqrt{1-\alpha^2}\sqrt{1-\beta^2}, \tag{4}$$

with $\Delta_0 = \sqrt{\Delta_{SAS}^2\alpha^2 + \Delta_Z^2(1-\alpha^2)(1-\beta^2)}$ and all the rest components are zero. The parameters α and β, satisfy the condition, $|\alpha| \le 1$ and $|\beta| \le 1$. As a physical variable it is convenient to use the imbalance parameter $\sigma_0 \equiv \mathcal{P}_z^0$, instead of the bias voltage Δ_{bias}. This is possible in the pseudospin and CAF phases. The bilayer system is balanced at $\sigma_0 = 0$, while all electrons are in the front layer at $\sigma_0 = 1$, and in the back layer at $\sigma_0 = -1$. There are three phases, and we discuss them in terms of α and β.

First, when $\alpha = 0$, it follows that

$$\mathcal{S}_z^0 = 1, \tag{5}$$

and all others being zero. This is the spin phase, which is characterized by the fact that the isospin is fully polarized into the spin direction. The spins in both layers point to the positive z axis due to the Zeeman effect.

Second, when $\alpha = 1$, we have

$$\mathcal{P}_x^0 = \sqrt{1-\beta^2}, \qquad \mathcal{P}_z^0 = \beta = \sigma_0, \tag{6}$$

and all the others being zero. This is the pseudospin phase, which is characterized by the fact that the isospin is fully polarized into the pseudospin direction as given by (6).

For intermediate values of α ($0 < \alpha < 1$), not only the spin and pseudospin but also some residual components are nonvanishing, where we may control the density imbalance by applying a bias voltage as in the pseudospin phase. It follows from (4) that, as the system goes away from the spin phase ($\alpha = 0$), the spins begin to cant coherently and make antiferromagnetic correlations between the two layers.

The interlayer phase coherence is an intriguing phenomenon in the bilayer QH system[1]. Since it is enhanced in the limit $\Delta_{SAS} \to 0$, it is worthwhile to investigate the effective Hamiltonian at $\nu = 2$ in this limit. We need to know how the parameters α and β are expressed in terms of the physical variables. Up to the order $O(\Delta_{SAS}^2)$, the solutions are

$$\beta = \pm\sqrt{1-\left(\frac{\Delta_{SAS}}{\Delta_Z}\right)^2} + O(\Delta_{SAS}^4), \tag{7}$$

with $\Delta_0 \to \Delta_{SAS} + O(\Delta_{SAS}^3)$. Then we have $\mathcal{P}_z^0 = \sigma_0 = \pm\alpha^2 + O(\Delta_{SAS}^2)$. The parameters α and β are simple functions of the physical variables Δ_{SAS}/Δ_Z and σ_0 in the limit $\Delta_{SAS} \to 0$. We might expect novel phenomena associated with the interlayer phase coherence in the CAF phase.

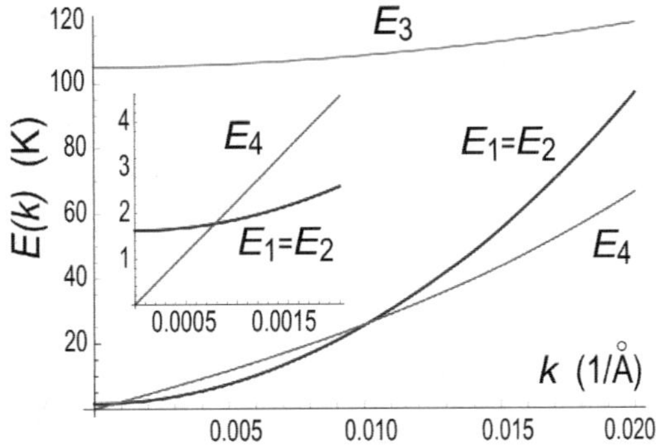

Figure 1. Dispersion relations (10) for the four Goldstone modes E_i. The sample parameters are $d = 231\text{Å}$, B=5.5T, $\rho_0 = 2.7 \times 10^{15}\text{m}^{-2}$, and $\alpha = 1/\sqrt{2}$. Inset: Dispersion relations near $k = 0$. It is clear that $E_4(k)$ is linear.

3. Effective Hamiltonian for Goldstone Modes

Having reviewed the three phases in the bilayer system at $\nu = 2$, we proceed to discuss the symmetry breaking pattern and construct the effective Hamiltonian for the Goldstone modes in each phase. There is a systematic method for this purpose[6].

We analyze excitations around the classical ground state (4). It is convenient to introduce the SU(4) isospin notation such that, $\mathcal{I}^{(0)}_{a0} = \mathcal{S}^0_a$, $\mathcal{I}^{(0)}_{0a} = \mathcal{P}^0_a$, $\mathcal{I}^{(0)}_{ab} = \mathcal{R}^0_{ab}$. We set all of them into one 15-dimensional vector $\mathcal{I}^{(0)}_{\mu\nu}$ with the index $\mu\nu$: Note that there is no component $\mathcal{I}^{(0)}_{00}$. We parametrize the SU(4) isospin operators as

$$\mathcal{I}_{\mu\nu}(x) = \left[\exp\left(i\sum_{\gamma\delta}\pi_{\gamma\delta}T_{\gamma\delta}\right)\right]^{\mu'\nu'}_{\mu\nu} \mathcal{I}^0_{\mu'\nu'}, \tag{8}$$

where $T_{\gamma\delta}$ are the matrices of the broken SU(4) generators in the adjoint representation of SU(4), each of which is a 15×15 matrix. The greek indices run over $0, x, y, z$. The phase field $\pi_{\gamma\delta}(x)$ are the eight Goldstone modes associated with the broken generators, and hence, only eight generating matrices $T_{\gamma\delta}$ are involved in the formula (8). Then we may identify $\mathcal{S}_a = \mathcal{I}_{a0}$, $\mathcal{P}_a = \mathcal{I}_{0a}$, $\mathcal{R}_{ab} = \mathcal{I}_{ab}$, and express various physical variables in terms of the Goldstone modes $\pi_{\gamma\delta}(x)$. We then expand the formula (8) in $\pi_{\gamma\delta}$ as, $\mathcal{I}_{\mu\nu}(x) = \mathcal{I}^0_{\mu\nu} + \mathcal{I}^{(1)}_{\mu\nu}(x) + \mathcal{I}^{(2)}_{\mu\nu}(x) + \cdots$, where $\mathcal{I}^{(n)}_{\mu\nu}(x)$ is the nth order term in the Goldstone mode $\pi_{\gamma\delta}$. Each phase is characterized by the order parameter $\mathcal{I}^0_{\mu\nu}$, which are nothing but (4). The key observation is that the first order term $\mathcal{I}^{(1)}_{\mu\nu}(x)$ contains all informations about the symmetry breaking pattern and the associated Goldstone modes, yielding their kinematic terms. On the other hand, the second order term $\mathcal{I}^{(2)}_{\mu\nu}(x)$ provides them with gaps. For the detailed discussions, see Ref.[7].

3.1. CAF Phase in $\Delta_{SAS} \to 0$

The effective Hamiltonian in the CAF phase is too complicated. We take the limit $\Delta_{\text{SAS}} \to 0$ to examine if some simplified formulas are obtained. In particular we would like to seek for gapless modes. Such gapless modes will play an important role to drive the interlayer coherence in the

CAF phase. In this limit we have the Hamiltonian

$$H = \sum_{i=1}^{4} \int d^2k E_i \eta_{i,\boldsymbol{k}}^{\dagger} \eta_{i,\boldsymbol{k}}, \tag{9}$$

together with the dispersion relations (Figure 1)

$$E_1 = E_2 = \frac{4}{\rho_0}((1-\alpha^2)J_s + \alpha^2 J_s^d)\boldsymbol{k}^2 + \Delta_Z, \quad E_3 = \frac{4\boldsymbol{k}^2}{\rho_0} J_s^d + 2\Delta_Z + 8\cos^2\theta_\alpha \epsilon_X^-,$$
$$E_4 = |\boldsymbol{k}| \sqrt{\frac{8J_s^d}{\rho_0}\left(\frac{2\boldsymbol{k}^2}{\rho_0}(\cos^2 2\theta_\alpha J_s^d + \sin^2 2\theta_\alpha J_s) + 2\sin^2 2\theta_\alpha(\epsilon_D^- - \epsilon_X^-)\right)}, \tag{10}$$

where $\cos\theta_\alpha = \sqrt{1-\alpha^2}$, $\sin\theta_\alpha = \alpha$, and $\eta_{i,\boldsymbol{k}}$ $(i = 1,2,3,4)$ are the annihilation operators satisfying the commutation relation $\left[\eta_{i,\boldsymbol{k}}, \eta_{j,\boldsymbol{k}'}^{\dagger}\right] = \delta_{ij}\delta(\boldsymbol{k}-\boldsymbol{k}')$.

We summarize the Goldstone modes in the CAF phase in the limit $\Delta_{\text{SAS}} \to 0$. It is to be emphasized that there emerges one gapless mode, $\eta_{4,\boldsymbol{k}}$, reflecting the realization of an exact and its spontaneous breaking of a U(1) part of the SU(4) rotational symmetry. Furthermore, it has the linear dispersion relation as in (10), as leads to a superfluidity associated with this gapless mode. All other modes have gaps.

4. Discussion

We have presented a systematic method based on the formula (8) to investigate the symmetry breaking pattern and to derive the effective Hamiltonian for the Goldstone modes in the ν=2 bilayer QH system, particularly, for the CAF phase. Eight Goldstone modes emerge in each phase, which are shown to be smoothly transformed one to another across the phase boundary.

The interlayer phase coherence and the Josephson effect are among the most intriguing phenomena in the $\nu = 1$ bilayer QH system[1]. They are enhanced in the limit $\Delta_{\text{SAS}} \to 0$. It is natural to seek for similar phenomena in the $\nu = 2$ bilayer QH system in the CAF phase, where the electron densities can be controlled arbitrarily in both layers. Having investigated the dispersion relations and the coherence length in the limit $\Delta_{\text{SAS}} \to 0$, remarkably, we have found one coherent mode whose coherence length diverges. Furthermore it has the linear dispersion relation. It might be responsible to the interlayer phase coherence.

Acknowledgments

Y. Hama thanks Norio Kumada, Koji Muraki, Takahiro Morimoto, Takeo Kato, and Yoshiro Hirayama for useful discussions and comments. This research was supported in part by JSPS Research Fellowships for Young Scientists, and a Grant-in-Aid for Scientific Research from the Ministry of Education, Culture, Sports, Science and Technology (MEXT) of Japan (Nos. 23340067, 24740184, 21540254).

References

[1] Z. F. Ezawa, *Quantum Hall effects: Field theoretical approach and related topics, Second Edition* (World Scientific, Singapore, 2008).
[2] L. Zheng *et al.*, Phys. Rev. Lett. **78**, 2453 (1997); V. Pellegrini *et al.*, Phys. Rev. Lett. **78**, 310 (1997); V. S. Khrapai *et al.*, Phys. Rev. Lett. **84**, 725 (2000); A. Fukuda *et al.*, Phys. Rev. B **73**, 165304 (2006).
[3] Z. F. Ezawa *et al.*, Phys. Rev. B **70**, 125304 (2004).
[4] Z. F. Ezawa *et al.*, Phys. Rev. B **71**, 125318 (2005).
[5] N. Kumada *et al.*, Science **313**, 329 (2006); Phys. Rev. Lett. **99**, 076805 (2007).
[6] S. R. Coleman *et al.*, Phys. Rev. **177**, 2239 (1969).
[7] Y. Hama *et al.*, Eur. Phys. J. B (2012) 85: 368.

Journal of Physics: Conference Series **456** (2013) 012013 doi:10.1088/1742-6596/456/1/012013

Chiral Symmetry and Many-Body Effect in Multilayer Graphene

Yuji Hamamoto
Institute of Physics, University of Tsukuba, Tsukuba 305-8571, Japan
E-mail: hamamoto.yuji.fp@u.tsukuba.ac.jp

Tohru Kawarabayashi
Department of Physics, Toho University, Funabashi 274-8510 Japan

Hideo Aoki
Department of Physics, University of Tokyo, Hongo, Tokyo 113-0033, Japan

Yasuhiro Hatsugai
Institute of Physics, University of Tsukuba, Tsukuba 305-8571, Japan
Tsukuba Research Center for Interdisciplinary Materials Science (TIMS), University of Tsukuba, Tsukuba 305-8571, Japan

Abstract. Influence of the chiral symmetry on the many-body problem in multilayer graphene in magnetic fields is investigated. For a spinless electron model on the honeycomb lattice the many-body ground state is shown to be a doubly-degenerate chiral condensate *irrespective of the number of layers.* The energy spectrum calculated numerically with the exact diagonalization method reveals for ABC-stacked multilayer graphenes that the many-body gap decreases monotonically with the number of layers.

1. Introduction

Graphene's intriguing physics [1, 2, 3, 4] is intimately related to the chiral symmetry, a fundamental feature of the honeycomb lattice. The symmetry is defined by the anticommutation relation, $\{\mathcal{H}, \Gamma\} = 0$, with the chiral operator Γ acts as

$$\Gamma c_i \Gamma = +c_i(-c_i) \quad \text{for} \quad i \in \bullet(\circ), \qquad \Gamma^2 = 1, \tag{1}$$

where $\bullet$ and $\circ$ respectively denote the two sublattices. The symmetry guarantees [5, 6] the topological stability of the doubled Dirac cones, an example of Nielsen-Ninomiya theorem [7, 8], as well as the degeneracy of the $n = 0$ Landau level (LL), which accommodates Aharonov-Casher's argument [9]. A remarkable phenomenon in the latter is the delta-function-like density of states (DOS) for the $n = 0$ LL even in the presence of ripples with wavelengths exceeding a few lattice constants [10]. These properties are expected to be inherited by multilayer graphene [11, 12, 13], since the leading hopping matrix elements, γ_0, γ_1 and γ_3, conserves the

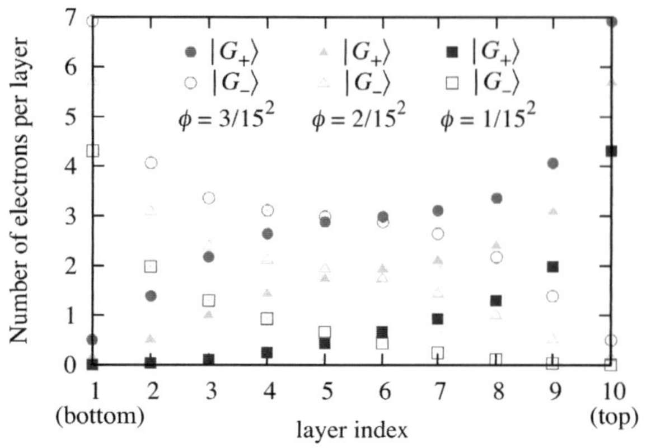

Figure 1. Layer-by-layer charge distribution in the chiral condensates for the ABC stack of 10 graphene sheets. $\gamma_1/\gamma_0 = 0.5$, and only the electrons in the zero-energy LL are included.

bipartite lattice structure, and hence the chiral symmetry. Indeed, a recent theory [14] predicts that AB-staked bilayer graphene retains the anomaly in DOS, which even extends to the case of a perpendicular electric field that in fact breaks the chiral symmetry.

Then an interesting question is how the chiral symmetry affects the many-body problem in multilayer graphene. Many-body effects have been extensively studied [4, 15, 16, 17, 18, 19, 20, 21, 22, 23, 24, 25, 26, 27, 28, 29, 30, 31, 32, 33, 34, 35, 36, 37, 38, 39, 40] from the beginning of the graphene research due to interests that include the gap opening in the central LL [41, 42, 43, 44] and the unconventional insulating $\nu = 0$ state [45]. So far, the role of the chiral symmetry in the many-body problem has been studied in the context of the lattice gauge theory, where monolayer graphene without magnetic fields is discussed, mainly based on the continuum model [46, 47, 48]. In the present work, we consider multilayer graphenes in magnetic fields, and investigate the many-body problem in terms of the chiral symmetry.

2. Ground states

We consider spinless electrons in multilayer graphenes in a magnetic field, for which the kinetic energy is

$$\mathcal{H}_{\text{kin}} = -\sum_{\langle ij \rangle} \gamma_{ij} e^{i\theta_{ij}} c_i^\dagger c_j + \text{H.c.}, \tag{2}$$

where γ_{ij} is the electron hopping between sites $\langle ij \rangle$. For the nearest-neighbor in-plane sites $\gamma_{ij} = \gamma_0$, while for the inter-layer vertical hopping $\gamma_{ij} = \gamma_1$. The magnetic field is included as the Peierls phase θ_{ij} such that magnetic flux per elementary hexagon equals to $\phi = \frac{1}{2\pi}\sum_{\hexagon} \theta_{ij}$ in units of the magnetic flux quantum h/e. In the string gauge [49], the magnetic flux reads $\phi = n/N_c$ with an integer n and the number of unit cells, N_c. Since the γ_0-γ_1 model preserves the chiral symmetry, one can construct a zero-energy multiplet, $\psi = (\psi_+, \psi_-)$ with $\psi_\pm = (\psi_{1\pm}, \cdots, \psi_{M_\pm \pm})$, where $\psi_{m\pm}$'s denote $M_\pm$-fold degenerate eigenstates of Γ corresponding to chirality ± 1, i.e., $\Gamma\psi_{m\pm} = \pm\psi_{m\pm}$. For an N_l-layer bulk system with a magnetic flux $\phi = n/N_c$ $(n = 1, 2, \cdots)$, the degeneracy reads $M_+ = M_- = nN_l$. It should be noted that the zero modes with chirality $+$ or $-$ are localized on sublattice • or ∘, respectively. Especially for ABC-stacked multilayers, the zero modes show sublattiece-selective charge accumulation towards the top or bottom layer, which decays rapidly away from the surface [50, 51, 52]. The surface states are analogous to the zigzag edge states in the ribbon structure [53] and should be observed experimentally as local density of states near zero energy with a scanning tunneling microscope.

For a half-filled multilayer graphene, the interaction for spinless electrons can be written in

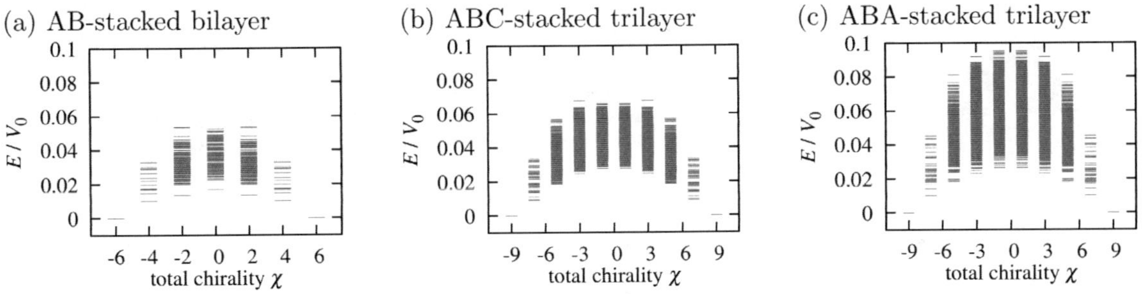

Figure 2. Energy spectra of AB-stacked bilayer graphene (a), ABC-stacked trilayer graphene (b) and ABA-stacked trilayer graphene (c), plotted against the total chirality, χ. Results for $N_c = 15^2$, $\phi = 3/15^2$, $\gamma_1/\gamma_0 = 0.5$, $V_1/V_0 = 0.5$ are shown.

a particle-hole symmetric form as

$$\mathcal{H}_{\rm int} = \sum_{i\neq j} V_{ij}\left(n_i - \frac{1}{2}\right)\left(n_j - \frac{1}{2}\right) = \frac{1}{2}\sum_{i\neq j} V_{ij}(c_i^\dagger c_j^\dagger c_j c_i + c_i c_j c_j^\dagger c_i^\dagger) + \text{const} \tag{3}$$

with interaction strength V_{ij}. Since it is hard to treat all the many-body states exactly, we focus on the many-body problem within the zero-energy LL. Such a treatment is adequate as long as the interaction energy is sufficiently smaller than the Landau gap between the central and the adjacent LLs. The projection onto the zero-energy LL is performed as $\tilde{c} = (\psi\psi^\dagger)c$, where $c \equiv (c_1, \cdots, c_N)^T$ with the number of sites N, and $\psi\psi^\dagger$ denotes the projection matrix. Note that the projected creation and annihilation operators are no longer fermionic; they obey anticommutation relations $\{\tilde{c}_i, \tilde{c}_j^\dagger\} = (\psi\psi^\dagger)_{ij}$ and $\{\tilde{c}_i, \tilde{c}_j\} = \{\tilde{c}_i^\dagger, \tilde{c}_j^\dagger\} = 0$. In what follows, we assume that the electron-electron repulsion only acts between electrons on different sublattices [$V_{ij} > 0$ for $(i,j) = (\bullet, \circ)$]. The simplest example is the nearest-neighbor repulsion, which is the leading term for spinless electrons. The total Hamiltonian in the projected subspace reads, up to a constant,

$$\tilde{\mathcal{H}} = \sum_{i\in\bullet, j\in\circ} \frac{V_{ij}}{2}(\tilde{c}_i^\dagger \tilde{c}_j^\dagger \tilde{c}_j \tilde{c}_i + \tilde{c}_i \tilde{c}_j \tilde{c}_j^\dagger \tilde{c}_i^\dagger) = \frac{1}{2}\sum_{klmn}(V_{klmn} d_{k+}^\dagger d_{l-}^\dagger d_{m-} d_{n+} + V_{klmn}^* d_{k+} d_{l-} d_{m-}^\dagger d_{n+}^\dagger), \tag{4}$$

where $d_{m\pm}^\dagger \equiv c^\dagger \psi_{m\pm}$ is the creation operator for the zero mode $\psi_{m\pm}$, while the pseudopotential appearing for $d, d^\dagger$ is defined as $V_{klmn} = \sum_{i\in\bullet, j\in\circ} V_{ij}(\psi_k)_i^*(\psi_l)_j^*(\psi_m)_j(\psi_n)_i$. Clearly, $\tilde{\mathcal{H}}$ is semi-positive definite; for an arbitrary many-body state $|\Phi\rangle$, $\langle\Phi|\tilde{\mathcal{H}}|\Phi\rangle = \frac{1}{2}\sum_{i\in\bullet, j\in\circ} V_{ij}\left|[\tilde{c}_i\tilde{c}_j + (\tilde{c}_i\tilde{c}_j)^\dagger]|\Phi\rangle\right|^2 \geq 0$. Since $\tilde{\mathcal{H}}$ commutes with $\mathcal{G} \equiv \sum_m (d_{m+}^\dagger d_{m+} - d_{m-}^\dagger d_{m-})$, eigenstates of $\tilde{\mathcal{H}}$ are classified according to the total chirality, $\chi = \langle\mathcal{G}\rangle$. The highest and lowest sectors in χ correspond to *chiral condensates*,

$$|G_\pm\rangle = d_{1\pm}^\dagger \cdots d_{M_\pm \pm}^\dagger |D_<\rangle, \qquad \mathcal{G}|G_\pm\rangle = \pm M_\pm |G_\pm\rangle, \tag{5}$$

where all the zero modes with chirality + (or −) are occupied, while others are empty. $|D_<\rangle$ denotes the Dirac sea of completely filled negative energy states. The chiral condensates are in fact the many-body ground states, since they are annihilated by both of $d_{k\pm}^\dagger d_{l\mp}^\dagger$ and $d_{m\mp} d_{n\pm}$, hence $\langle G_\pm|\tilde{\mathcal{H}}|G_\pm\rangle = 0$. Thus the chiral condensates form a ground-state *doublet*, $\Psi = (|G_+\rangle, |G_-\rangle)$, so that any unitary-mixture, $\Psi \mapsto \Psi^\omega \omega$ with $\omega \in \mathrm{U}(2)$, is again an eigenstate. We expect that the chiral condensate remains the ground state for e.g. Coulomb interaction

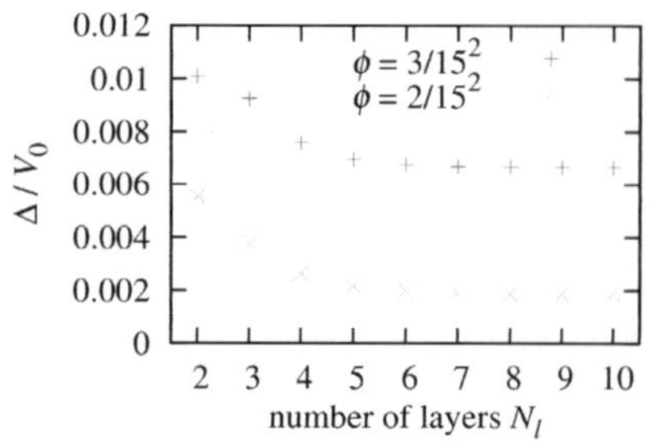

Figure 3. Energy gap Δ of ABC-stacked multilayer graphene composed of N_l graphene sheets as a function of N_l. Here a result for $\gamma_1/\gamma_0 = 0.5, V_1/V_0 = 0.5$ is shown.

because its leading part is the nearest-neighbor repulsion. Since the above discussion holds irrespective of the number of layers, it is expected that Ψ has properties similar to those in the monolayer case, such as vanishing Hall conductivity, or the Kekulé-type bond-order along armchair edges [54, 55]. Although the spinless model with interaction between opposite sublattices might seem rather artificial, the results obtained here should have an experimental relevance in e.g. cold atoms on an optical latteice.

As a physical property peculiar to the multilayer case, we plot in Figure 1 layer-by-layer charge distribution in $|G_\pm\rangle$ for an ABC-stacked multilayer with $N_c = 15^2$ and $N_l = 10$, where only the electrons in the zero-energy LL is shown. Since an experimental cyclotron frequency ω_c is too small to treat numerically based on our lattice model, we adopt a rather large parameter $\gamma_1/\gamma_0 = 0.5$ to access the realistic parameter region $\gamma_1 > \omega_c$. One can see charge accumulation towards the top or bottom layer, which reflects the property of the zero modes as the surface states [50, 51, 52]. Notably, a shoulder structure evolves on the intermediate layers with increasing ϕ, Although the non-monotonic behavior disagrees with the results of the continuum model, where low-energy wave functions decays exponentially away from a surface, the discrepancy seems to be due to the finite size effect of our lattice model.

3. Excited states

Next we investigate excited states at half filling by numerically diagonalizing the projected Hamiltonian (4). For simplicity, only the nearest-neighbor repulsions in the same layer V_0 and those between adjacent layers V_1 are taken into account. Fig. 2 shows typical examples of energy spectra plotted against the total chirality, χ, for AB-stacked bilayer graphene and for ABC- or ABA-stacked trilayer graphene. The spectra are symmetric with respect to χ, since now we have the degeneracies of the zero modes with chirality $+$ and $-$ are equal. From this total-chirality-resolved plot one can confirm that the ground states are indeed chiral condensates, as expected from the above discussion. More importantly, these data suggest that the first excited states are obtained by flipping a single-chirality of the chiral condensates as

$$|E_\pm\rangle = \sum_{mn} C_{mn} d^\dagger_{m\mp} d_{n\pm} |G_\pm\rangle \tag{6}$$

with coefficients C_{mn} determined numerically. Although the size of the Hilbert space blows up exponentially with the number of layers, the ansatz (6) enables us to calculate the energy gap Δ between the ground state and the first-excited state by restricting ourselves to the sector of total chirality $\chi = \pm(M_\pm - 2)$. Figure 3 plots thus obtained energy gap Δ of ABC-stacked multilayer graphene composed of N_l graphene sheets. The decrease in Δ as a function of N_l is consistent with the fact the overlap of the opposite-chirality zero modes, which are accumulated near the opposite surfaces, becomes small with increasing N_l. The saturation in Δ for large N_l suggests that the overlap does not completely vanish due to the finite size effect.

4. Summary

We have considered spinless fermions in multilayer graphene and investigated many-body states in the quantum Hall regime. For the electron-electron repulsion restricted to unlike sublattices, the ground state is exactly identified to be chiral condensates as in the case of monolayer graphene. Using the exact diagonalization method, we have numerically calculated the many-body gap, which decreases monotonically with increasing the number of layers.

Acknowledgments

The computation in this work has been done with the facilities of the Supercomputer Center, Institute for Solid State Physics, University of Tokyo. This work was supported in part by Grants-in-Aid for Scientific Research No. 23340112 and No. 23654128 from the JSPS.

References

[1] Castro Neto A H, Guinea F, Peres N M R, Novoselov K S and Geim A K 2009 *Rev. Mod. Phys.* **81** 109
[2] Das Sarma S, Adam S, Hwang E H and Rossi E 2011 *Rev. Mod. Phys.* **83** 407
[3] Goerbig M O 2011 *Rev. Mod. Phys.* **83** 1193
[4] Kotov V N, Uchoa B, Pereira V M, Guinea F and Castro Neto A H 2012 *Rev. Mod. Phys.* **84** 1067
[5] Hatsugai Y, Fukui T and Aoki H 2006 *Phys. Rev. B* **74** 205414
[6] Hatsugai Y 2011 *J. Phys. Conf. Ser.* **334** 012004
[7] Nielsen H B and Ninomiya M 1981 *Nuclear Physics B* **185** 20
[8] Nielsen H B and Ninomiya M 1981 *Nuclear Physics B* **193** 173
[9] Aharonov Y and Casher A 1979 *Phys. Rev. A* **19** 2461
[10] Kawarabayashi T, Hatsugai Y and Aoki H 2009 *Phys. Rev. Lett.* **103** 156804
[11] McCann E and Fal'ko V I 2006 *Phys. Rev. Lett.* **96** 086805
[12] Nilsson J, Neto A H C, Guinea F and Peres N M R 2006 *Phys. Rev. Lett.* **97** 266801
[13] Arovas D P and Guinea F 2008 *Phys. Rev. B* **78** 245416
[14] Kawarabayashi T, Hatsugai Y and Aoki H 2012 *Phys. Rev. B* **85** 165410
[15] Nomura K and MacDonald A H 2006 *Phys. Rev. Lett.* **96** 256602
[16] Alicea J and Fisher M P A 2006 *Phys. Rev. B* **74** 075422
[17] Goerbig M O, Moessner R and Douçot B 2006 *Phys. Rev. B* **74** 161407
[18] Gusynin V P, Miransky V A, Sharapov S G and Shovkovy I A 2006 *Phys. Rev. B* **74** 195429
[19] Fuchs J N and Lederer P 2007 *Phys. Rev. Lett.* **98** 016803
[20] Herbut I F 2007 *Phys. Rev. B* **75** 165411
[21] Sheng L, Sheng D N, Haldane F D M and Balents L 2007 *Phys. Rev. Lett.* **99** 196802
[22] Jung J and MacDonald A H 2009 *Phys. Rev. B* **80** 235417
[23] Nomura K, Ryu S and Lee D H 2009 *Phys. Rev. Lett.* **103** 216801
[24] McCann E 2006 *Phys. Rev. B* **74** 161403
[25] Min H, Borghi G, Polini M and MacDonald A H 2008 *Phys. Rev. B* **77** 041407
[26] Barlas Y, Côté R, Nomura K and MacDonald A H 2008 *Phys. Rev. Lett.* **101** 097601
[27] Shizuya K 2009 *Phys. Rev. B* **79** 165402
[28] Nakamura M, Castro E V and Dóra B 2009 *Phys. Rev. Lett.* **103** 266804
[29] Viola Kusminskiy S, Campbell D K and Castro Neto A H 2009 *Europhys. Lett.* **85** 58005
[30] Vafek O and Yang K 2010 *Phys. Rev. B* **81** 041401
[31] Zhang F, Min H, Polini M and MacDonald A H 2010 *Phys. Rev. B* **81** 041402
[32] Shizuya K 2010 *Phys. Rev. B* **81** 075407
[33] Gorbar E V, Gusynin V P and Miransky V A 2010 *Phys. Rev. B* **81** 155451
[34] Nandkishore R and Levitov L 2010 *Phys. Rev. Lett.* **104** 156803
[35] Nandkishore R and Levitov L 2010 *Phys. Rev. B* **82** 115431
[36] Lemonik Y, Aleiner I L, Toke C and Fal'ko V I 2010 *Phys. Rev. B* **82** 201408
[37] Gorbar E V, Gusynin V P, Jia J and Miransky V A 2011 *Phys. Rev. B* **84** 235449
[38] Lemonik Y, Aleiner I and Fal'ko V I 2012 *Phys. Rev. B* **85** 245451
[39] Min H and MacDonald A H 2008 *Phys. Rev. B* **77** 155416
[40] Zhang F, Sahu B, Min H and MacDonald A H 2010 *Phys. Rev. B* **82** 035409
[41] Zhang Y, Jiang Z, Small J P, Purewal M S, Tan Y W, Fazlollahi M, Chudow J D, Jaszczak J A, Stormer H L and Kim P 2006 *Phys. Rev. Lett.* **96** 136806
[42] Jiang Z, Zhang Y, Stormer H L and Kim P 2007 *Phys. Rev. Lett.* **99** 106802
[43] Zhao Y, Cadden-Zimansky P, Ghahari F and Kim P 2012 *Phys. Rev. Lett.* **108** 106804

[44] Young A F, Dean C R, Wang L, Ren H, Cadden-Zimansky P, Watanabe K, Taniguchi T, Hone J, Shepard K L and Kim P 2012 *Nat. Phys.* **8** 550
[45] Checkelsky J G, Li L and Ong N P 2008 *Phys. Rev. Lett.* **100** 206801
[46] Drut J E and Lähde T A 2009 *Phys. Rev. Lett.* **102** 026802
[47] Drut J E and Lähde T A 2009 *Phys. Rev. B* **79** 165425
[48] Araki Y and Hatsuda T 2010 *Phys. Rev. B* **82** 121403
[49] Hatsugai Y, Ishibashi K and Morita Y 1999 *Phys. Rev. Lett.* **83** 2246
[50] Guinea F, Castro Neto A H and Peres N M R 2006 *Phys. Rev. B* **73** 245426
[51] Koshino M 2010 *Phys. Rev. B* **81** 125304
[52] Hatsugai Y, Morimoto T, Kawarabayashi T, Hamamoto Y and Aoki H *Preprint* arXiv:1210.0714
[53] Fujita M, Wakabayashi K, Nakada K and Kusakabe K 1996 *J. Phys. Soc. of Jpn.* **65** 1920
[54] Hamamoto Y, Hatsugai Y and Aoki H *Preprint* arXiv:1108.1638
[55] Hamamoto Y, Aoki H and Hatsugai Y 2012 *Phys. Rev. B* **86** 205424

HMF-20 IOP Publishing
Journal of Physics: Conference Series **456** (2013) 012014 doi:10.1088/1742-6596/456/1/012014

Exchange interaction in chirally coupled quantum dots

Daniel Tutuc[1], Alexander W. Heine[1], Dieter Schuh[2], Werner Wegscheider[2,3], and Rolf J. Haug[1]

[1]Institut für Festkörperphysik, Leibniz Universität Hannover, Appelstraße 2, D-30167 Hannover, Germany

[2]Institut für Experimentelle und Angewandte Physik, Universtät Regensburg, Universitätsstraße 31, D-93053 Regensburg, Germany

[3]Laboratorium für Festkörperphysik, ETH Zürich, Schafmattstraße 16, CH-8093 Zürich, Switzerland

E-mail: heine@nano.uni-hannover.de

Abstract. We present transport measurements on a system of two lateral quantum dots in a perpendicular magnetic field. Due to edge channel formation in an open conducting region, the quantum dots are chirally coupled. When both quantum dots are tuned into the Kondo regime simultaneously, we observe a change in the temperature dependence of the differential conductance. This is explained by the RKKY exchange interaction between the two dots. As a function of bias the differential conductance shows a splitting of the Kondo resonance which changes in the presence of RKKY interaction.

1. Introduction

Semiconductor quantum dots (QDs) are highly tunable devices and therefore have attracted much interest in the scientific community. Using the spin of the QD as a qubit in quantum information processes is one of the proposed applications [1]. For this application it is necessary to provide non-local manipulation and readout of the spin information of a QD. One possible mechanism to entangle the spins of different QDs beyond the nearest neighbour approach is the Ruderman-Kittel-Kasuya-Yosida (RKKY) interaction between magnetic moments. This indirect non-local exchange interaction is mediated by charge carriers and locks the magnetic moments in a new ground state. The latter is either ferromagnetic or antiferromagnetic depending on the distance between the magnetic moments and the Fermi wave vector of the charge carriers [2].

Another entanglement mechanism of a single QD spin to a many-body electron system is the Kondo effect [3, 4, 5]. At very low temperatures the spin of a QD is screened by the electrons in the leads and forms a new singlet ground state with the binding energy T_k. In an otherwise Coulomb-blocked situation this singlet ground state allows the electron of the highest occupied energy level to leave the dot and to be replaced by an electron from the leads with opposite spin. Therefore the dot becomes transparent at zero bias (the so-called zero bias anomaly) because its spin is constantly flipped.

In a system where the Kondo effect as well as the RKKY exchange interaction are present, one expects a competition between these two mechanisms which can be observed in transport properties [6, 7].

2. Experimental Setup

Our sample is prepared on a GaAs/AlGaAs heterostructure containing a two dimensional electron system 37 nm below the surface with an electron density of 3.95×10^{15} m^{-2}. Two lateral QDs (QD1 and QD2) are defined by local anodic oxidation using an AFM [8, 9]. Figure 1 (a) shows an AFM picture of the structure. The QDs are tunnel coupled to individual drains (D1 and D2) and to an open conducting region with a length of $\sim$ 600 nm. This region connects the QDs and is used as source contact (S). The tunnel barriers and the energy levels of the QDs can be tuned by six in-plane gates and an individual bias can be applied to each QD. The measurements were performed in a ^{3}He/^{4}He dilution refrigerator with an electron temperature of about 80 mK. Standard lock-in technique was used and the differential conductance $G_\mathrm{i} = dI_\mathrm{i}/dV$ (i = 1, 2) of each dot was measured individually. A magnetic field was applied perpendicular to the sample. More detailed information concerning the experimental setup can be found in Ref. [10].

3. Kondo chessboard and RKKY interaction

The measurement of the differential conductance G_1 of QD1 at zero bias as a function of gate voltage V_{G6} and perpendicular magnetic field leads to a pattern of alternating tiles of low and high differential conductance (Fig. 1 (c)). This pattern originates from the formation of Landau levels (LLs) in the QD in perpendicular magnetic field when two LLs are observable. In the presented situation the filling factor is $2 < \nu < 4$. Kondo transport then only involves energy states of the lowest LL. The number of electrons in this LL0 can be controlled by varying the number of flux quanta on the dot which redistributes the electrons between the LLs. This is achieved by a change in the magnetic field. A modulation of the total number of electrons on the dot by sweeping the gate voltage also changes the occupation of LL0. These two mechanisms lead to the so-called *Kondo chessboard* [11, 12, 13, 14], seen in Fig. 1 (c).

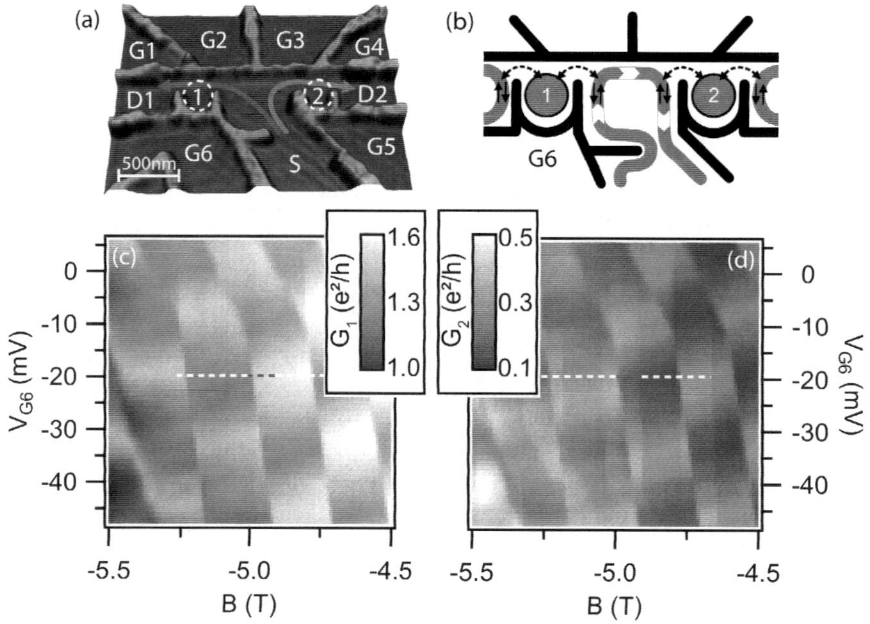

Figure 1. (a) AFM picture of the sample. The oxide lines in bright colour define the structure. (b) Schematic of the edge channels for perpendicular magnetic field with negative polarity. (c)+(d) Differential conductance of QD1 (c) and QD2 (d) as function of magnetic field and gate voltage.

The differential conductance G_2 of QD2 in Fig. 1 (d) shows a more complicated pattern: a negative of the Kondo chessboard in QD1 is visible that is superimposed by a stripe pattern. The reason for that is the formation of edge channels in the open central region. For the given magnetic field polarity a schematic of transport through the edge channels is shown in Fig. 1 (b). QD2 lies "downstream" and the changes of the electrostatic potential at the site of QD1, e.g. by onset of the Kondo effect, can be seen in the transport properties of QD2. Thus the coupling between the QDs is chiral and the chirality can be changed by reversing the magnetic field polarity [10].

The intervals where both QDs are tuned into the Kondo region simultaneously can be easily identified in the superimposed pattern exhibited by QD2. In QD1, Kondo transport can be observed between -4.97 T and -4.77 T, while QD2 exhibits the Kondo effect between -5.05 T and -4.92 T. Therefore between -4.97 T and -4.92 T transport through both QDs is Kondo enhanced (marked in red). A detailed temperature analysis in this region yields a strong change in the behaviour of the Kondo temperatures compared to the case when only one QD is tuned into the Kondo regime (see [10]). This change indicates a modulation in the ground state of the system. Since this effect can be observed in both QDs for both magnetic field polarities, it is not caused by any electrostatic effect. Instead it is attributed to the quantum mechanical interaction mediated by the electrons in the edge states, the RKKY exchange interaction.

4. Bias dependence

To probe the influence of the RKKY interaction on the zero bias anomaly, a bias V_{sd1} was applied at drain 1 and the differential conductance G_1 was measured along the dashed line in Fig. 1 (c). The result is shown in Fig. 2 (a). The region of Kondo transport can be seen clearly and between -4.9 T and -4.8 T a splitting can be observed in Fig. 2 (b). In traces taken along the coloured lines in Fig. 2 (a) the detailed evolution of this splitting is revealed. For the trace of -4.83 T two peaks at about ± 0.24 mV are clearly visible. When QD2 enters the Kondo regime at -4.92 T, this splitting decreases to ± 0.12 mV and the peaks are considerably smaller. In the centre of the overlap region at -4.96 T the peaks vanish completely.

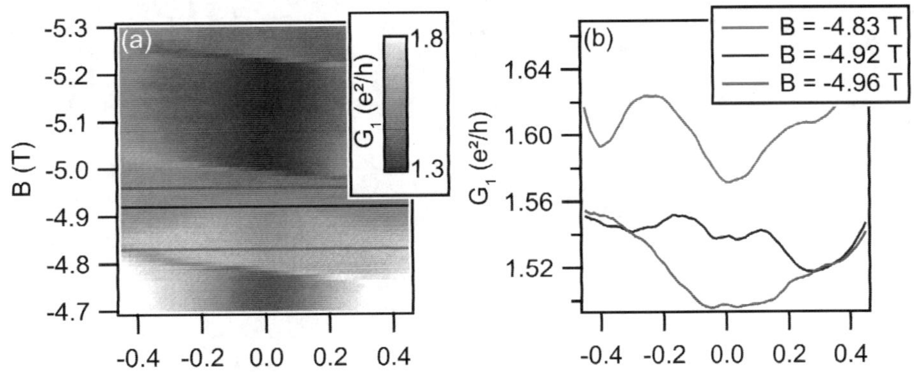

Figure 2. (a) Differential conductance G_1 as function of magnetic field and bias voltage for fixed gate voltage $V_{G6} = -20$mV (dashed line in Fig. 1 (c)). (b) Traces measured along the coloured lines in (a).

This clear change in the zero bias anomaly splitting when both QDs are tuned into the Kondo regime is another hint for the presence of the RKKY exchange interaction between the two QDs.

When the polarity of magnetic field is changed and QD2 is "upstream", a similar effect can be seen as function of the bias on drain 2.

Note that the observed splitting at -4.83 T is a factor of 2 larger than the expected Zeeman splitting for bulk GaAs ($|g^*| = 0.44$) of about 130 μeV at 5 T.

5. Conclusion

We have shown the chiral coupling of two QDs in a perpendicular magnetic field. The RKKY exchange interaction between the magnetic moments of the QD can be probed by using the Kondo effect as a spectroscopic tool. By analysing the Kondo temperature a change in the ground state of the system can be confirmed which is present regardless of chirality. In bias dependent measurements a splitting of the zero bias anomaly of QD1 can be observed, when only one QD shows the Kondo effect. It is strongly suppressed, when both QDs exhibit Kondo transport and the RKKY interaction is present.

Acknoledgments

We acknowledge financial support by the German Cluster of Excellence QUEST and the NTH School for Contacts in Nanosystems.

References

[1] Loss D and DiVincenzo D P 1998 *Phys. Rev. A* **57**(1) 120–126
[2] Ruderman M A and Kittel C 1954 *Phys. Rev.* **96**(1) 99–102
[3] Goldhaber-Gordon D, Göres J, Kastner M A, Shtrikman H, Mahalu D and Meirav U 1998 *Phys. Rev. Lett.* **81** 5225–5228
[4] Goldhaber-Gordon D, Shtrikman H, Mahalu D, Abusch-Magder D, Meirav U and Kastner M A 1998 *Nature* **391** 156–159
[5] Cronenwett S M, Oosterkamp T H and Kouwenhoven L P 1998 *Science* **281** 540–544
[6] Craig N J, Taylor J M, Lester E A, Marcus C M, Hanson M P and Gossard A C 2004 *Science* **304** 565–567
[7] Simon P, López R and Oreg Y 2005 *Phys. Rev. Lett.* **94** 086602
[8] Held R, Vancura T, Heinzel T, Ensslin K, Holland M and Wegscheider W 1998 *Applied Physics Letters* **73** 262–264
[9] Keyser U F, Schumacher H W, Zeitler U, Haug R J and Eberl K 2000 *Applied Physics Letters* **76** 457–459
[10] Tutuc D, Popescu B, Schuh D, Wegscheider W and Haug R J 2011 *Phys. Rev. B* **83**(24) 241308
[11] Keller M, Wilhelm U, Schmid J, Weis J, v Klitzing K and Eberl K 2001 *Phys. Rev. B* **64** 033302
[12] Fuehner C, Keyser U F, Haug R J, Reuter D and Wieck A D 2002 *Phys. Rev. B* **66** 161305
[13] Stopa M, van der Wiel W G, Franceschi S D, Tarucha S and Kouwenhoven L P 2003 *Phys. Rev. Lett.* **91** 046601
[14] Rogge M C, Fühner C and Haug R J 2006 *Phys. Rev. Lett.* **97** 176801

HMF-20 IOP Publishing
Journal of Physics: Conference Series **456** (2013) 012015 doi:10.1088/1742-6596/456/1/012015

Tuning of the Landé g-factor in $Al_xGa_{1-x}As/AlAs$ single and double quantum wells

F. G. G. Hernandez[1], **G. M. Gusev**[1], **and A. K. Bakarov**[2]
[1] Instituto de Física, Universidade de São Paulo, Caixa Postal 66318 - CEP 05315-970, São Paulo, SP, Brazil
[2] Institute of Semiconductor Physics, Novosibirsk 630090, Russia

E-mail: felixggh@if.usp.br

Abstract. We report on the spin dynamics of a high mobility two-dimensional electron gas in a $Al_xGa_{1-x}As/AlAs$ double quantum well structure. For high electron density samples, the g-factor was measured using time-resolved Kerr rotation technique. The g-factor tuning capability was observed by changing the aluminum content x independently in each well. Experiments demonstrated an unusual spin dephasing time robustness for high excitation power. The effect of the interaction between wells was analyzed in samples with different tunneling barriers. Results were compared with experiments on single well systems demonstrating higher spin polarization generation, longer spin dephasing time, and coupling for the double structures.

1. Introduction

The influence of a dense two-dimensional gas in the trion and exciton optical transitions of a single quantum well has been extensively studied [1, 2, 3, 4]. Determining the electronic structure of such semiconductor nanostructures requires a detailed study of their spin properties [5]. An important tool for such investigation is the Landé g-factor as it depends on the confinement potential details and chemical composition.

For example, the electron g-factor of a $Al_xGa_{1-x}As$ quantum well depends strongly on the Al content x. In a single parabolic $Al_xGa_{1-x}As/Al_{0.4}Ga_{0.6}As$ well, a gate-voltage-mediated control of coherent spin precession has already been demonstrated including g-factor sign change and vanishing values [6]. Also, the spin-beat spectroscopy in $GaAs/Al_xGa_{1-x}As$ quantum wells reveals a g-factor universal dependence on the optical transition energies [7].

Adding an extra control parameter, bilayer systems can be constructed by two parallel quantum wells with a high mobility electron gas separated by the tunneling barrier [8]. When a magnetic field is applied perpendicular to well plane, the quantum Hall states are formed. For double wells with vanishing g-factor, experiments reveal unsual magneto-transport features in the quantum Hall effect [9].

The possibility to generate direct and indirect excitonic states in double quantum wells focused great attention in optical experiments [10, 11, 12]. Furthermore, the g-factor variation with an external electric field has been observed in coupled quantum wells with unequal widths [13]. In this reference, time-resolved Faraday rotation showed a continuous g-factor tune or switch between the Larmor frequencies of the individual wells.

Here, time-resolved Kerr rotation data displays a strong tunability of the g-factor by independently adjusting the Al content in a pair of square quantum wells with equal width.

Furthermore, pump-probe spectroscopy shows that the spin polarization generation and dephasing time strongly dependent on this chemical composition control parameter. For double quantum wells, the tunneling barrier width plays the major role and allows additional tuning capability with coupling.

2. Samples

The investigated samples consist of single and double quantum wells containing a dense high-mobility two-dimensional electron gas (2DEG). Several samples were grown using the Al content (x<16%) as a g-factor engineering parameter in every individual well. The mobility and electron density were characterized through transport measurements. The sample specifications are presented in table 1. For the double quantum well structures, we denoted the sample configuration following a convention similar to reference [13].

Table 1. Single and double quantum well samples where t_b is the AlAs barrier width, n is the electron density, and μ is the mobility. The well width is 14 nm equal for all samples.

Sample	Structure	x% QW1	t_b (nm)	x% QW2	$n(\times 10^{11} cm^{-2})$	$\mu(\times 10^3 cm^2/Vs)$
SQW	QW	10	-	-	4.79	35.8
10-1.4-10	DQW	10	1.4	10	4.76	12
8-5-14	DQW	8	5	14	6.63	37.9
11-5-16	DQW	11	5	16	5.0	39

3. Experimental Results and Discussions

In order to obtain the effective g-factor, time-resolved Kerr rotation (TRKR) studies were performed. For optical excitation a mode-locked Ti:sapphire laser was used emitting pulses with 100 fs duration at a rate of 75.6 MHz. The pump/probe lasers were tuned to each quantum well trion optical transitions to excite/monitor the 2DEG spin polarization. The pump beam was circularly polarized by an photo-elastic modulator with a frequency of 50 kHz. The probe beam polarization was not modulated and its change induced by the spin dynamics was detected with a brigde using coupled photodiodes. The Kerr signal was robust under high excitation power with pump-probe intensities of 10 mW and 5 mW respectively. The sample was immersed in the variable temperature insert of a superconductor magnet for fields B up to 6 T aligned perpendicular to the optical axis.

The electron spin precession was observed under an in-plane magnetic field by the time-resolved Kerr angle Θ_K oscillations with a frequency determined by the g-factor according to [6, 13]:

$$\Theta_K = A \exp(-\Delta t/T_2^*) \cos(2\pi f_L \Delta t + \phi) \quad (1)$$

where A is the amplitude proportional to the injected spin polarization along the optical axis, Δt is the pump-probe delay time, T_2^* is the spin dephasing time, $f_L = g\mu_B B/h$ is the Larmor frequency at a given magnetic field B, and ϕ is a phase offset. Only the magnitude of the g value will be consider in this report and not the sign ($g = |g|$).

Figure 1a shows the time evolution of the Kerr signal Θ_K for a single quantum well with Al content x=10% and for a double quantum well with x=10% in both wells separated by a narrow tunneling barrier. First, the SQW electron g-factor displays a large shift (Δg~44%) considering the calculated value of g~0.27 from reference [14]. From the SQW-DQW comparison, three

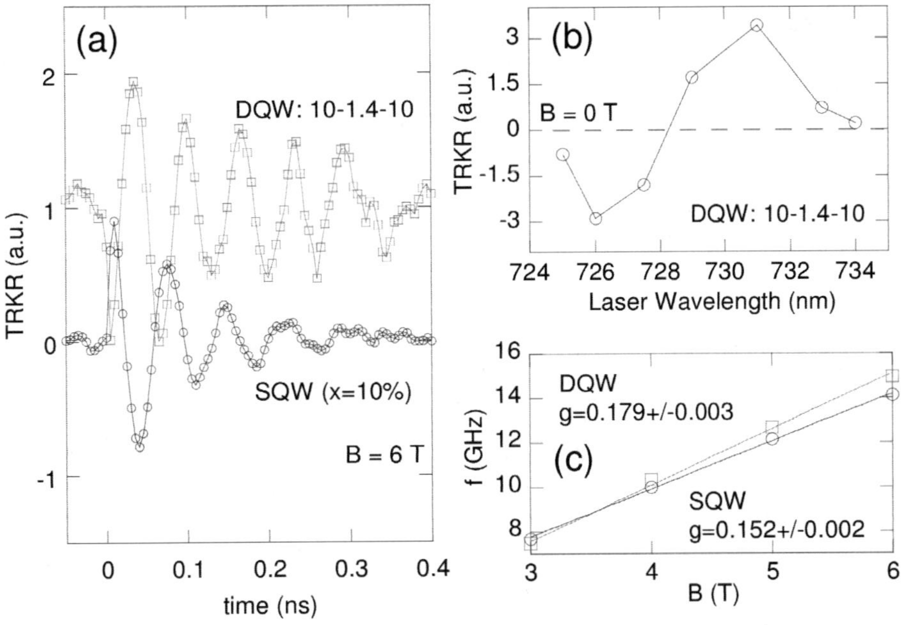

Figure 1. Single and double quantum wells with x=10%. (a) Kerr rotation at 6 Tesla. (b) Zero-field signal amplitude as function of the laser detuning at a fixed delay of 0.26 ns for sample 10-1.4-10. (c) Larmor frequency as function of the magnetic field strength. The solid line are linear fits with the g-factor as parameter.

principal features can be directly extracted: (i) the spin dephasing time appears to be longer in the double well system, (ii) the Larmor frequency is slightly larger in the double well sample, and (iii) Kerr signal oscillations display opposite phases.

To explore the opposite phase between the curves in figure 1(a), the Kerr signal amplitude was measured as function of the laser detuning for sample 10-1.4-10. Figure 1b shows that a phase change occurs leading to a maximum negative and positive amplitude a fixed time delay. In a single GaAs quantum well with the same width, the appearance of a phase change was assigned to the hole character (light/heavy) in the trion formation [4]. The electron g-factor measured for the excitation wavelength at both transitions was exactly the same ($g=0.179\pm0.003$ and $g=0.178\pm0.002$ at 726 nm and 731 nm respectively). This result indicates that this measured value is related to the electrons in the 2DEG also in accordance to this previous report. Given the fact that the larger spin polarization amplitude was obtained with excitation at the higher energy transition (light-hole trion), we plotted this data for a g-factor comparison between the SQW and the DQW 10-1.4-10 in figure 1c. The data shows that the quantum wells in the double structure with with narrow barrier must be coupled producing a considerably large g-factor magnitude enhancement from the SQW value $\Delta g\sim15\%$.

Figure 2a and b shows double quantum well samples with different Al composition in each well. Since the x parameter changes the energy gap, we can identify the emission peaks associated with the individual quantum wells in the photoluminescence spectra. These peaks are separated by the large x difference. The laser tuning near every peak position allow to observe the two opposite phase components in the Kerr signal as in figure 1b. In a similar manner, both data plots (figure 2a and b) display a larger g-factor and longer spin dephasing time for increasing x.

A possible coupling between the wells is indicated by the presence of the higher g-factor into

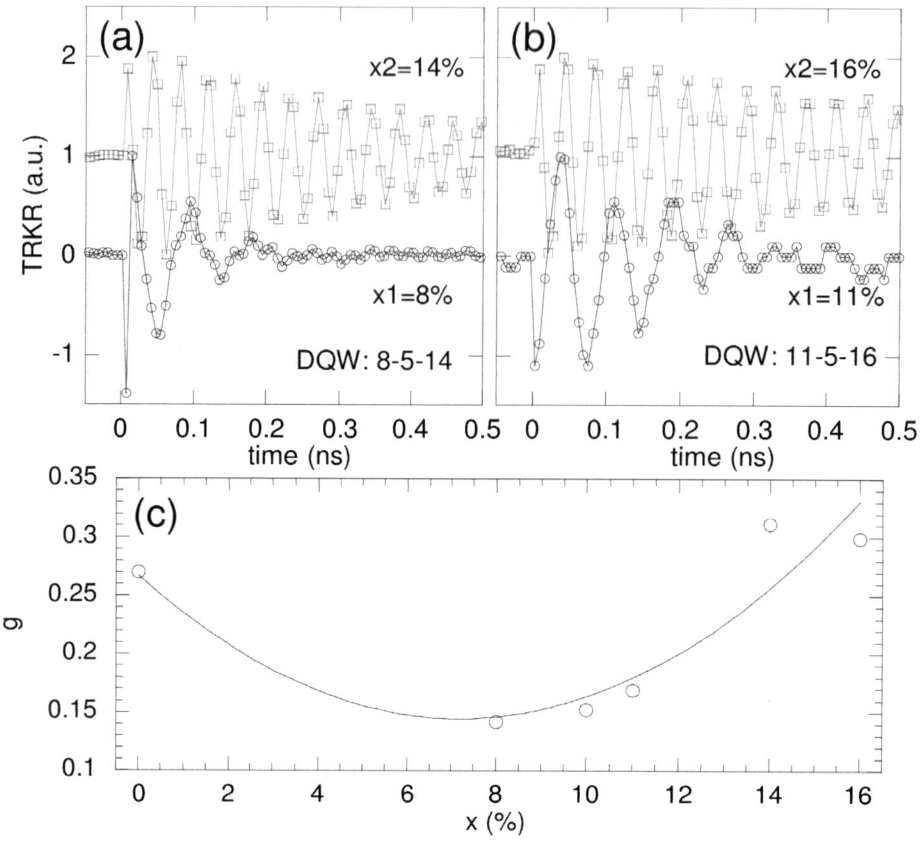

Figure 2. Kerr rotation signal in double structures using resonant excitation with each single quantum well for sample (a) DQW: 8-5-14 and (b) DQW: 11-5-16. (c) g-factor dependence on the Al content. The data point at x=0 was extracted for a 14 nm quantum well width from reference [14]. The solid line is a fitted parabolic function.

the other well signal in the same pair. This faster frequency appears with lower amplitude, as it is far from resonance, at a long delay time ($\sim$ 0.3 - 0.5 ns in the black curves). We can conclude that the double wells with a hard AlAs tunneling barrier of 5 nm still present coupling. However, this coupling results in the mixing of the individual g-factors and may not produce a shift in the individual values. Following this assumption, the g-factor associated to an individual well was plotted in figure 2c including a parabolic fitting as trend line. We extracted the value for a GaAs well (x=0) with the same width from reference [14] and we use the SQW data for x=10%. From the fitted function, it is expected a lower magnitude of the g-factor around 7%. We must note that the g-factor sign is considered in our present analysis. Nevertheless, if we take the know negative value at x=0, a vanishing g-factor if expected near x=7% as obtained in reference [6].

We conclude that the composition engineering in $Al_xGa_{1-x}As$ single and double square quantum wells allows to tuned the g-factor in large range as well as the spin polarization efficiency and dephasing time. For the double wells, the tunneling barrier adds another control parameter in a coupling regime. A detailed study of the indirect exciton in double quantum wells with unequal aluminum content as well as the excitation power dependence study will be published

Journal of Physics: Conference Series **456** (2013) 012015 doi:10.1088/1742-6596/456/1/012015

elsewhere [15].

4. Acknowledgments

This work was supported by FAPESP (Contracts 2009/15007-5 and 2010/09880-5) and CNPq (Brazilian agencies).

References

[1] Astakhov G V, Kochereshko V P, Yakovlev D R, Ossau W, Nürnberger J, Faschinger W, Landwehr G, Wojtowicz T, Karczewski G,and Kossut J 2002 *Phys. Rev. B* **65** 115310

[2] Suris R A, Kochereshko V P, Astakhov G V, Yakovlev D R, Ossau W, Nürnberger J, Faschinger W, Landwehr G, Wojtowicz T, Karczewski G,and Kossut J 2001 *Phys. Stat. Sol. b* **227** 343

[3] Zhukov E A, Yakovlev D R, Bayer M, Karczewski G, Wojtowicz T, and Kossut J 2006 *Phys. Stat. Sol. b* **243** 878

[4] Kennedy T A, Shabaev A, Scheibner M, Efros Al L, Bracker A S, and Gammon D 2006 *Phys. Rev. B* **73** 045307

[5] Mayer Alegre T P, Hernandez F G G, Pereira A L C, Medeiros-Ribeiro G 2006 *Phys. Rev. Lett.* **97** 236402

[6] Salis G, Kato Y, Ensslin K, Driscoll D C, Gossard A C, and Awschalom D D 2001 *Nature* **414**, 619

[7] Yugova I A, Greilich A, Yakovlev D R, Kiselev A A, Bayer M, Petrov V V, Dolgikh Yu K, Reuter D, and Wieck A D 2007 *Phys. Rev. B* **75** 245302

[8] Girvin S M and MacDonald A H 1996 *Perspectives in Quantum Hall Effects* ed S. Das Sama and A. Pinczuk (New York: Wiley)

[9] Armas L E G, Gusev G M, Lamas T E, Bakarov A K, and Portal J C 2009 *Int. J. Mod. Phys. B* **23** 2933

[10] Orlita M, Byszewski M, Döhler G H, Grill R, Hlídek P, Malzer S, Zvára M 2006 *Physica E* **34** 284

[11] Orlita M, Grill R, Hlídek P, Zvára M, Döhler G H, Malzer S, and Byszewski M 2005 *Phys. Rev. B* **72** 165314

[12] Butov L V, Levitov L S, Mintsev A V, Simons B D, Gossard A C, and Chemla D S 2004 *Phys. Rev. Lett.* **92** 117404

[13] Poggio M, Steeves G M, Myers R C, Stern N P, Gossard A C, and Awschalom D D 2004 *Phys. Rev. B* **70** 121305(R)

[14] Winkler R 2003 *Springer Tracts in Modern Physics* vol 191 (Berlin: Springer-Verlag) p 133

[15] Hernandez F G G, Gusev G M, Bakarov A K to be published

Journal of Physics: Conference Series **456** (2013) 012016 doi:10.1088/1742-6596/456/1/012016

Weak Localisation in Clean and Highly Disordered Graphene

Michael Hilke, Mathieu Massicotte, Eric Whiteway, Victor Yu
Department of Physics, McGill University, Montréal, Canada H3A 2T8
E-mail: hilke@physics.mcgill.ca

Abstract. We look at the magnetic field induced weak localisation peak of graphene samples with different mobilities. At very low temperatures, low mobility samples exhibit a very broad peak as a function of the magnetic field, in contrast to higher mobility samples, where the weak localisation peak is very sharp. We analyze the experimental data in the context of the localisation length, which allows us to extract, both the localisation length and the phase coherence length of the samples, regardless of their mobilities. This analysis is made possible by the observation that the localisation length undergoes a generic weak localisation dependence with striking universal properties.

1. Introduction:

Weak localisation (WL), which is the enhanced backscattering of coherent electrons in a disordered media [1, 2], has been observed in many materials, including in thin metals and in two dimensional electron gases [3]. Since WL strongly depends on phase coherence, it is often used to probe the phase coherence properties of a device and as a tool for extracting information on various scattering mechanisms. More recently, WL was discussed and observed in graphene [4, 5, 6], where the existence of two degenerate Dirac valleys leads to interesting new WL properties. For instance, the strength of the WL effect strongly depends on inter-valley scattering, and WL is expected to vanish in its absence [7]. Perturbative expressions for the conductivity corrections in graphene due to WL have been worked out in great detail by McCann et al. [5] and successfully fitted to the magnetic field dependence of the resistance at low temperatures [6]. The dominant feature is a peak in the resistance at zero magnetic field. Most of the works were done in a regime of weak disorder, where the perturbative approach by McCann et al. is applicable and where strong localisation is absent.

The aim of this paper is to discuss an alternate point of view to WL, which is not restricted to weak disorder. This was largely driven by the desire to understand experimental observations of WL in a large range of graphene samples, of which some were highly disordered. At low temperatures, the sharpness of the WL peak in clean large scale graphene is quite striking [8], whereas in low mobility graphene samples, a wide negative magnetoresistance behaviour is observed [9]. What we show in this work, is that WL can be naturally understood within the framework of strong localisation and applied to the experimental case of strong disorder as well as weak disorder.

2. Localisation length:

Our results follow from the observation shown in figure 1, where the localisation length (L_c) exhibits a dip at zero magnetic field, which in turn leads to a peak in the resistance. L_c was extracted by computing the resistance of a disordered graphene flake using non-equilibrium Green's functions. The two terminal resistance is obtained by computing the Green's function of the honeycomb tight binding lattice iteratively as a function of the length of the system. L_c is then determined from the exponential increase in resistance with the length of the graphene flake, when the length exceeds L_c. Disorder is assumed to be on-site and uncorrelated. Figure 1 shows the dip at zero field of L_c for different strengths of the disorder V, where each carbon site has a random potential $-V/2 < v_i < V/2$. V is in units of the hopping element, which is $\sim 3eV$ in graphene.

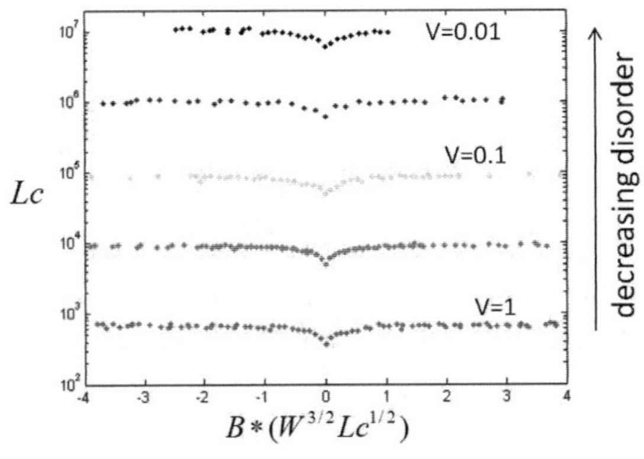

Figure 1. The magnetic field dependence of the extracted localisation length (in units of lattice sites) for different values of the disorder V.

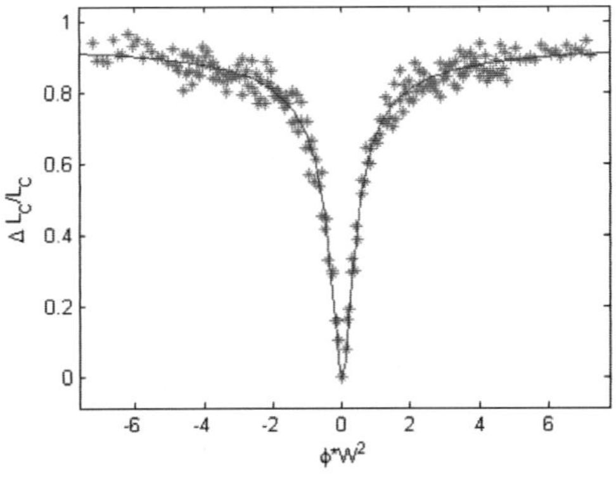

Figure 2. The magnetic field dependence of the relative localisation length for the disorder strength $V = 0.3$ and for different ribbon widths (40 to 120 lattice sites). The line is fitted according to equation (1).

L_c follows the same behavior regardless of the amount of disorder or the width of the device. This is nicely illustrated by rescaling the magnetic field by $B\sqrt{W^3Lc}$ and then by plotting the relative $\Delta L_c/L_c = (L_c(B) - L_c(0))/L_c(0)$ as a function of field. In this case all curves overlap and are described by the same function within numerical accuracy. The edge plays an important role here because, in the weak disorder limit, we always have $L_c \gg W$, where W is the width of the system. This is because L_c scales as $L_c \sim W$ for a given disorder strength, hence no matter how wide the device, we will always have $L_c \gg W$ in the low disorder limit.

In the presence of edge scattering, it was shown that the correction to the conductivity is given by $\Delta\sigma \sim (1 - 1/\sqrt{1 + B^2/B_c^2})$, where B_c is a characteristic field determined by the device width and the scattering strength [5]. Since Anderson localisation is also a measure of the return probability, it is not surprising that the same functional form also determines L_c, hence we find that the expression

$$\Delta L_c/L_c = \frac{L_c(B) - L_c(0)}{L_c(0)} \simeq 0.95 \left(1 - 1/\sqrt{1 + B^2/B_c^2}\right) \tag{1}$$

convincingly fits the numerical data shown in figure 2. However, when the disorder becomes very strong as measured by $L_c \ll W$, edge scattering is no longer relevant, since in this case W always exceeds L_c. This becomes then a two-dimensional problem, where the shortest length scale is given by L_c. In this case expression (1) is not valid anymore and we have to use a two-dimensional analogue. Numerically, this is illustrated in figure 3, where we show again the relative L_c, as a function of magnetic field, but this time for stronger disorder. When comparing figure 3 to figure 2, the shape has changed but not the general behavior, namely a dip at zero field.

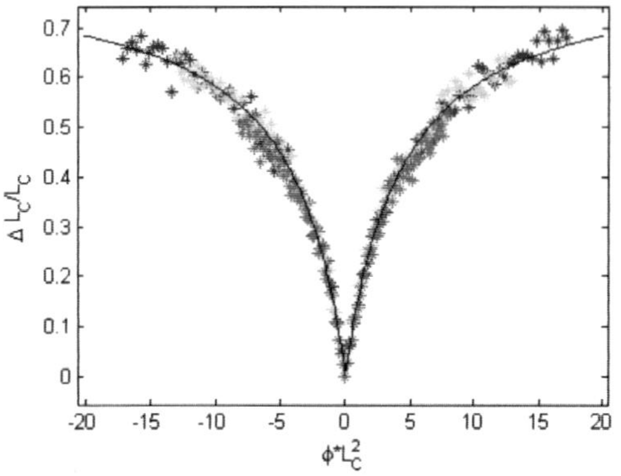

Figure 3. The magnetic field dependence of the relative L_c for the disorder strength $V = 2$ and for different widths (40 to 120). The full line is obtained using equation (2).

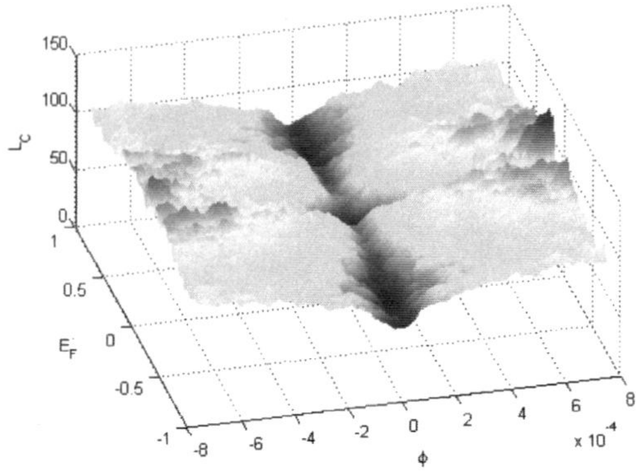

Figure 4. The density (or Fermi energy) dependence, as well as the magnetic field dependence of L_c is plotted for $V = 2$ and the width $W = 120$. The Dirac point corresponds to $E_F = 0$.

The functional form of the field dependence has to be modified according to the two-dimensional nature of the problem and can be obtained by following, for instance, Beenakker's approach to WL [10]. The field correction to the conductivity can be written as $\Delta\sigma \sim \ln(t_\phi/\tau_e + 1)$, where $t_\phi \sim 1/(1/\tau_\phi + 1/\tau_B)$, τ_e is the elastic scattering time, τ_ϕ the phase coherence time and $\tau_B \sim B^{-1}$ the field induced destruction of coherent backscattering. For the numerical simulations, we effectively work at zero temperature, which leads to an infinite phase coherence time. In this case, the coherent backscattering time is not bound by τ_ϕ but is bound by the localisation time, τ_c, where $L_c = \sqrt{D\tau_c}$ and D is the diffusion constant. Hence, for the simulations, τ_ϕ has to be replaced by τ_c in the expression for the field dependence. Now, assuming again that the relative change in L_c is proportional to the coherent return probability, i.e., the relative change of the conductivity and further assuming that $\tau_e \simeq \tau_c$, we obtain

$$\Delta L_c/L_c \simeq 1.4 \cdot \ln\left(\frac{1 + B/B_c}{1 + B/2B_c}\right), \tag{2}$$

which nicely fits the numerical data in figure 3 (the prefactor is $\sim 1/\ln(2)$ which normalizes expression (2) at high fields). We used $\tau_c/\tau_B = B/B_c$, where B_c is again the characteristic magnetic field. In the limit, where $\tau_e \ll \tau_c$, we obtain simply $\Delta L_c/L_c \sim \ln(1 + B/B_c)$, which leads to a renormalization of B_c by a factor of 2 at small fields. At low temperatures and in the presence of strong disorder, L_c is the shortest length scale and the characteristic field is $B_c \simeq \hbar/eL_c^2$, which defines the area of a phase coherent loop, before coherent backscattering is suppressed when penetrated by a flux quantum. Indeed, equation (2) fits the numerical results for the relative L_c, when the field is rescaled by $B \cdot L_c^2$, as shown in figure 3. For small disorder, on the other hand, where $L_c \gg L_\phi$, the characteristic field is instead determined by L_ϕ and given by $B_c \simeq \hbar/eL_\phi^2$. Hence, using equation (2) we can describe the generic behaviour of both the low disorder ($L_c \gg L_\phi$) and high disorder ($L_c \ll L_\phi$) limits. To make the connection to localisation as above, it is important to distinguish between the behaviour of L_c and the behaviour of the relative L_c. Indeed, L_c does not follow a generic dependence on field or density as shown in figure 4. It is only the relative L_c which shows this generic and an almost universal behaviour as graphed in figure 8.

3. Experiments:

The next step is to apply these concepts to the experimental data. We performed experiments on large scale graphene as well as lithographically defined Hall bars and graphene nano-ribbons, in addition to large (over 100 μm) single crystal grains, of which some show strong dendritic structures and are dubbed graphlocons. Monolayers of graphene were grown by chemical vapor deposition (CVD) of hydrocarbons on 25 μm-thick commercial Cu foils. The CVD process used was similar to those described in previous works [8, 11, 12].

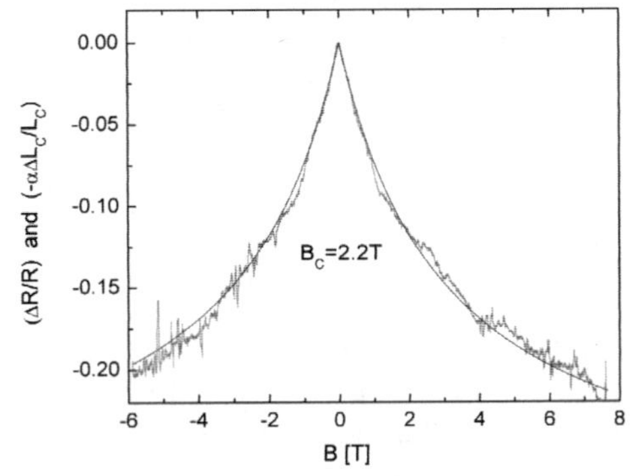

Figure 5. The measured magneto-resistance of a low mobility ($\mu \simeq 100cm^2/Vs$) graphene Hall bar device at 100mK. Shown is the relative resistance and the fit to the relative L_c. The fitting parameters are $B_c = 2.2$T and $\alpha = 0.4$.

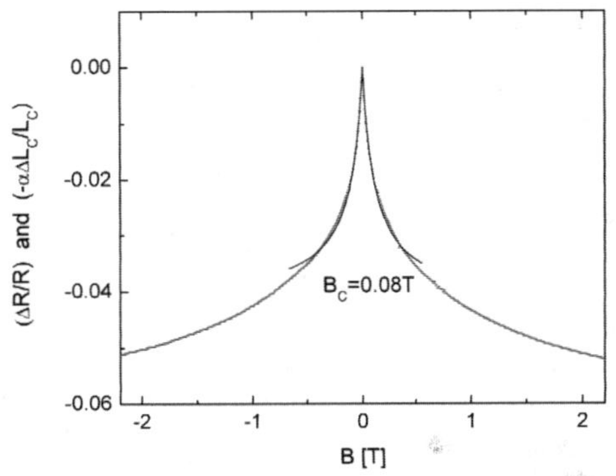

Figure 6. This sample had a medium mobility of $\mu \simeq 700cm^2/Vs$) and the fitting parameters are $B_c = 0.08$T and $\alpha = 0.06$.

The overall experimental behavior can be summarized as follows: at low temperatures, low mobility samples, typically below 500 cm^2/V·s show a very wide peak in the resistance at zero field (as shown in figure 5). With increasing mobility, the peak becomes increasingly sharper (figures 6 and 7). All the samples we measured show a WL peak at zero field. The relative change in resistance varies between 0.5% for high mobilities to 20% for low mobilities. This is also expected from McCann's expression, since it yields $\Delta\rho/\rho \sim \rho\Delta F(B)$, where $\Delta F(B)$ describes the magnetic field dependence and ρ is the resistivity [5]. The main difference is that equation (2) leads to an increased sensitivity on the sample geometry, since for a phase coherent sample the resistance will depend exponentially on the length at lengths longer than the localization length.

To connect the data with L_c, we can simply assume that the resistance is given by $R \sim e^{L/L_c}$ where L is the length of the sample. If L_ϕ is smaller than the length of the sample, then L needs to be replaced by L_ϕ. For small changes in the relative resistance, we can then write

$$\frac{\Delta R}{R} \simeq -\frac{\Delta L_c}{L_c} \cdot \frac{L_\phi}{L_c}. \tag{3}$$

Using equations (2) and (3) we can now fit the relative resistance as a function of the magnetic field for different samples. The fits are shown in figures 5 to 7 for samples with increasing mobilities. The fits are quite good and require only two fitting parameters, B_c and $\alpha \simeq 1.4L_\phi/L_c$, which are given in the corresponding figure captions. Recalling that for large devices, the limits are given by $B_c \simeq \hbar/e\min\{L_\phi^2, L_c^2\}$, which can be interpolated to $B_c \simeq \hbar(L_\phi^{-2} + L_c^{-2})/e$ and which allows us to obtain L_c and L_ϕ. Plugging in the numbers we find that $L_\phi \simeq 20$nm and $L_c \simeq 60$nm for the sample in figure 5, whereas $L_\phi \simeq 90$nm and $L_c \simeq 2\mu$m for the sample in figure 6, and finally $L_\phi \simeq 0.5\mu$m and $L_c \simeq 65\mu$m for the sample in figure 7.

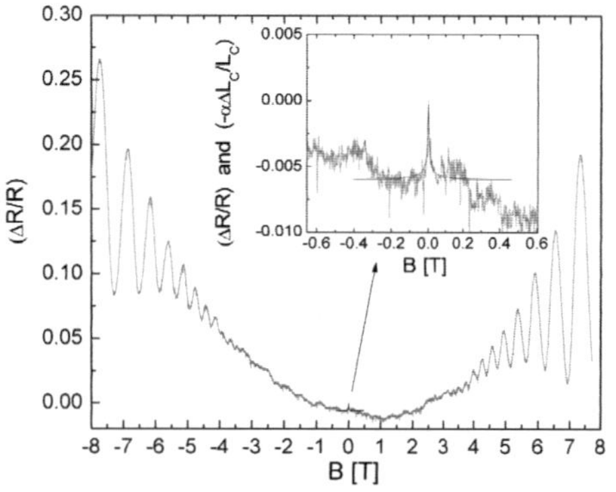

Figure 7. The relative resistance change of a high mobility sample ($\mu \simeq 5000cm^2/Vs$) with fitting parameters $B_c = 0.003$T and $\alpha = 0.01$. The inset is a zoom-in of the low field region.

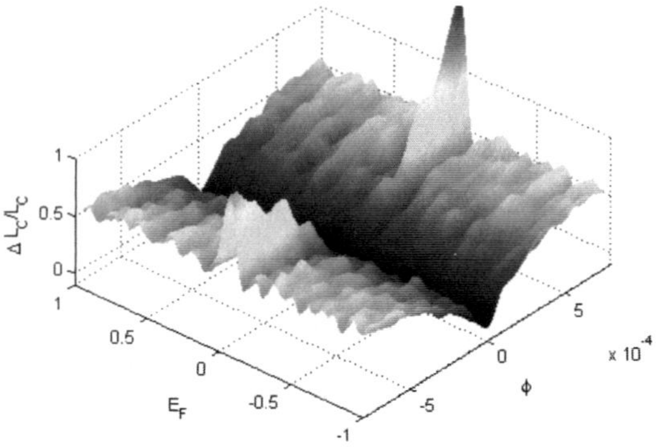

Figure 8. The relative L_c as a function of the Fermi energy and magnetic field for $V = 2$ and $W = 120$. The absence of a systematic dependence on the Fermi energy outside of the Dirac peak is quite striking and determines a close to universal dependence, irrespective of density and disorder strength.

4. Conclusion:
We presented an alternate way of looking at the experimental weak localisation peak, by realizing that the localisation length also undergoes a weak localisation effect, in the sense that there is a dip at zero field, which implies a peak in resistance. This allows us to connect weak localisation to strong localisation in a natural way. Moreover, we can associate a characteristic field, B_c, which determines the magnetic field dependence. B_c is determined by the square of the inverse localization length for strong disorder and by the square of the inverse phase coherence length for weak disorder. Indeed, we find that the magnetic field dependence of the relative localisation length becomes close to universal, and that all curves can be rescaled on top of each other, by rescaling the field dependence with B_c. We used this generic localisation length dependence to fit our experimental data for graphene samples of very different mobilities, which not only allowed us to obtain a good fit, but also to extract the localisation length and phase coherence length independently. This method is very robust and is applicable to high and low mobility samples. We expect this analysis to be very generic and applicable to a large class of two dimensional systems and not only to graphene.

5. Acknowledgments:
We thank R. Gagnon and J. Lefebvre for technical assistance and NSERC and FQRNT for financial assistance.

6. References

[1] S. Hikami, A.I. Larkin and Y. Nagaoka, Prog. Theor. Phys. Vol. **63**, 707 (1980)
[2] B.L. Altshuler, D. Khmel'nitzkii and A.I. Larkin, P. A. Lee, Phys. Rev. B, **22**, 5142 (1980).
[3] G. Bergmann, Phys. Reports, **107**, 1 (1984); D.J. Bishop, R.C. Dynes, and D.C. Tsui, Phys. Rev. B, **22**, 773 (1982); G.M. Gusev, et al., J.Phys.:Cond. Matter **4**, L269 (1992).
[4] H. Suzuura and T. Ando, Phys. Rev. Lett. **89**, 266603 (2002).
[5] E. McCann et al., Phys. Rev. Lett. **97**, 146805 (2006)
[6] S. V. Morozov et al. Phys. Rev. Lett. **100**, 016602 (2008); F. V. Tikhonenko et al. Phys. Rev. Lett. **100**, 056802 (2008); D.-K. Ki, et al. Phys. Rev. B **78**, 125409 (2008).
[7] D.R. Cooper et al., ISRN Cond. Mat. Phys., 501686 (2012).
[8] E. Whiteway, V. Yu, J. Lefebvre, R. Gagnon, and M. Hilke, arXiv:1011.5712
[9] X. Hong, S.-H. Cheng, C. Herding, and J. Zhu, Phys. Rev. B, **83**, 085410 (2011).
[10] C.W.J. Beenakker and H. van Houten, Phys. Rev. B, **38**, 3232 (1988).
[11] V. Yu, E. Whiteway, J. Maassen, and M. Hilke, Phys. Rev. B, **81**, 205407 (2011).
[12] S. Bernard, E. Whiteway, V. Yu, D. G. Austing, and M. Hilke, Phys. Rev. B, **86**, 085409 (2012).

Journal of Physics: Conference Series **456** (2013) 012017 doi:10.1088/1742-6596/456/1/012017

Charge conversion of nearly free and impurity bound magneto-trions immersed in 2D electron or hole gas with optically tunable concentration

J Jadczak[1], L Bryja[1], J Misiewicz[1], A Wójs[1], M Potemski[2], F Liu[3], D R Yakovlev[3], M Bayer[3], D Reuter[4], A D Wieck[4], C A Nicoll[5], I Farrer[5], D A Ritchie[5]

[1]Institute of Physics, Wroclaw University of Technology, 50-370 Wroclaw, Poland
[2]Laboratoire National des Champs Magnétiques Intenses, CNRS-UJF-UPS-INSA, 38042 Grenoble, France
[3]Experimentelle Physik 2, Technische Universität Dortmund, 44221 Dortmund, Germany
[4]Lehrstuhl für Angewandte Festkörperphysik, Ruhr-Universität Bochum, 44780 Bochum, Germany
[5]Cavendish Laboratory, University of Cambridge, Cambridge CB3 0HE, UK

joanna. jadczak@pwr.wroc.pl

Abstract. We report collection of precise photoluminescence maps (emission intensity versus energy and magnetic field) of superior-quality two-dimensional hole gases. The maps reveal field evolution of both direct and cyclotron-assisted recombination lines attributed to various excitonic complexes, either moving nearly freely in the plane or bound to the acceptors placed inside or outside the quantum well. Under two-beam illumination (with photon energies below and above the band-gap in the barrier) we were able to control hole concentration (in the same well), and in particular to decrease it beyond the point of p- to n-type conversion. Our results demonstrate contrast between charge conversion of free and acceptor-bound positive trions resulting from the breaking of charge reflection symmetry by a fixed impurity.

1. Introduction

Two-dimensional (2D) electron gas subject to quantizing magnetic field is a paradigmatic system for the investigation of many-body interaction effects and as well as single-particle localization [1]. Most previous experiments were devoted to the electron systems due to their much higher mobility. However, recent advances in the growth of high purity semiconductor structures have also refocused attention onto the 2D hole systems. The magneto-optical spectra of the 2D holes are not equivalent to those of the 2D electrons mainly due to their much larger effective mass, which results in different relations between the binding energies of neutral and charged complexes. In presented studies we have recorded extremely precise photoluminescence maps as a function of excitation intensity and energy vs. magnetic field of high quality 2D hole gases, revealing field evolution of both direct and cyclotron-assisted recombination lines attributed to various excitonic complexes moving nearly freely in the plane or bound to residual acceptors positioned inside or outside the quantum well. Combining red and green laser excitation power, with photon energies below and above the band-gap of the barrier, we were able to control the hole concentration and even decrease it beyond the point of conversion from p- to n-type (in the same structure) [2].

2. Experiment and results

The investigated samples were superior quality 15 nm wide GaAs/$Ga_{0.66}Al_{0.34}As$ quantum wells, grown by molecular beam epitaxy on a (001) semi-insulating GaAs substrate. The 2D hole gas was introduced into the quantum well by Carbon δ-doping in the barriers. Symmetric and asymmetric structures were fabricated with either the same or different doping dose in both barriers. In all samples, the hole mobility measured at $T = 4.2$ K was in the range of $\mu = 10^5$ cm^2/Vs and the hole concentration measured in the dark varied in the range of $p = (1.1 - 2.4) \times 10^{11}$ cm^{-2}. Photoluminescence (PL) was excited by a combination of two beams: a "red" 720-nm line of a titanium sapphire tunable laser and a "green" 514-nm line of an ion argon laser with the photon energy below and above the band gap of the barrier respectively. As it was demonstrated previously [1,2], the selective photoexcitation inside and outside of the quantum well allows control of the dynamical carrier population and even of the sign of majority carriers inside the well. The experiments were performed in the Faraday configuration, in magnetic fields up to $B = 23$ T changed with a small step $\Delta B = 0.05$ T at temperatures varied in the range of $T = 1.8$ to 30 K. Fiber optics geometry was used. The switch between the σ^- and σ^+ circular polarizations was achieved by reversing the direction of the field (current). The PL spectra were analyzed in a 0.7 m long monochromator, equipped with a Nitrogen-cooled charge coupled device.

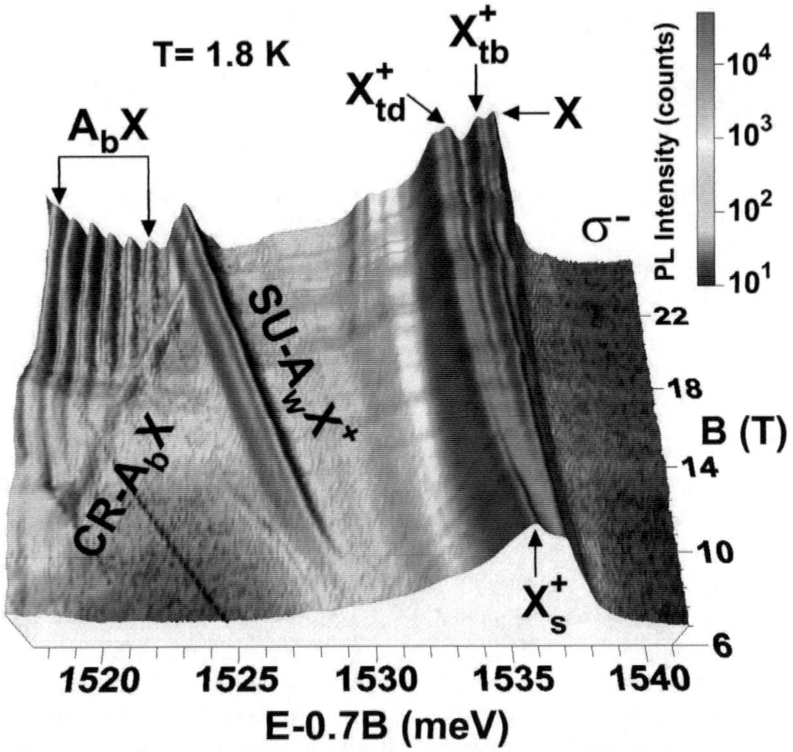

Figure 1. (color online) Evolution of the PL spectrum of the symmetric GaAs quantum well of (dark) concentration $p = 1.15 \times 10^{11}$ cm^{-2}, in magnetic field B varied from 0 to 23 T, in a strong intensity σ^- polarization, at $T = 1{:}8$ K, under power density $P = 30$ mW/cm^2 of the red laser. Photoluminescence intensity is presented in color logarithmic scale.

In the Fig. 1 evolution of PL spectra in σ^- polarization at temperature $T = 1.8$ K under excitation of red laser only are presented in magnetic field varied from $B = 6$ to 23 T.

In the Fig. 2 variations of the spectrum as a function of the power density of an additional green laser (decreasing 2D hole concentration), in magnetic field $B = 23$ T in σ+ polarization is presented.

Aided with the realistic numerical calculations [3,4] detailed analysis of the experimental data (field evolution of the emission spectra detected for different concentrations or holes or electrons), allowed us to interpret almost all observed emission lines related to the 2D gas. We divided them into two main groups.

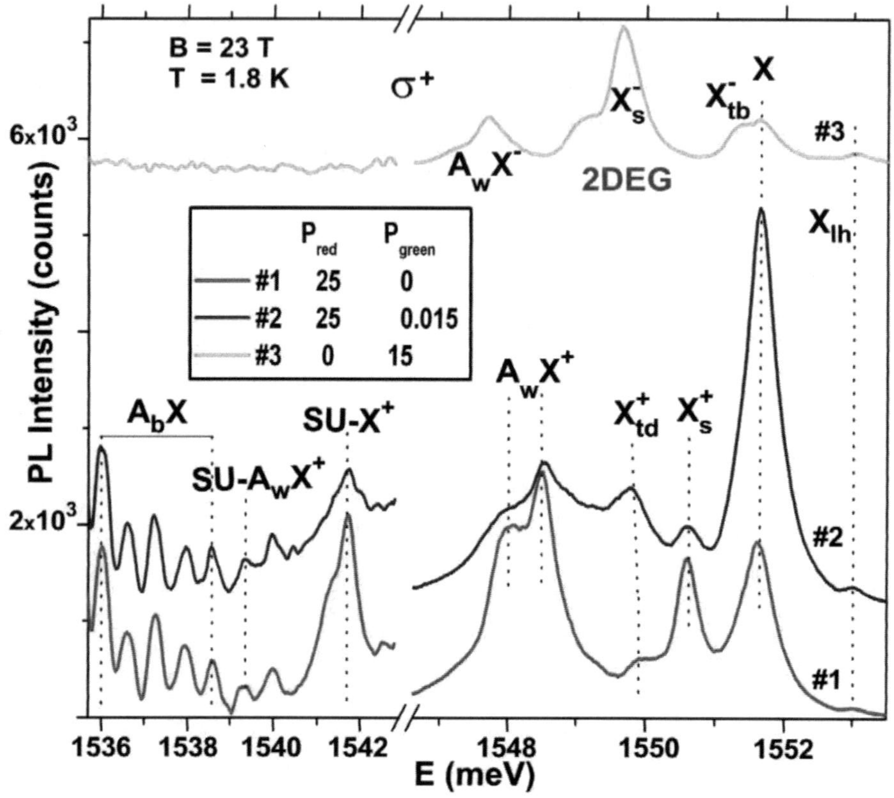

Figure 2. (color online) The PL spectra at B=23T, under different excitation of red and green lasers, in the symmetric GaAs quantum well.

The first group of line, detected in the high energy region of the PL spectra, are strongly dependent on the quantum well width [5]. In this group we detected emission (highest in energy and intensity) from all theoretically predicted trions: the singlet X^+_s and two triplet states (bright X^+_{tb} and dark X^+_{td}) as well as the neutral excitons ($X_{hh} \equiv X$ and X_{lh} containing a heavy or light hole, respectively). Below, in energy sequence, to these lines, we detected the following lines in PL spectra: doublet of acceptor bound trion A_wX^+ (Fig. 2), hole cyclotron replicas of almost free and bound trions (SU – X^+ and SU – A_wX^+) related to well known "shake-up" processes [6]. We interpreted all above lines as radiative recombination of the electron-hole complexes whose wave functions are predominately located inside the quantum well. With increase of the power density of green laser the structure converts from p to n type. All positively charged complexes disappear from the PL spectra and are replaced by negatively charged ones: X^- and A_wX^- (Fig. 2). The former transition: charge conversion between the positive and negative excitonic complexes bound to acceptors ($A_wX^+ \rightarrow A_wX^-$) are the most spectacular observation in our investigation. This effect is essentially distinct from the previously observed charge conversion of free trions [7]. In the $A_wX^+ \rightarrow A_wX^-$ conversion two holes are subtracted whereas in $X^+ \rightarrow X^-$ conversion one hole in trion complex is substituted by one electron.

The second group of lines are detected in the lower energy sector of the PL spectrum. Their energy positions are insensitive to the width of the quantum well. We have attributed them to the radiative recombination of positive trions strongly localized by Coulomb potentials of remote ionized acceptors located in the barriers: $A_bX=A^-+X^+$. The multiple, parallel lines correspond to the A_bX complexes with acceptors placed in the barrier at subsequent crystalline layers. Above interpretation is supported with the following observation. In the studied the high-quality structure the residual bulk acceptor density in the barrier is on the order of 10^{15} cm^{-3}. This gives the average acceptor density per crystalline layer of 0.28×10^8 cm^{-2},which is over three orders of magnitude lower then 2D hole concentration in the well yielding considerably weaker intensity of each single A_bX^- line in comparison to X^+.

3. Conclusions

In polarization-resolved two-beam magneto-photoluminescence experiments performed on high-mobility 15 nm wide GaAs/$Ga_{0.66}Al_{0.34}As$ quantum wells we have identified and understood several optical transitions, including the nearly free excitons and trions, excitonic complexes bound to charged impurities placed in the well and in the barriers, and also their cyclotron replicas. In particular we have observed coexistence of nearly free and strongly bound trions as well as charge conversion between the positive and negative exciton complexes (both free and bound to acceptors in the well).

Acknowledgements

This work has been supported by the Polish NCN Grant No. 2011/01/B/ST3/04504 and by EU Marie Curie Grant No. PCIG09-GA-2011-294186.

References

[1] Jain J K, *Composite Fermions* (Cambridge University Press, Cambridge, 2007); Ihn T, *Semiconductor Nanostructures: Quantum States and Electronic* Transport (Oxford University Press, Oxford, 2010)
[2] Bartsch G et al., *Phys. Rev.* B **83,** 235317 (2011).
[3] Bryja L, Wójs A, Misiewicz J, Potemski M, Reuter D, Wieck A, *Phys. Rev.* B **75**, 035308 (2007).
[4] Wójs A, Quinn J J, Hawrylak P, *Phys. Rev.* B **62**, 4630 (2000); Wójs A, *Phys. Rev.* B **76**, 085344 (2007).
[5] Bryja L et al., *Phys. Rev.* B **85**, 165308 (2012)
[6] Glasberg S, Finkelstein G, Shtrikman H and Bar-Joseph I, *Phys. Rev.* B **59**, 10425R (1999).
[7] Glasberg S, Shtrikman H and Bar-Joseph I, *Phys. Rev.* B **63** 201308(R) (2001).

Journal of Physics: Conference Series **456** (2013) 012018 doi:10.1088/1742-6596/456/1/012018

Theory of Defect-Induced Kondo Effect in Graphene: Effect of Zeeman Field

Taro Kanao, Hiroyasu Matsuura and Masao Ogata

Department of Physics, the University of Tokyo, Bunkyo, Tokyo 113-0033, Japan

E-mail: kanao@hosi.phys.s.u-tokyo.ac.jp

Abstract. The effect of the Zeeman field on the defect-induced Kondo effect in graphene is investigated. The effective model of the Kondo effect is derived, and is analyzed on the basis of the numerical renormalization group method. It is found that, under the Zeeman field, at the border of the spin polarized state and the Kondo-Yosida singlet state, there is an unusual fixed point where the entropy of the defect is $k_B \ln 2$ and the expectation value of S_z^2 is 1/8.

1. Introduction

The effects of defects on the properties of graphene are of scientific and practical importance [1]. For studying the effects of defects experimentally, ion irradiation has been used, which introduces point defects in graphene. Recently, in such an ion-irradiated graphene, the Kondo effect is observed [2]. Usually, the Kondo effect indicates the presence of localized moments in the sample. Therefore the above result suggests that a point defect in graphene causes a localized moment. The magnetic behavior of the defects in graphene is also observed in the recent magnetization measurement of ion-irradiated graphene [3].

As a possible origin of this Kondo effect, we proposed a localized moment on a sp^2 orbital around the defect [4]. There are three sp^2 orbitals around a point defect. With Jahn-Teller distortion, two of them form a covalent bond and remaining one behaves as a localized moment with spin-half. Furthermore, with this lattice distortion, there is a hybridization between the localized sp^2 orbital and conduction π electrons, and the Kondo effect can occur. On this scenario, we introduced an effective model and analyzed it by the numerical renormalization group (NRG) method [5, 6]. As a result, we showed that the experimentally observed gate voltage dependence of Kondo temperature can be understood by assuming that the hybridization between the sp^2 and π orbitals is strong enough [4].

In the above experiment, the negative magnetoresistance (MR), which is the evidence of the Kondo effect, was also observed. Although the characteristic magnetic field for this negative MR is found to be smaller by one order than the usual Kondo-effect case, its origin has not been understood. So far, the studies on the effect of magnetic field on the Kondo effect in graphene are limited, and there are few studies on the effect of Zeeman field [7].

Motivated by this negative MR, in this paper, we construct an effective model of graphene with a point defect and study the magnetic field effect on the thermodynamics of the model on the basis of the NRG method. Here, as a first step, we consider only the Zeeman field and neglect the orbital magnetic effect on the conduction electrons.

2. Model Hamiltonian with the Zeeman field

The effective Hamiltonian of the graphene with a point defect has been discussed in ref. [4], and the one-dimensional energy representation for the conduction electrons [6, 8, 9] which is necessary to apply the NRG method has been derived. In this representation, the effective Hamiltonian is written as

$$H = H_{\rm gra} + H_{\rm def} + H_{\rm hyb}, \tag{1}$$

where $H_{\rm gra}$, $H_{\rm def}$, and $H_{\rm hyb}$ are Hamiltonians for the conduction electrons, the defect sp^2 orbital, and the hybridization, respectively, with

$$H_{\rm gra} = \sum_{\sigma=\uparrow,\downarrow} \int_{-D}^{D} {\rm d}\varepsilon\,(\varepsilon-\mu)\,c^\dagger_{\varepsilon\sigma}c_{\varepsilon\sigma}, \tag{2}$$

$$H_{\rm def} = \sum_\sigma (\epsilon_{sp^2}-\mu)d^\dagger_\sigma d_\sigma + U d^\dagger_\uparrow d_\uparrow d^\dagger_\downarrow d_\downarrow - \sum_\sigma h\sigma d^\dagger_\sigma d_\sigma, \tag{3}$$

$$H_{\rm hyb} = \frac{V}{D}\sum_\sigma \int_{-D}^{D} {\rm d}\varepsilon|\varepsilon|^{1/2}\left(c^\dagger_{\varepsilon\sigma}d_\sigma + {\rm h.c.}\right). \tag{4}$$

Here, $c_{\varepsilon\sigma}$ is an annihilation operator of the conduction electrons with energy ε and spin $\sigma =\uparrow$, $\downarrow$, which satisfies the anti-commutation relation $\left\{c_{\varepsilon\sigma}, c^\dagger_{\varepsilon'\sigma'}\right\} = \delta_{\sigma\sigma'}\delta(\varepsilon-\varepsilon')$ [4, 6, 8, 9]. $\delta(x)$ is the Dirac delta function. D is a cut-off energy that is the same order as the band width of the conduction electrons. d_σ is an annihilation operator of an electron on the defect sp^2 orbital with spin $\sigma =\uparrow$, $\downarrow$, which satisfies $\left\{d_\sigma, d^\dagger_{\sigma'}\right\} = \delta_{\sigma\sigma'}$. μ is the chemical potential of electrons. ϵ_{sp^2}, U, and h are the energy level of the defect sp^2 orbital measured from that of p_z orbital on carbon atom, Coulomb interaction, and Zeeman field on the defect sp^2 orbital, respectively. V is the amplitude of the hybridization between conduction electrons and the sp^2 orbital.

This model of eqs. (1)-(4) is known as the single-channel pseudogap Anderson model [10], where the density of states (DOS) of the conduction electron is proportional to $|\varepsilon|$. The Kondo problem in this pseudogap system was studied in detail for the case with $\mu = 0$ and $h = 0$ [10]. The main feature in this case is as follows. In the particle-hole symmetric ($2\epsilon_{sp^2} + U = 0$) case, no Kondo screening occurs, i.e. the localized moment is not screened by conduction electrons. On the other hand, in the particle-hole asymmetric ($2\epsilon_{sp^2} + U \neq 0$) case, the localized moment is screened when the coupling constant V is larger than a critical value V_c. In this paper, we assume the particle-hole asymmetry and study the case with $h \neq 0$ and both $\mu = 0$ and $\mu \neq 0$.

3. Numerical renormalization group method

We apply the NRG method for the model (1)-(4), following the formalism of ref. [10] to investigate the electronic state at low temperatures. The NRG method can include the effects of the pseudogap DOS, a finite chemical potential, and Zeeman field. In the NRG method, the effective Hamiltonian is transformed into the one-dimensional form with a chain length N,

$$H = \lim_{N\to\infty} \alpha\Lambda^{-(N-1)/2}H_N, \tag{5}$$

where

$$H_{N+1} = \Lambda^{1/2}H_N + \sum_\sigma \left[\Lambda^{-1/2}\epsilon_{N+1}c^\dagger_{N+1\sigma}c_{N+1\sigma} + t_N\left(c^\dagger_{N\sigma}c_{N+1\sigma} + c^\dagger_{N+1\sigma}c_{N\sigma}\right)\right], \tag{6}$$

$$H_0 = \Lambda^{-1/2}\left[\sum_\sigma \epsilon_0 c^\dagger_{0\sigma}c_{0\sigma} + \frac{1}{\alpha}(H_{\rm def} + H_{\rm hyb})\right], \tag{7}$$

$$H_{\rm hyb} = VF\sum_\sigma \left(c^\dagger_{0\sigma}d_\sigma + {\rm h.c.}\right). \tag{8}$$

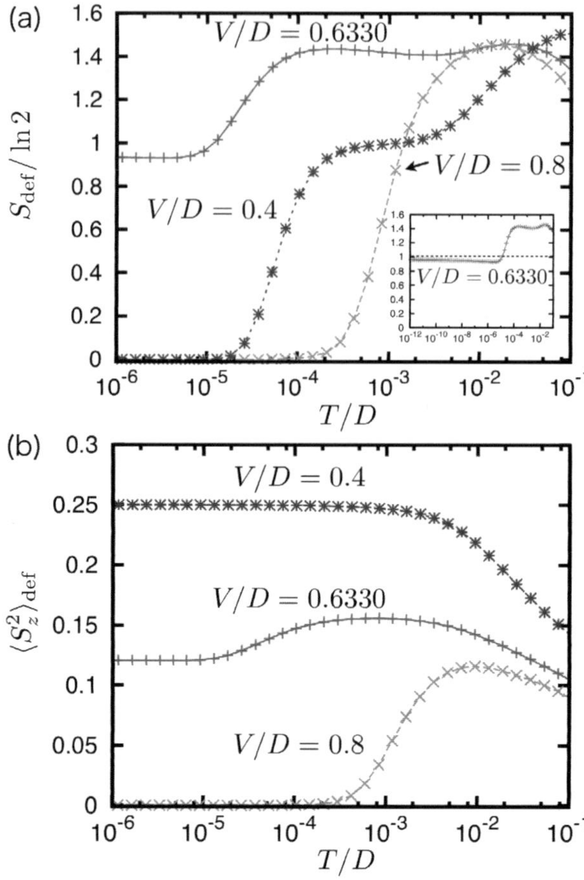

Figure 1. (a) Temperature dependence of entropy of the electron on the defect sp^2 orbital, $S_{\rm def}$, under the Zeeman field $h/D = 10^{-4}$ at chemical potential $\mu = 0$ for the hybridization $V/D = 0.4$, 0.6330, and 0.8. Here, Boltzmann constant $k_B = 1$. Inset shows $S_{\rm def}$ at $V/D = 0.6330$ for $10^{-12} < T/D < 10^{-1}$. (b) Temperature dependence of expectation value of S_z^2, $\langle S_z^2\rangle_{\rm def}$, for the same parameters as in Fig. 1(a).

Here, $c_{n\sigma}$ is an annihilation operator of the conduction electrons on the nth site of the chain. $\Lambda(>1)$ is a logarithmic discretization parameter and $\alpha = (1+\Lambda^{-1})/2$. In the following, Λ is set to be $\Lambda = 2$. t_N is the "hopping parameter", and ϵ_N is the "on-site potential", and F is a factor. These parameters reflect the DOS of the conduction electrons, and have numerical values of order of unity [10].

By diagonalizing this Hamiltonian iteratively N times and by eliminating higher energy states, we obtain the energy eigenvalues and the eigenstates. Here 300 states are retained. By using these eigenvalues and eigenstates, thermodynamic quantities are calculated [6]. In this paper, the localized moment contributions to the entropy, $S_{\rm def}$, and the expectation value of square of the z component of localized moment, $\langle S_z^2\rangle_{\rm def}$, are calculated.

4. Numerical results

In order to clarify the low-temperature states of this model, we calculate the temperature dependence of $S_{\rm def}$ and $\langle S_z^2\rangle_{\rm def}$ in the NRG method. The parameters D, ϵ_{sp^2}, and U are assumed to be $D \simeq 8$ eV, $\epsilon_{sp^2} \simeq -1$ eV, and $U \simeq 3$ eV. Thus we use the parameters $\epsilon_{sp^2}/D = -0.125$, and $U/D = 0.375$, in the following. With these parameters, in the case of no Zeeman field, $h/D = 0$, it has been shown [4] that the critical hybridization is $V_c/D = 0.6311$. At $\mu = 0$, the entropy of the defect is $S_{\rm def} = \ln 2$, $\ln 3$, and 0, for $V < V_c, V = V_c$, and $V > V_c$, respectively at low temperatures. These three cases correspond to the local moment state, the valence-fluctuation fixed point, and the Kondo-Yosida singlet state, respectively. Here, the Boltzmann constant is set to be $k_B = 1$.

Figure 1(a) shows $S_{\rm def}$ as a function of temperature for the hybridization $V/D = 0.4$, 0.6330, and 0.8 under the Zeeman field of $h/D = 10^{-4}$ at $\mu = 0$. For the case with $V/D = 0.4$ and 0.8, $S_{\rm def} \simeq 0$ at low temperatures. However, their temperature dependences are different: At $V/D = 0.4$, for $T \simeq 10^{-4}D(\simeq h)$, $S_{\rm def}$ changes from $\ln 2$ to $\ln 1 = 0$. Thus, it is suggested that the ground state is a spin polarized state. At $V/D = 0.8$, on the other hand, $S_{\rm def}$ changes from $\ln 3$ to $\ln 1$ for $T/D \simeq 10^{-3}$. This temperature is much larger than the Zeeman field ($h/D = 10^{-4}$). Thus, the ground state is considered to be the Kondo-Yosida singlet state. At a critical point $V/D = 0.6330$, we find that $S_{\rm def}$ approaches $\ln 2$ at the lowest temperature as shown in the inset of Fig. 1(a).

In order to obtain more information on the electronic states at low temperatures, we calculate $\langle S_z^2 \rangle_{\rm def}$. Figure 1(b) shows $\langle S_z^2 \rangle_{\rm def}$ as a function of temperature for the same parameters. For the case with $V/D = 0.4$ and 0.8, we find $\langle S_z^2 \rangle_{\rm def} \simeq 1/4$ and 0, respectively at low temperature, which confirms the ground states suggested above: the spin polarized state for $V/D = 0.4$, and the Kondo-Yosida singlet state for $V/D = 0.8$. It is also found that $\langle S_z^2 \rangle_{\rm def}$ suddenly changes from 1/4 to 0 at $V/D = 0.6330$ when we change V. Thus, $V/D = 0.6330 \simeq \tilde{V}_c/D$ is a critical value which separates the spin polarized state and the Kondo-Yosida singlet state. At $V \simeq \tilde{V}_c$, $\langle S_z^2 \rangle_{\rm def} \simeq 1/8$.

The critical value in the Zeeman field, $\tilde{V}_c$, is larger than the critical value V_c at $h = 0$, because the Kondo-Yosida singlet state is suppressed by the Zeeman field. We suppose that $\tilde{V}_c$ is determined by the relation $T_K \sim h$, where T_K is a Kondo temperature. The values of $S_{\rm def} = \ln 2$ and $\langle S_z^2 \rangle_{\rm def} = 1/8$ at $\tilde{V}_c$ imply that there are two degenerate states at this point. We expect that these two states are up-spin state and empty (unoccupied) state, since $\langle S_z^2 \rangle_{\rm def} = 1/8$ can be interpreted as $[(1/2)^2 + 0]/2$ [11]. The detail of this unusual fixed point will be discussed elsewhere.

We also calculate the finite chemical potential case, $\mu \neq 0$, under the Zeeman field. $S_{\rm def}$ is always zero at low temperatures. At low temperatures, $\langle S_z^2 \rangle_{\rm def}$, shows the change from $\langle S_z^2 \rangle_{\rm def} = 1/4$ to 0, as V increases. We do not find any unusual fixed point in this case.

5. Conclusion

In summary, the effect of the Zeeman field on the defect-induced Kondo effect in graphene has been investigated. The effective model with the Zeeman field has been derived and has been analyzed on the basis of numerical renormalization group method. It has been found that, at $\mu = 0$, there is an unusual fixed point with the entropy of $k_B \ln 2$ and the expectation value of S_z^2 of 1/8 at the border of the spin polarized state and the Kondo-Yosida singlet state.

Acknowledgments

T. K. is supported by the Global COE program "the Physical Sciences Frontier" of the Ministry of Education, Culture, Sports, Science and Technology, Japan.

References

[1] Terrones H, Lv R, Terrones M and Dresselhaus M S 2012 *Rep. Prog. Phys.* **75** 062501
[2] Chen J-H, Li L, Cullen W G, Williams E D and Fuhrer M S 2011 *Nat. Phys.* **7** 535

[3] Nair R R, Sepioni M, Tsai I-L, Lehtinen O, Keinonen J, Krasheninnikov A V, Thomson T, Geim A K and Grigorieva I V 2012 *Nat. Phys.* **8** 199
[4] Kanao T, Matsuura H and Ogata M 2012 *J. Phys. Soc. Jpn.* **81** 063709
[5] Wilson K G 1975 *Rev. Mod. Phys.* **47** 773
[6] Bulla R, Costi T A and Pruschke T 2008 *Rev. Mod. Phys.* **80** 395
[7] Cornaglia P S, and Usaj G and Balseiro C A 2009 *Phys. Rev. Lett.* **102** 046801
[8] Jones B A and Varma C M 1987 *Phys. Rev. Lett.* **58** 843
[9] Sakai O and Shimizu Y 1992 *J. Phys. Soc. Jpn.* **61** 2333
[10] Gonzalez-Buxton C and Ingersent K 1998 *Phys. Rev.* B **57** 14254
[11] After the manuscript of this paper had been accepted, we became aware that this situation was discussed with perturbative renormalization group methods in Fritz L and Vojta M 2004 *Phys. Rev.* B **70** 214427

Towards an Electronic Interferometer based on Spin-Resolved Quantum Hall Edge States

B. Karmakar[1,*], D. Venturelli[1], L. Chirolli[1], F. Taddei[1], V. Giovannetti[1], R. Fazio[1], S. Roddaro[1], G. Biasiol[2], L. Sorba[1], L. N. Pfeiffer[3], K. W. West[3], V. Pellegrini[1] and F. Beltram[1]

[1]*NEST, Scuola Normale Superiore and Instituto Nanoscienrze-CNR, Pisa, Italy*

[2]*Istituto Officina dei Materiali CNR, Laboratorio TASC, Basovizza (TS), Italy*

[3]*School of Engineering and Applied Science, Princeton University, New Jersey, USA*

*b.karmakar@sns.it

Abstract. Spin resolved edge states are ideal candidates for the implementation of dual-rail quantum computation architectures by encoding the qubit in the spin degree of freedom of the co-propagating edge states. An important element for the realization of such architectures is a coherent beam splitter that controllably mixes the two co-propagating spin-resolved edge channels. Coupling of the spin resolved edge states is demonstrated recently by spin-flip scattering event that is induced by in-plane spatially-dependent periodic magnetic field of the nano-magnet array placed at the boundary of the mesa. In this paper we discuss the nanofabrication and our preliminary transport analysis of an electronic interferometer device made of two nano-magnetic arrays placed in close proximity. The impact of temperature in the coherent properties of the devices is addressed.

1. Introduction

Observation of quantum interference effect in electronic Mach-Zehnder interferometer (MZI) using counter propagating quantum Hall (QH) edge channels[1] proved that coherent QH edge channel circuit is a promising solid state system for quantum information processing. Very large visibility ~62% of the quantum interference in MZI is the signature of long coherence length[1] of the QH edge channels at low temperature ~10mK. The edge channels are orthogonal Eigen states and they are immune to impurity scattering and geometrical defects[2]. As a result, the edge channels are topologically protected that is characterized by a bulk insulating gap[3] created under application of a high magnetic field B on a two dimensional electron gas (2DEG). Based on these promising properties of the QH edge states, various coherent electronic circuits have been investigated[4,5,6].

Implementation of dual-rail[7] quantum computation architectures[8] by encoding the qubit in the two distinct co-propagating energy-degenerate orthogonal QH edge states is very promising for testing quantum algorithms. A key element for the realization of such architectures[8,9,10] is a coherent beam splitter that makes it possible to prepare any superposition of the two logic states, thus realizing one-qubit gate transformations. Recently we have demonstrated resonant coupling[11] of the spin resolved edge states (SRESs) at filling factor $\nu = 2$ (number of filled Landau level) by spin-flip scattering events that is induced by in-plane spetially-dependent periodic magnetic field of the nano-magnet array placed at the boundary of the mesa. In this paper we introduce an electronic interferometer device based on such resonant nano-magnetic coupling of the SRESs and present initial characterization of the device down to $T = 250$mK.

2. Dual-rail architecture

The dual-rail architectures[8] proposed by Giovannetti *et al.* are analogues to linear quantum optics architectures[12,13] for quantum information processing. The proposed architectures are based on coherent electronic transport in the topologically protected QH edge states, where superposition of states are formed by coherent beam mixing of the two orthogonal edge states and flying qubit gate transformation is performed by inducing Aharonov–Bohm flux in the circuit. The main advantage of this architecture[8] is the availability of the two beams after qubit transformation for further processing. On top of that the devices will suffer less flux instability due to small Aharonov–Bohm loop area defined by a small metal gate (see fig.1) in the circuit. In this topologically protected edge channel circuit quantum information is preserved and does not interact with the environment. Moreover, development of single electron source[14] for single electron interferometer circuits is easy because of Fermionic nature of the electrons. Importantly, the proposed architecture entirely relies on the coherent mixing of the two edge channels. Recently, controlled coupling[11] of the SRESs was demonstrated by our group by triggering a resonant charge transfer by spin-flip events with a nano-magnet array placed at the boundary of the mesa. Based on the exploitation of this resonant nano-magnetic coupling, our further aim is to investigate an electronic interferometer following dual-rail architectures[8].

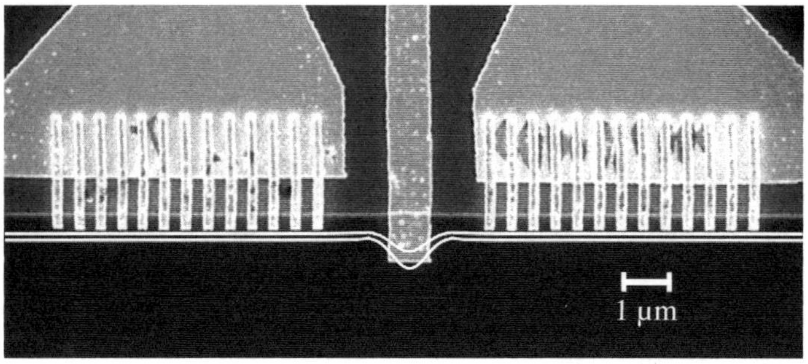

Figure 1: SEM micrograph of the interferometer device with nano-magnet array of periodicity λ = 500nm. Two sets of nano-magnet arrays are used as beam mixer and the metal gate at the middle opens up the loop for inducing Aharonov–Bohm flux under application of a negative voltage bias on it. Two white lines schematically represent SRESs.

3. Device fabrication

The interferometric device is fabricated on a one-sided modulation-doped AlGaAs/GaAs heterostructure grown by molecular beam epitaxy. The 2DEG resides at the AlGaAs/GaAs hetero-interface located 100 nm below the top surface. A spacer layer of 42 nm separates the 2DEG from the Si δ-doping layer above it. The 2DEG has nominal electron density of $1.3 \times 10^{11}/\text{cm}^2$ and low-temperature mobility nearly $4 \times 10^6\text{cm}^2/\text{V-s}$. The mesa is defined by e-beam lithography and wet chemical etching. Ohmic contacts to the mesa are defined by photolithography, thermal evaporation of Ni/Au-Ge/Au and then rapid thermal annealing. The Cobalt nano-magnet arrays of periodicity λ = 500nm are defined at the mesa boundary of the 2DEG using e-beam lithography and thermal evaporation of 10nm Ti followed by 140nm Co and 10nm Au. Evaporation of Ti enhances adhesion of Co and removes remnant O_2 from the evaporation chamber and top 10nm Au protects forming of CoO, which is anti-ferromagnetic in nature. The nano-magnet arrays have overlap on the mesa of about 300nm and are extended over 6.2μm along the mesa boundary. For inducing Aharonov–Bohm flux a metal gate in between the two nano-magnet arrays (fig.1) is defined by e-beam lithography and thermal evaporation of 10nm Ti and 150nm of Au. The SRESs are separately contacted at the source and drain sides (not shown in fig.1) by suitable metal gating near the Ohmic contacts (see ref.[11]).

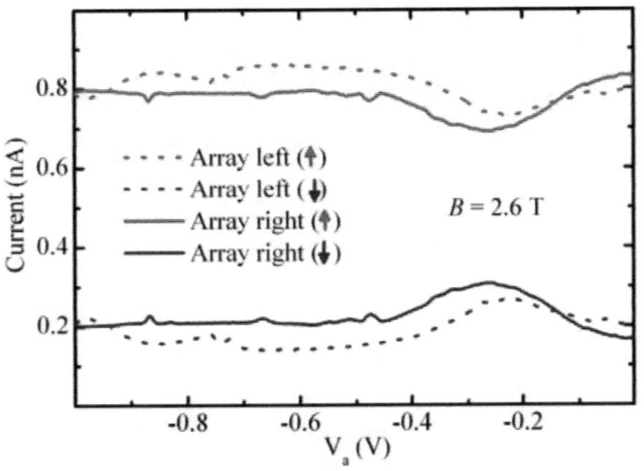

Figure 2: (Color online) Plot of the transmitted spin up (↑) and spin down (↓) edge channel current with the nano-magnet voltage V_a for the two left and right arrays. The working magnetic field B = 2.6T is chosen nearly at the center of the ν = 2 plateau. The spin up edge channel is excited with a voltage bias of 25.8μV. Spin-flip process enhances the current in the spin down channel and decreases spin up channel current, while the total current remain constant nearly 1nA.

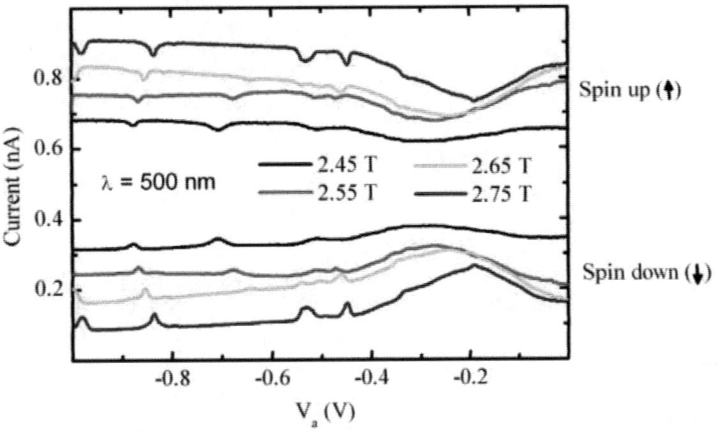

Figure 3: (Color online) Plot of the transmitted spin up (↑) and spin down (↓) edge channel current with the nano-magnet voltage V_a for the right array at different magnetic field B within the ν = 2 plateau. The multiple resonant peaks move towards higher voltage bias V_a with increasing magnetic field B.

4. Experimental results

At the beginning of the experiments, the working points like magnetic field B, gate voltage for separately contacting the edge channels etc. are needed to set in the device. At first a two-terminal magneto-current (2TMR) measurement (see ref.[11]) is carried out at temperature T = 250mK with a voltage excitation of 25.8μV in one terminal while monitoring the transmitted current in the other terminal. From the 2TMR measurement, the extension of the ν = 2 plateau is found out in the magnetic field range from 2.45T to 2.75T. To separately contacting the SRESs (see ref.[11]), a voltage bias V_g = 0.068V is found out to be suitable on the metal gates near the Ohmic contacts in this device. Then the nano-magnetic coupling is tested in the interferometer device (fig.1) by changing the voltage bias V_a

on the nano-magnet arrays. The traces of the voltage V_a dependence of the transmitted spin up (↑) and spin down (↓) edge channel currents at a magnetic field of B = 2.6T are plotted in fig.2. At large negative bias V_a = -1V on both the arrays, where the nano-magnetic coupling is expected to be negligible, we see a spin-mixing of nearly 20%. On top of this inherent spin-mixing (as background), an enhancement of spin-mixing of about 10% is observed for both the arrays at resonance with low applied voltage bias V_a nearly at -0.22V and -0.27V. The voltage bias V_a at the resonant spin-mixing for the two arrays are found to be little different, probably due to difference in local electrostatic conditions.

In fig.3 the traces of the nano-magnet voltage V_a dependence of the transmitted spin up (↑) and spin down (↓) edge channel current at different magnetic fields B within the $\nu = 2$ plateau are plotted for the right array. The multiple resonance peaks move towards higher voltage V_a of the nano-magnet array with increasing magnetic field B. We note that such a behavior was already seen in our previous work[11]. It should be also stressed that the inherent background mixing also reduces with increasing magnetic field, but does not become negligible. This inherent background spin-mixing[15] suggests an incoherent spin-mixing process in the system, which might prevent quantum interference oscillation in the device. Indeed, quantum interference oscillation is not observed in the device at the lowest available temperature 250mK.

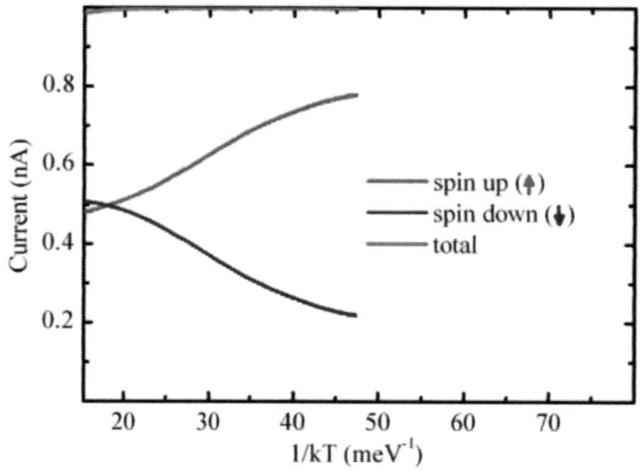

Figure 4: (Color online) Temperature dependence of the spin-resolved edge channel current at magnetic field B = 2.62T with nano-magnet voltage bias V_a = -1V. SRESs completely relax over the copropagating length of 100μm by thermally activated spin-flip process above temperature 0.7K.

Temperature dependence of the spin-resolved edge channel current at magnetic filed B = 2.62T is plotted in fig.4. In this experiment, the nano-magnet voltage bias is kept at V_a = -1V, such that the edge channels are deflected far away from the nano-magnet arrays and the nano-magnetic coupling become negligible. In this condition, thermally activated spin-relaxation processes completely equilibrate the SRESs over the copropagating distance of 100μm above temperature 0.7K. This experiment is carried out at a magnetic field B = 2.62T at filling factor $\nu = 2$, where the Zeeman gap is nearly E_z = 0.064meV (0.8K). We believe that such a low Zeeman gap E_z is responsible for the significant spin-relaxation observed at the temperature 250mK. The evolution of the experimental curves in fig.4 suggests that the spin-relaxation can be further suppressed by lowering the sample temperature below 250mK.

5. Discussions

In this low-density 2DEG, resonant nano-magnetic coupling is observed on top of the thermally activated spin-relaxation process. The behavior of the resonant coupling by changing the magnetic field B within the $\nu = 2$ plateau is similar to our previous work[11]. The resonant coupling is not sufficient to observe the quantum interference oscillation in the device at the available temperature of 250mK. At this temperature coexisting thermally-activated spin-relaxation incoherent processes take place over the co-propagating distances. Therefore, suppression of the thermally activated spin-relaxation events is a key requirement and that can be reached by further lowering the sample temperature. After achieving the suppression of such incoherent processes at low temperature, the coherence of the resonant nano-magnetic coupling can be tested. We plan to carry out such experiments in the near future by using a dilution fridge with a base temperature of 10mK.

6. Conclusions

Resonant nano-magnetic coupling is observed in a low-density sample (n_s = $1.3\times10^{11}/cm^2$) with a nano-magnet periodicity of λ = 500nm at a temperature of 250mK on top of a thermally activated spin-relaxation background. The existence of such incoherent spin-relaxation processes is detrimental to the expected coherent behavior of our device. Suppression of the thermally activated spin-relaxation process is an essential step to observe the coherent transport in our interferometric device. This condition can be reached at dilution fridge temperatures.

7. Acknowledgement

This work was supported by MIUR through FIRBIDEAS Project No. RBID08B3FM and by EU through Projects SOLID and NANOCTM.

References

[1] Y. Ji et al., Nature (London) **422**, 415 (2003).
[2] B. I. Halperin, Phys. Rev. B **25**, 2185 (1982); M. Bu¨ttiker, Phys. Rev. B **46**, 12 485 (1992).
[3] C. L. Kane and E. J. Mele, Phys. Rev. Lett. **95**, 226801 (2005); B.A. Bernevig, T. L. Hughes, and S. C. Zhang, Science **314**, 1757 (2006); D. Hsieh et al., Nature (London) **452**, 970 (2008); S. Das Sarma, M. Freedman, and C. Nayak, Phys. Rev. Lett. **94**, 166802 (2005).
[4] I. Neder et al., Nature (London) **448**, 333 (2007); Phys. Rev. Lett. **96**, 016804 (2006); Phys. Rev. Lett. **98**, 036803 (2007).
[5] P. Roulleau et al., Phys. Rev. B **76**, 161309(R) (2007); Phys. Rev. Lett., **100**, 126802 (2008); Phys. Rev. Lett., **102**, 236802 (2009). E. Bieri et al., Phys. Rev. B **79**, 245324 (2009). S. Roddaro et al., Phys. Rev. Lett. **103**, 016802 (2009).
[6] R. L. Willett, L. N. Pfeiffer and K. W. West., PNAS **106**, 8853 (2009).
[7] Isaac L. Chuang and Yoshihisa Yamamoto. Phys. Rev. A **52**, 3489–3496 (1995).
[8] V. Giovannetti, F. Taddei, D. Frustaglia, and R. Fazio, Phys. Rev. B **77**, 155320 (2008).
[9] T. M. Stace, C. H.W. Barnes, and G. J. Milburn, Phys. Rev. Lett. **93**, 126804 (2004).
[10] C. Nayak et al., Rev. Mod. Phys. **80**, 1083 (2008).
[11] B. Karmakar et al., Phys. Rev. Lett. **107**, 236804 (2011).
[12] E. Knill, R. Laflamme and G. J. Milburn. Nature **409,** 46 (2001).
[13] Alberto Politi, Martin J. Cryan, John G. Rarity, Siyuan Yu and Jeremy L. O'Brien. Science **320,** 646 (2008); Jonathan C. F. Matthews, Alberto Politi, Andre´ Stefanov and Jeremy L. O'Brien. Nature Photonics **3,** 346 (2009).
[14] G. Feve, et al., Science **316**, 1169-1172 (2007).
[15] G. Muller et al., Phys. Rev. B **45**, 3932 (1992); S. Komiyama, H. Hirai, M. Ohsawa, and Y. Matsuda, Phys. Rev. B. **45**, 11085 (1992).

Stability of zero-mode Landau levels in bilayer graphene against disorder in the presence of the trigonal warping

Tohru Kawarabayashi[1], **Yasuhiro Hasugai**[2] **and Hideo Aoki**[3]

[1]Department of Physics, Toho University, Funabashi 274-8510, Japan
[2]Institute of Physics, University of Tsukuba, Tsukuba 305-8571, Japan
[3]Department of Physics, University of Tokyo, Hongo, Tokyo 113-0033, Japan

E-mail: tkawa@ph.sci.toho-u.ac.jp

Abstract. The stability of the zero-energy Landau levels in bilayer graphene against the chiral symmetric disorder is examined in the presence of the trigonal warping. Based on the tight-binding lattice model with a bond disorder correlated over several lattice constants, it is shown that among the four Landau levels per spin and per valley, two Landau levels exhibit the anomalous sharpness as in the absence of the trigonal warping, while the other two are broadened, yielding split peaks in the density of states. This can be attributed to the fact that the total chirality in each valley is ± 2, which is protected topologically even in the presence of an intra-valley scattering due to disorder.

1. Introduction

In bilayer graphene, the Landau level structure in the vicinity of the contact point is drastically changed by the trigonal warping. In the presence of the trigonal warping, the four Dirac cones appear in low energies out of the parabolic bands touching at K and K' points [1]. Due to this modification, we have four-fold zero-energy Landau levels per spin and per valley in weak magnetic fields. So the trigonal warping doubles the degeneracy of the zero-energy Landau levels from the situation in the absence of the trigonal warping. Now an interesting question is how the four Landau levels react to disorder.

Bilayer graphene, in the absence of the trigonal warping, has two-fold degenerated zero-energy Landau levels, and the stability of such zero-modes against disorder in gauge degrees of freedom (ripples) has been shown to be a consequence of the index theorem [2]. We have also demonstrated with an explicit numerical approach based on a tight-binding model as well as with Aharonov-Casher argument that the doubly-degenerated zero-energy Landau levels ($n = 0$ and $n = 1$) exhibit an anomalous sharpness for a bond disorder spatially correlated over more than several lattice constants and respecting the chiral symmetry of the system [3]. So the question now is whether the additional two zero-energy Landau levels in the presence of the trigonal warping in a weak magnetic field are all robust against the bond disorder as well. It is to be noted that the chiral symmetry itself is respected by the trigonal warping.

To clarify the stability of these zero-energy Landau levels in the presence of the trigonal warping, we have performed numerical calculations with a tight-binding lattice model in weak

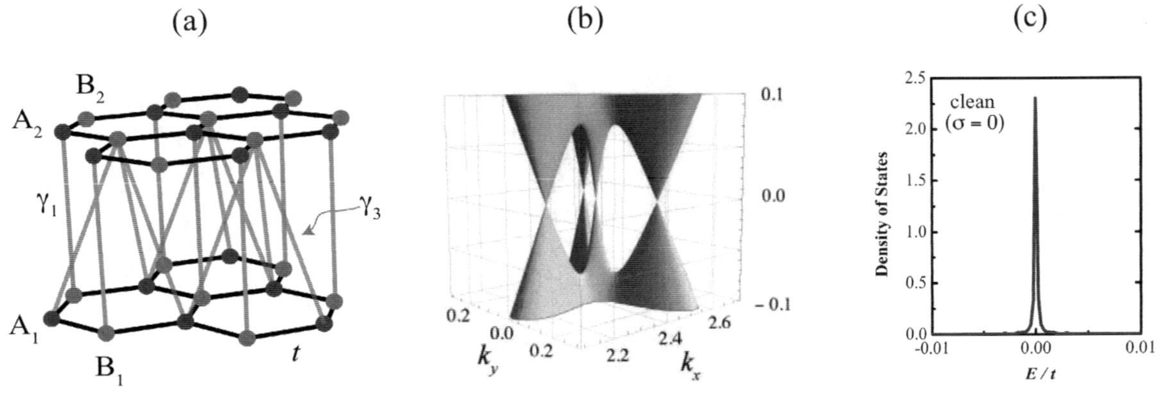

Figure 1. (a) Lattice model for bilayer graphene with the largest interlayer coupling γ_1 and another, γ_3, that gives rise to the trigonal warping. (b) Energy dispersion $E(k_x, k_y)/t$ around $(k_x, k_y) = (4\ \ /3\sqrt{3}, 0)$ for the interlayer coupling $\gamma_1/t = 0.2$ and $\gamma_3/t = 1.5$. (c) Density of states around $E = 0$ for $\phi/(h/e) = 1/5000$ evaluated by the Green function $G(E + i\varepsilon)$ with $\varepsilon/t = 1.0 \times 10^{-4}$, which can be fit as $(4\phi/\ \ \phi_0)(\varepsilon/(E^2 + \varepsilon^2))$.

magnetic fields to examine the robustness against the spatially-correlated bond disorder. We shall find that, among the four Landau levels per spin and per valley, two levels exhibit the anomalous sharpness against the finite-ranged bond disorder as in the absence of the trigonal warping, while the other two are broadened. This can be attributed to the fact that the *total chirality of the four Dirac cones* at K(K') is +2(−2), and that, remarkably enough, these are topologically protected even in the presence of the intra-valley scattering arising from the disorder.

2. Model

We adopt a tight-binding lattice model for bilayer graphene with the Bernal (A-B) stacking having the inter-layer couplings γ_1 and γ_3 (Fig.1(a)). The coupling γ_1 connecting the A_2 site in the top layer and the B_1 in the bottom determines the overall parabolic dispersion, while the coupling γ_3 between the B_2 and the A_1 introduces the trigonal warping of the Fermi surface and produces the four Dirac cones at K(K') point in a low-energy scale [1]. We introduce the bond disorder $\ \ t$ in the nearest-neighbor hopping t in each layer as in Ref.[3]. The disorder is assumed to have a gaussian distribution with variance $\ \ $ and have a spatial correlation as $\langle\ \ t(\boldsymbol{r})\ \ t(\boldsymbol{r}')\rangle = \ \ ^2 \exp(-|\boldsymbol{r} - \boldsymbol{r}'|^2/4\eta^2)$ with the correlation length η. In order to access to the low-energy regime in weak magnetic fields, we consider a system as large as 4×10^6 sites which is 4 times larger than those in our previous studies [3, 4] to evaluate the density of states from the Green function $\rho = -\mathrm{Im}\langle G_{r,r}(E + i\varepsilon)\rangle_r/\ \ $ [5]. The magnetic field applied perpendicular to the graphene sheet is taken into account by the Peierls phases for the hopping amplitude, where the magnetic flux per hexagon of the honeycomb lattice is denoted by ϕ in units of the flux quantum $\phi_0 = h/e$. Spin degrees of freedom are neglected for simplicity, and all the lengths are measured in units of the nearest-neighbor distance, a, of the honeycomb lattice.

3. Numerical Results

In order to study numerically the low-energy physics where the trigonal warping (coming from γ_3) is relevant, we assume $\gamma_1/t = 0.2$ and $\gamma_3/t = 1.5$ in the above lattice model. Although the present value of γ_3 is much larger than the value estimated for the bulk graphite $\gamma_3/t\ \ \ 0.1$ [6],

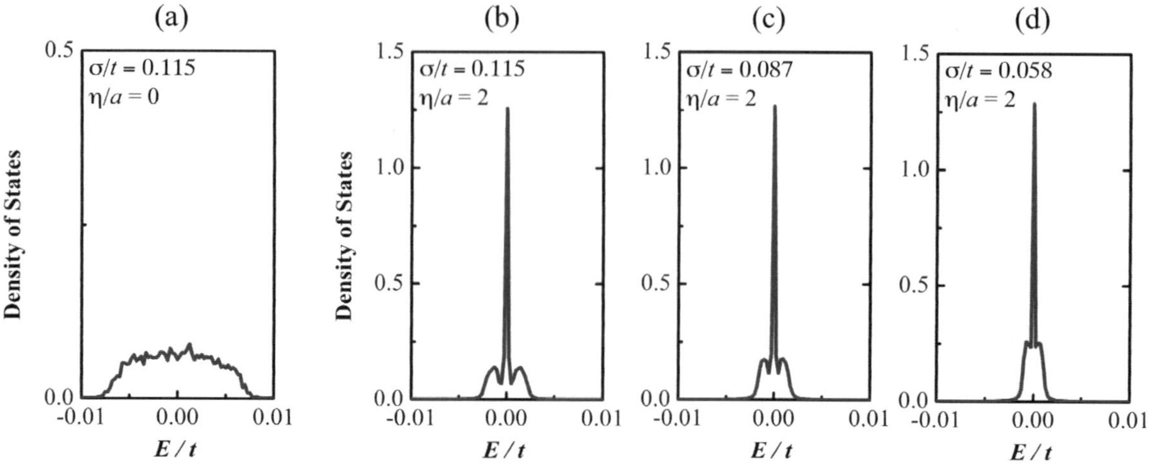

Figure 2. Density of states at zero energy for an uncorrelated disorder with $\eta = 0$ (a) and for a correlated disorder with $\eta/a = 2$ ((b),(c),(d)). Disorder strength is assumed to be $/t = 0.115$ for (a) and (b), while $/t = 0.087$ for (c) and $/t = 0.058$ for (d). The magnetic flux per hexagon is $\phi/\phi_0 = 1/5000$ and the imaginary part of energy is $\varepsilon/t = 1.0 \times 10^{-4}$.

the dispersion at low energies is still described by the four Dirac cones as shown in Fig.1(b), which should be adiabatically connected to the realistic situation with $\gamma_1/t = 0.12$ and $\gamma_3/t = 0.1$. It is expected therefore that the stability of the zero-mode Landau levels against the disorder scattering, which mixes the four Dirac cones, can be discussed within the present lattice model by assuming an appropriate (see below) value of the correlation length η of the bond disorder.

First, we show in Fig.1(c) the density of states for $\phi/\phi_0 = 1/5000$ in the absence of disorder ($= 0$), in which we clearly see that the zero-energy Landau level is 4-fold degenerated per valley. It is to be noted that the zero-energy Landau levels in a tight-binding lattice model with a finite system-size have small but finite widths. Exact zero-energy states with the -function density of states are realized only in the thermodynamic and continuum limit. In the present calculation, such a small width, however, is much smaller than the energy resolution determined by the imaginary part ε of the energy, since the sharp peak in the absence of disorder can be fitted fairly accurately by the Lorentzian form $\varepsilon/(E^2 + \varepsilon^2)$ with a pre-factor consistent with the four-fold degeneracy of the Landau levels.

We then introduce the disorder in Fig.2. For the case where the bond disorder is uncorrelated in space ($\eta = 0$), we confirm that the usual broadening occurs for the zero-energy Landau levels (Fig.2(a)). In contrast, when the disorder is correlated over a few lattice constants ($\eta/a = 2$), we have one central sharp Landau levels accompanied by two broadened ones (Fig.2(b),(c),(d)). The peak height of the central peak is one-half of that in the absence of disorder (Fig.1(c)), and its shape is insensitive to the disorder strength . The splitting of the broadened satellite peaks becomes larger for stronger disorder. The satellite peaks indicate the splitting of the critical energy for the quantum Hall transition, which has been discussed for the monolayer graphene with a short-ranged bond disorder [7].

All these results suggest that among four-fold zero energy Landau levels, two Landau levels remain anomalously sharp against the spatially correlated bond disorder as in the absence of the trigonal warping ($\gamma_3 = 0$), while the other two are broadened and split as two satellite peaks. If one notes that the separation Δk between Dirac cones for the present model with $\gamma_3/t = 1.5$ and $\gamma_1/t = 0.2$ is order of $0.2a^{-1}$ (Fig.1(b)), one might expect that there should be a considerable disorder scattering among these Dirac cones, since the correlation length $\eta/a = 2$ of disorder is smaller than Δk^{-1} $5a$. In actual bilayer graphene, where γ_3 is much smaller, the separation Δk is estimated to be $\Delta k = 2\gamma_1\gamma_3/(3t^2 a)$ [1], which gives Δk^{-1} $120a$ 17nm for the bulk values of γ_1 and γ_3. The scale of ripples, on the other hand, is estimated to be 10 15nm [8, 9], which is again smaller than Δk^{-1} 17nm. We thus expect that the mixing of the four Dirac cones due to ripples can be also relevant in actual bilayer graphene.

4. Summary and discussions

We have investigated the stability of zero-mode Landau levels of bilayer graphene in small magnetic fields where the trigonal warping is relevant. We have considered a bond disorder that respects the chiral symmetry and is correlated over a few lattice constants, which suppresses the inter-valley scattering but causes the intra-valley scattering. We have found that, among four-fold Landau levels per spin and per valley, two Landau levels are stable against such a disorder with the anomalously sharp density of states at $E = 0$. On the other hand, the other two levels are broadened and yield split satellite peaks.

This result can be attributed to the fact that the total chirality of the four Dirac cones at K (K') is $2(-2)$, which is topologically protected even in the presence of intra-valley scattering due to disorder. When the inter-valley scattering is switched on by making the the disorder uncorrelated spatially ($\eta = 0$), the anomalously sharp peak is washed away (Fig. 2(a)). The present results then suggest that the mixing of the two Dirac cones with opposite chirality by the disorder scattering generally destroys the anomalous stability of zero-mode Landau levels of each Dirac cone.

Acknowledgments

This work is partly supported by the Grants-in-Aid for Scientific Research (No. 22540336 and No. 23340112) from JSPS.

References

[1] McCann E and Falko V I 2006 *Phys. Rev. Lett.* **96** 086805
[2] Katsnelson M I and Prokhorova M F 2008 *Phys. Rev.* **B77** 205424
[3] Kawarabayashi T, Hatsugai Y and Aoki H 2012 *Phys. Rev.* **B85** 165410
[4] Kawarabayashi T, Hatsugai Y., and Aoki H 2009 *Phys. Rev. Lett.* **103** 156804; 2010 *Physica* **E42** 759.
[5] Schweitzer L, Kramer B and MacKinnon A 1984 *J. Phys. C* **17** 4111
[6] Dresselhaus M S and Dresselhaus G 2002 *Adv. Phys.* **51** 1
[7] Koshino M and Ando T 2007 *Phys. Rev.* **B75** 033412
[8] Meyer J C, Geim A K, Katsnelson M I, Novoselov K S, Booth T J, and Roth S 2007 *Nature* **446** 60
[9] Geringer V, Liebmann M, Echtermeyer T, Runte S, Schmidt M, Ruckamp R, Lemme M C, and Morgenstern M 2009 *Phys. Rev. Lett.* **102** 076102

Journal of Physics: Conference Series **456** (2013) 012021 doi:10.1088/1742-6596/456/1/012021

Effects of Rashba spin splitting and exchange interaction in electron spin resonance in narrow-gap quantum well heterostructures

S S Krishtopenko[1,2], K V Maremyanin[1], A V Malyzhenkov[1], K P Kalinin[1], A V Ikonnikov[1], V I Gavrilenko[1], M Goiran[2] and Yu G Sadofyev[3]

[1] Institute for Physics of Microstructures, Russian Academy of Sciences, GSP-105, 603950, Nizhny Novgorod, Russia

[2] Laboratoire National des Champs Magnétiques Intenses (LNCMI-T), CNRS UPR 3228 Université de Toulouse, 143 Avenue de Rangueil, F-31400 Toulouse, France

[3] P.N. Lebedev Physical Institute, Russian Academy of Sciences, 119991 Moscow, Russia

E-mail: gavr@imp.sci-nnov.ru

Abstract. We report an electron spin resonance (ESR) study in *n*-type narrow-gap quantum well (QW) heterostructures. By using the Hartree-Fock approximation based on the 8 $\mathbf{k}\cdot\mathbf{p}$ Hamiltonian the many-body theory of the ESR in narrow-gap QWs is developed. We have discovered significant enhancement of the ESR g-factor and its low-magnetic-field divergence in both in asymmetric and symmetric QWs which is caused by the exchange interaction in 2D electron gas (2DEG). The ESR energies estimated for 2DEG in asymmetrical InAs/AlSb QWs are compared with the experimental results obtained by electrically detected ESR technique.

1. Introduction

The electron-electron (*e-e*) and Zeeman interaction in the two-dimensional (2D) electron system in a strong perpendicular magnetic field at low temperatures leads to a rich variety of many-body phenomena that do not have analogues in three-dimensional systems [1]. The experimental observation of the effects of an *e-e* interaction in these states is complicated due to a number of symmetry restrictions. The latter include the Kohn theorem, which states that the *e-e* interaction in an ideal translationally invariant system does not affect the cyclotron resonance (CR) energy in 2D electron gas (2DEG) [2]. According to a similar theorem for 2D systems with a rotational invariance in spin space (the Larmor theorem) the *e-e* interaction does not enter into the energy of ESR in such systems [3].

Since typical narrow-gap QW heterostructures based on InSb, HgTe and InAs are characterized by pronounced spin-orbit interaction (SOI), which leads to breaking of rotational symmetry in spin space, and nonparabolicity of the electronic subbands, resulted from perturbation of translational invariance by electric potential of crystal lattice, the both theorems should be violated in these structures [4, 5]. Our earlier studies [5-10] have shown the SOI and mixing between $|S\rangle$- and $|P\rangle$-states in the Γ_6, Γ_7 and Γ_8 bands to largely influence the *e-e* interaction effects such as quasiparticle and ESR g-factor enhancement in 2DEG. In this paper we extend the limits of our previous work [5] to a theoretical and experimental study of ESR in symmetric and asymmetric InAs/AlSb QWs.

2. Theory

To take into consideration SOI and the mixing of the conduction band (Γ_6) with the light- and heavy-hole bands (Γ_8) and with the split-off band (Γ_7) exactly, the 8-band **k·p** Hamiltonian [6] as a single-particle kinetic energy operator is used. It directly takes into consideration the influence of nonparabolicity, lattice-mismatch deformation and SOI on electron energy spectrum. Many-body corrections to ESR energies were obtained in the Hartree-Fock approximation (HFA) [5]. The Coulomb Green function was obtained by solving the electrostatic problem in three-layered system with different permittivity values in QW and in barriers [6]. As all calculations in this paper are performed for InAs/AlSb QW heterostructures [7, 11], we take into account the 'built-in' electric field of ionized donors in single-particle Hamiltonian.

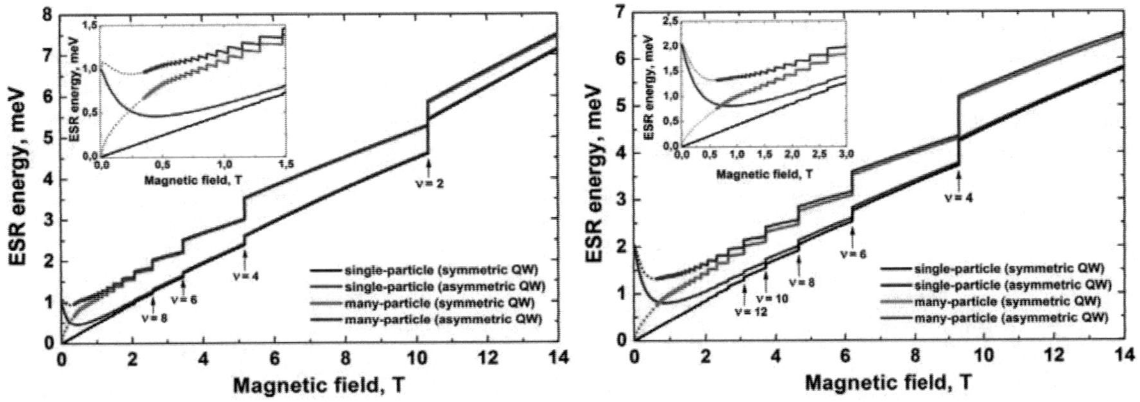

Figure 1: ESR energy in a 15 nm wide InAs/AlSb QW with symmetric and asymmetric profile versus magnetic field at different values of 2DEG concentration: $5\cdot10^{11}$ cm^{-2} (right panel) and $9\cdot10^{11}$ cm^{-2} (left panel). The insets show the results of calculations in weak magnetic fields. Red and blue dotted curves correspond to the zero-field values interpolated by using (1) and (2).

Fig. 1 illustrates the ESR energy calculations versus magnetic field in InAs/AlSb heterostructures with a 15 nm wide QW for different values of the 2DEG concentration. Black and brown curves correspond to the single-particle approximation in a symmetric and asymmetric QW. Red and blue curves are for the ESR energies calculated with allowance for the *e-e* interaction within HFA in QW with a symmetric and asymmetric profile, respectively. The arrows indicate the magnetic field values corresponding to the even-valued filling factors of Landau levels (LLs). The features observed at even-valued LL filling factors appear due to the electronic subband nonparabolicity in InAs/AlSb QW [5-10]. As the magnetic field increases, the Fermi level 'jumps' from one pair of spin-split Landau levels to the lower-lying pair with a greater spin-splitting. This causes a sharp rise of the spin-splitting energy at the Fermi level, followed by an abrupt change in the ESR energy. It is seen that the single-particle ESR energy in a symmetric QW is determined by the Zeeman splitting of LLs, demonstrating a linear dependence on magnetic field in the region of weak magnetic fields.

The Rashba spin splitting in an asymmetric QW leads to distortion of the monotonic dependence of single-particle ESR energy on magnetic field and appearance of a pronounced minimum in the low magnetic fields region. In this case the spin splitting at the Fermi level is determined by two contributors: the Rashba- and the Zeeman splitting [11]. As the magnetic field increases from zero, the ESR energy quickly drops, going smoothly over to the Zeeman splitting. Note that for high LL indices n (in weak magnetic fields) the single-particle ESR energies can be obtained analytically [12]:

$$E_{ESR} \approx \left[\left(\hbar\omega_C - g_{(1e)}\mu_B B\right)^2 + \Delta_R^2\right]^{1/2} - \hbar\omega_C, \qquad (1)$$

where $g_{(1e)}$ is a single-particle g-factor [5, 6], $\omega_C=eB/m^*c$, m^* and Δ_R are the Rashba splitting and the electron effective mass at the Fermi wavevector in a zero magnetic field, respectively. Thus, the 'single-electron' ESR energy decreases linearly with B in weak magnetic fields.

Strong SOI and mixing between $|S>$- and $|P>$-states in the conduction and valence bands in the InAs/AlSb QWs leads to violation of Larmor theorem applicability conditions in symmetric and asymmetric QW. It is clear that *e-e* interaction in narrow-gap QW causes considerable enhancement of ESR energy that at low magnetic fields is more than twice its single-electron value. The Rashba spin splitting in the asymmetric QWs resulted in a noticeable effect on the many-body ESR energy in weak magnetic fields only. The behavior of many-body ESR energy in asymmetric QWs at low magnetic fields strongly depends on a magnitude of the Rashba spin splitting. We note that when the latter is weak in a zero magnetic field, the field dependence of ESR energy is monotonically increasing, whereas at high values of Rashba splitting we observe a monotonically decreasing curve in the region of low magnetic fields [10]. It can be shown that HFA corrections to ESR energy at *integer* LL filling factors in symmetric [5] and asymmetric [10] narrow-gap QWs have a square-root dependence on the magnetic field:

$$\Delta_{ESR}^{(e-e)} = C\sqrt{B}\,, \qquad (2)$$

where constant C is independent of magnetic field, it is defined by quantum numbers of the upper fully filled LL [5, 10].

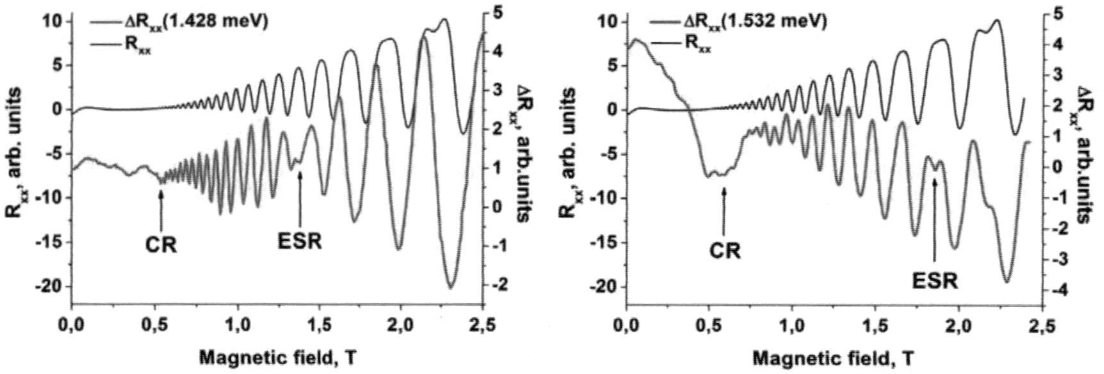

Figure 2: Shubnikov-de Haas oscillations and change in the resistance due to microwave-illumination as a function of applied magnetic field at different BWT photon energy: 1.428 meV (left panel) and 1.532 meV (right panel). The magnitude of current in the measurements was 50 μA. Features attributed to CR and ESR are labeled.

3. Experiment

ESR measurements have been carried out at 4.2 K using electrical detection technique [3]. As a radiation source the backward-wave tube (BWT) which covers the range of 250-500 GHz is used. The measurements were performed on a sample T340 with a QW width of 15 nm. The 2DEG concentration and mobility in the sample at $T = 4.2$ K were $7.65\cdot10^{11}$ cm^{-2} and $4.5\cdot10^{5}$ cm^2/V·s, respectively. Fig. 2 is a typical change in magneto-resistance in the sample T340 induced by BWT radiation. Magnetic field was oriented perpendicular to the sample plane. Features attributed to CR and ESR, are clear observed. The experimental values are presented in Fig. 3 by symbols. One can see a good agreement between theoretical predictions and experimental data if the asymmetric built-in electric field of the surface donors in GaSb cap layer [7, 13, 14] resulting in the Rashba spin splitting [15] is taken into account. A small discrepancy between calculated curve and experimental data is attributed, in our opinion, to the influence of dynamic polarization (Overhauser shift [16]) on the ESR line shape and position. In addition, we demonstrate that exchange interaction leads to the low-magnetic-field divergence of many-body ESR g-factor in symmetric QWs. In asymmetric QW it increases the Rashba splitting induced divergence of the g-factor under low magnetic fields. In high magnetic fields the contribution of the exchange enhancement to ESR energy and g-factor decreases with a growing magnetic field (see Fig. 3).

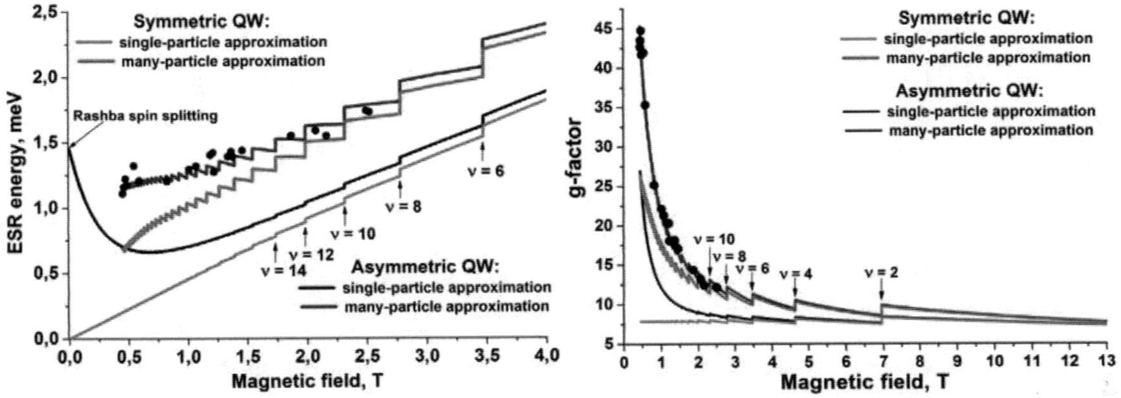

Figure 3: ESR energy (left panel) and g-factor (right panel) in a 15 nm wide InAs/AlSb QW with symmetric and asymmetric profile calculated for 2DEG concentration of $7.65 \cdot 10^{11}$ cm^{-2}. The symbols are experimentally data in T340 sample. Magnetic fields corresponding to even-valued LL filling factors are indicated by arrows.

This work demonstrates that ESR can serve as a tool to probe the *e-e* interaction effects in narrow-gap QWs. We hope that our results will stimulate further experimental studies of ESR in narrow-gap QW heterostructures and will provide an in-depth understanding of the physical mechanisms responsible for the exchange enhancement of ESR energy in 2DEG.

4. Acknowledgements

This work is partially supported by the non-profit Dynasty Foundation, by the RFBR (Grants 11-02-93111, 11-02-00952, 12-02-00940, 13-02-00894), by the Russian Academy of Sciences, and by Russian Ministry of Education and Science (Grant НШ-4756.2012.2).

References

[1] *Perspectives in Quantum Hall Effect*, Ed. by S. Das Sarma and A. Pinczuk (Wiley, New York, 1997).
[2] W Kohn, *Phys. Rev.* **123**, 1242 (1961).
[3] M Dobers, K von Klitzing and G Weimann, *Phys. Rev. B* **38**, 5453 (1988).
[4] A V Ikonnikov, S S Krishtopenko, V I Gavrilenko et al, *J. Low Temp. Phys.* **159**, 197. (2010).
[5] S S Krishtopenko, V I Gavrilenko and M Goiran, *J. Phys.: Condens. Matter* **24**, 252201 (2012).
[6] S S Krishtopenko, V I Gavrilenko and M Goiran, *J. Phys.: Condens. Matter* **23**, 385601 (2011).
[7] S S Krishtopenko, K P Kalinin, V I Gavrilenko et al, *Semiconductors* **46**, 1163 (2012).
[8] S S Krishtopenko, V I Gavrilenko and M Goiran, *J. Phys.: Condens. Matter* **24**, 135601 (2012).
[9] S S Krishtopenko, V I Gavrilenko and M Goiran, *Solid State Phenomena* **190**, 554 (2012).
[10] S S Krishtopenko, V I Gavrilenko and M Goiran (to be published).
[11] P Pfeffer and W Zawadzki, *Semicond. Sci. Technol.* **19**, R1 (2004).
[12] Yu A Bychkov and E I Rashba, *J. Phys. C* **17**, 6039 (1984).
[13] V I Gavrilenko, A V Ikonnikov, S S Krishtopenko et al, *Semiconductors* **44**, 616 (2010).
[14] K E Spirin, K P Kalinin, S S Krishtopenko et al, *Semiconductors* **46**, 1396 (2012).
[15] S S Krishtopenko, V I Gavrilenko and M Goiran, *Semiconductors* **45**, 110 (2011).
[16] A Berg, M Dobers, P R Gerhardts and K von Klitzing, *Phys. Rev. Lett.* **64**, 2563 (1990).

Weak localization and electron-electron interactions in polycrystalline tin dioxide films

V K Ksenevich[1], T A Dauzhenka[1] and J Galibert[2]

[1] Department of Physics, Belarusian State University, Nezalezhnastsi Ave. 4, 220030, Minsk, Belarus

[2] Laboratoire National des Champs Magnétiques Intenses CNRS, Université de Toulouse, UPS, INSA Toulouse, UJF, 143 Avenue de Rangueil F-31400 Toulouse, France

e-mail address: jean.galibert@lncmi.cnrs.fr

Abstract. Electrical and magnetotransport properties of polycrystalline tin dioxide films were investigated in the temperature range 2-300 K and in high magnetic fields up to 27 T. The experimental data were analyzed using models inherent both for 2D and 3D disordered systems. A crossover from 2D to 3D behaviour was observed as the temperature was increased.

1. Introduction

Besides a number of perspective applications in optoelectronics due to the coexistence of high optical transparency in visible and infrared range of electromagnetic waves' spectra and high electrical conductivity [1], tin dioxide films provide the possibility to study different charge transport mechanisms, such as hopping, tunneling, diffusion transport. Quantum corrections to conductivity in the vicinity of metal-insulator transition as a result of weak localization (WL) and electron-electron interactions (EEI) phenomena can be also observed in tin dioxide films [2].

We recently reported the observation of negative magnetoresistance (MR) due to WL and EEI in thin polycrystalline tin dioxide films fabricated by means of combination of ion-exchange processes in polymer films with successive gas transport reaction [3].

In this work we focus our efforts on investigations of WL and EEI in polycrystalline tin dioxide (obtained from $SnCl_2$ solution) by means of MR measurements in high magnetic fields. Temperature dependencies of the resistance *R(T)* and MR of polycrystalline tin dioxide films were measured within temperature range 2-300 K and in pulsed magnetic fields up to 27 T. The experimental data were analysed using models inherent both for 2D and 3D disordered systems [4]. A crossover from 2D to 3D behaviour was observed as the temperature was increased.

2. Fabrication of samples

The samples of SnO_2 polycrystalline films were fabricated following a three-stage procedure [5]: (a) formation (on either Al_2O_3 or Si/SiO_2 substrates) of a Sn-containing polymer networks, (b) gas transport of tin chloride vapour [as a result of the heating of the tin chloride hydrate ($SnCl_2 \cdot 2H_2O$)] on the substrates with previously deposited Sn-containing networks. After the deposition of tin chloride on the substrate, it was decomposed to the tin (II) oxide during the thermohydrolysis, (c) during the subsequent heat treatment in air at temperature 550 ^{o}C, SnO was transformed to SnO_2. Electron diffraction results confirmed the polycrystallinity (type cassiterite) of obtained films. Typical SEM image of tin dioxide film is shown in figure 1.

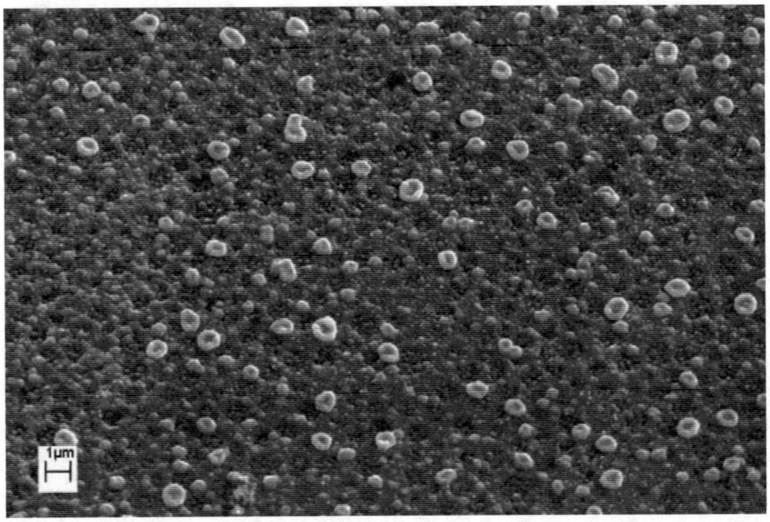

Figure 1. SEM image of tin dioxide films (after gas-transport reaction).

3. Experimental results and discussion

3.1. Temperature dependence

Typical *R(T)* dependence for SnO_2 film is presented in figure 2.

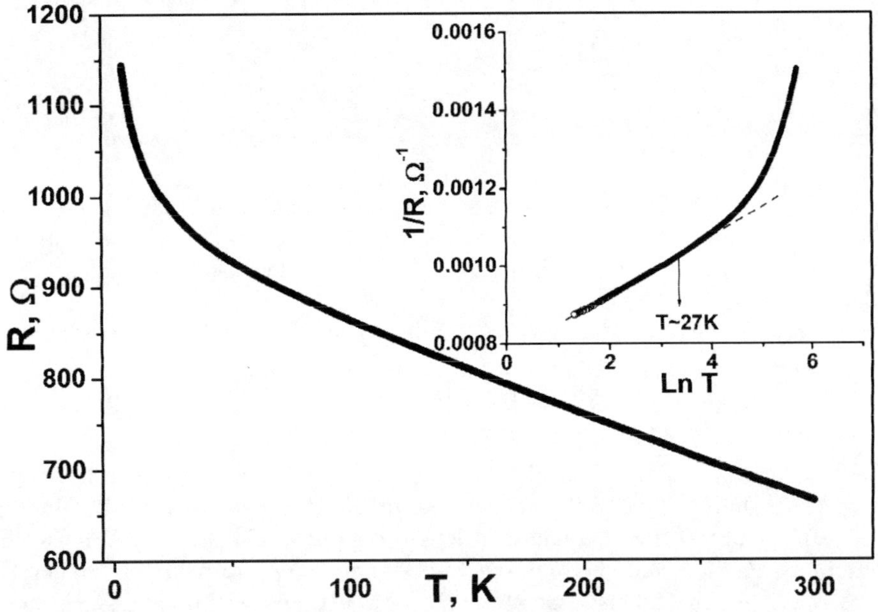

Figure 2. Temperature dependence of resistance of SnO_2 film. In the inset figure temperature dependence of conductance in the scale *1/R-Ln T* is shown. Dashed line corresponds to linear approximation of experimental data at low T.

Such types of *R(T)* dependencies (with negative temperature coefficient of the resistance ($dR/dT<0$) and ratio *R(2 K)/R(300 K)~1-2*) are inherent for different disordered systems in the vicinity of metal-insulator transition. WL and EEI effects define the temperature dependence of conductivity of these systems [4].

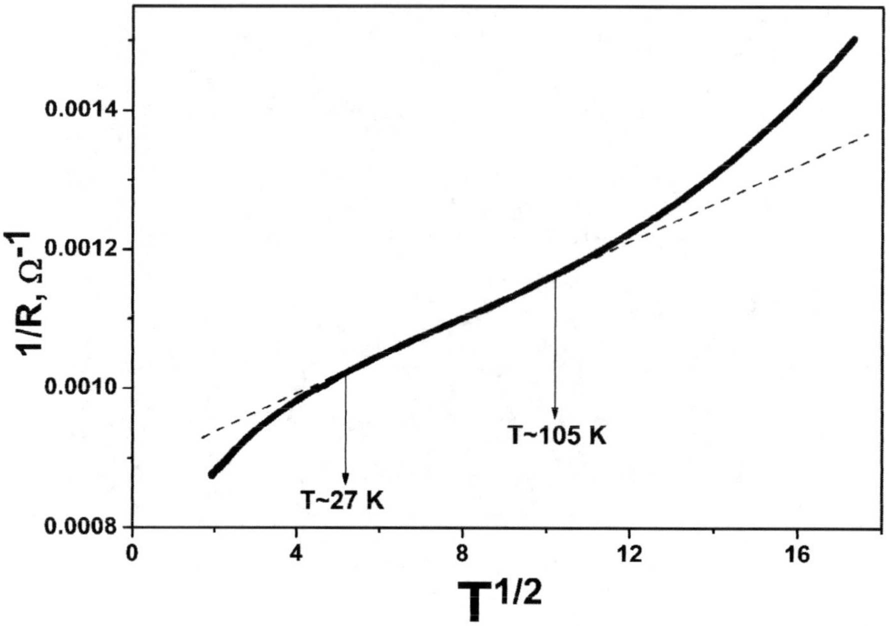

Figure 3. Temperature dependence of resistance of SnO_2 film in the scale $1/R$- $T^{1/2}$ is shown. Dashed line corresponds to linear approximation of experimental data in the intermediate temperature range.

At low temperatures temperature dependence of conductivity for 2D disordered systems can be expressed as [6]:

$$\sigma_{2D} = \sigma_0 + aG_0 \ln\frac{T}{T_0} + K_{ee}G_0 \ln\frac{kT\tau}{\hbar}, \qquad (1)$$

where first term is a constant related to the residual conductivity of the system at T=0 K, the second and the third term are the temperature dependence of conductivity due to effects WL and EEI, respectively. Thus, according Eq. (1) both the WL and EEI produce $\sim lnT$ dependence of conductivity of 2D disordered systems.

Temperature dependence of conductivity for 3D disordered systems follows the law[7]:

$$\sigma_{3D} = \sigma_0 + \alpha T^{1/2} + \beta T^{p/2}, \qquad (2)$$

taking into account both WL and EEI effects. The first term is a constant related to the residual conductivity of the system at T=0 K, the second and the third terms takes into account the effects of EEI and WL, respectively. If the characteristic length parameters, describing WL and EEI (inelastic scattering length $L_\varphi=(D\tau_\varphi)^{1/2}$ and thermal length $L_T=(D\hbar/k_BT)^{1/2}$, respectively, where D being the diffusion constant, τ_φ is the inelastic scattering time), exceeds the thickness of the system, the temperature dependence of the resistance follows a behaviour inherent for 2D disordered systems. As far as L_φ and L_T are temperature dependent parameters, which decrease when the temperature rises, a crossover from the $\sigma(T)$ behaviour typical for 2D systems to the behaviour inherent for 3D disordered materials is expected as the temperature increases. As one can see from inset to Fig.2, $\sim lnT$

dependence of conductivity is observed in the low temperature range (up to ~27 K). In the temperature range T ~ 27-105 K dependence of conductivity on temperature can be linearized in scale $\sigma - T^{1/2}$ as shown in Fig.3. Thus, we can assume existence of crossover from 2D to 3D behaviour for SnO_2 films at T ~ 27K with prevailing of quantum corrections to conductivity due to EEI.

3.2. Magnetic field dependences

MR data measured in pulsed magnetic fields up to 27 T and in the temperature range 2.5-80 K are shown in Fig.4. The negative MR in the whole range of available magnetic fields was observed. In the low-temperature range upturn from negative to positive MR with tendency to saturation in the high magnetic field range was observed. The minimum position of negative MR shifts to higher fields as the temperature rises. At temperatures T>15 K only negative MR was observed.

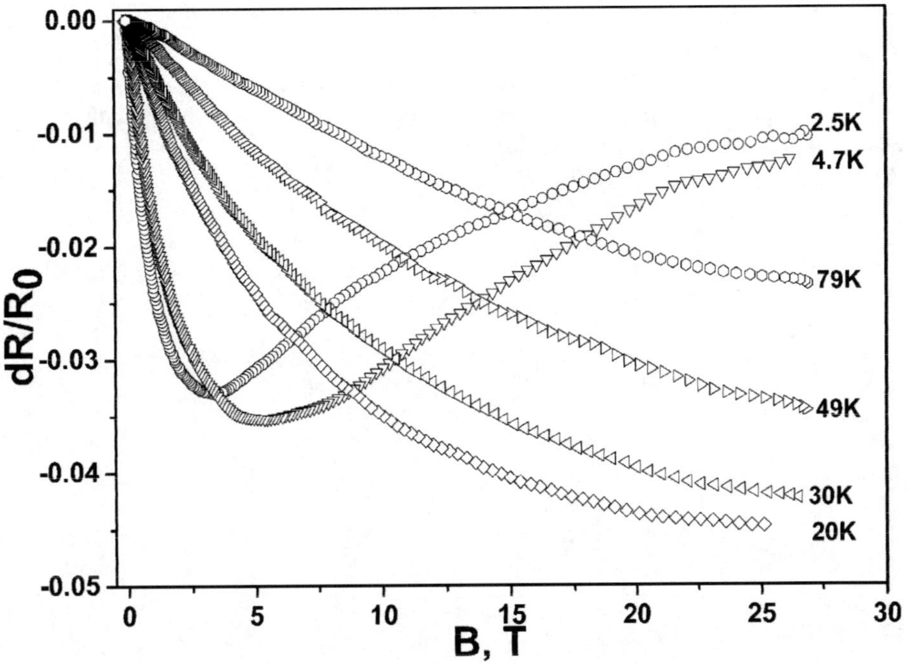

Figure 4. Magnetic field dependences of normalized magnetoresistance of SnO_2 films obtained from $SnCl_2$ solution, measured in the temperature range 2.5-80 K.

Negative MR due to WL effect usually dominates at higher temperatures and lower fields, while at lower temperatures and higher fields, such that the energy difference between the Zeeman split levels is much more than the thermal energy ($g\mu_B H >> k_B T$) electron-electron interactions should be taken into account, providing possibility to observe positive MR.

Existence of both negative and positive MR indicate on the possibility of interplay of EEI in addition to WL effects in polycrystalline tin dioxide films. Detailed analysis of MR data in order to distinguish influence of WL and EEI by means of method proposed in [8] will be reported elsewhere.

4. Conclusion

Quantum corrections to conductivity of thin polycrystalline tin dioxide films, fabricated by means of combination of ion-exchange processes in polymer films with successive gas transport reaction, were observed. Both weak localization (WL) and electron-electron interactions (EEI) effects can influence on the conductivity of films. A crossover from 2D to 3D behaviour was observed at temperature T~ 27 K.

Acknowledgments

We acknowledge support from by FR7 I3 EuroMagnet II and by CNRS/BRFFI grant No. F11F-007. The authors are grateful to I.A. Bashmakov for samples' fabrication, P.I. Gaiduk and S.V. Gusakova for implementation of TEM and SEM measurements.

One of us (T.D.) acknowledges the Embassy of France in Minsk for a French Foreign Affairs Ministry PhD grant.

References

[1] Batzill M and Diebold U 2005 *Progress in Surface Science* **79** 47
[2] Serin T *et al* 2010 *J. Electron. Mater.* **39** 1152
[3] Dauzhenka T A *et al* 2011 *Phys. Rev. B* **83** 165309
[4] Lee P A and Ramakrishnan T V 1985 *Rev. Mod. Phys.* **57** 287
[5] Bashmakov I A, Dauzhenka T A and Ksenevich V K 2009 *Proceedings of IV International Conference: Actual Problems of Solid State Physics (Minsk) [in russian]* **2** 43
[6] Minkov G M *et al* 2001 *Phys. Rev. B* **64** 235327
[7] Dai P, Zhangand Y and Sarachik M P 1992 *Phys. Rev. B.* **45** 3984
[8] Bogdanovich S *et al* 1995 *Phys. Rev. Lett.* **74** 2543

Journal of Physics: Conference Series **456** (2013) 012023 doi:10.1088/1742-6596/456/1/012023

"Magnetic Gradient" edge magnetoplasmons in non-uniform magnetic field

O G Balev[1] and I A Larkin[2]

Abstract. It is shown that laterally inhomogeneous strong magnetic field applied to otherwise spatially homogeneous two-dimensional electron system (2DES) allows "magnetic gradient" or special magnetic-edge magnetoplasmons (MEMPs). This mechanism is different from usual "density gradient" edge magnetoplasmons. Symmetric and antisymmetric families of MEMPs are obtained. They are localized at magnetic field inhomogeneity (magnetic-edge). Both symmetric and antisymmetric MEMPs have the modes of opposite chirality.

[1] Departamento de Física, Universidade Federal do Amazonas, 69077-000, Manaus, AM, Brazil
[2] Department of Physics, Minho University, Braga 4710-057, Portugal

E-mail: ogbalev@ufam.edu.br
vaniala2000@yahoo.co.uk

Different types of "density gradient" edge magnetoplasmons (EMPs) have been studied for 2DES subjected to a strong homogeneos magnetic fiels [1, 2, 3, 4, 5]. These EMPs appear due to a strong change of the stationary local electron density at the edge of the channel (in particular, in the vicinity of edge states) that induces a strong modulation of the local magnetoconductivity tensor (modulation of the nondiagonal components is especially important), e.g., see [1, 3, 5]. Here we present theoretical study of the chiral modes in 2DES induced by "magnetic gradient". I.e., of special magnetic-edge magnetoplasmons (MEMPs) localized in a vicinity of magnetic field inhomogeneity. MEMPs are absent if a strong magnetic field applied to 2DES is homogeneous.

We consider homogeneous 2DES, localized within $z = 0$ plane, that is embedded in GaAs based sample, with the dielectric constant ε, that occupies a half-space $z < d$. In addition, 2DES is subjected to a strong laterally inhomogeneos magnetic field $\mathbf{B}(y) = B(y)\hat{\mathbf{z}}$, which appears due to ferromagnetic semi-infinite film of a finite thickness ηd. Fig. 1 presents a model geometry under discussion; the x-axis points out of the figure plane, and the y-axis points to the right from the ferromagnetic film. A finite external spatially homogeneous magnetic field $\mathbf{B}_{ext} = B_{ext}\hat{\mathbf{z}}$ is applied as well; $B_{ext} > 0$. We assume that the surface of heterostructure crystal, at $z = d$, has pined potential [6], or covered by a nonmagnetic-metallic gate. In addition, we assume that at $z = 0$ plane the ions jellium background of the constant area density n_I is located. Let us assume that magnetic field $B(y)$ is a smooth function of y, with the characteristic scale Δy. Comparing the pertubation of the electron density of 2DES, without metallic gates, due to spatially inhomogeneous magnetic field shows that even at the characteristic filling factor $\nu \leq 2$ the density of 2DES is very weakly modified if the Bohr radius $a_B \ll 2\Delta y$ and the quantum magnetic length $\ell_0(y) \ll 2\Delta y$; $\ell_0(y) = \sqrt{\hbar c/|e|B(y)}$. Below we assume that these conditions are satisfied and, respectively, the static electron density will be taken as n_I.

We assume that the low-frequency, $\omega \ll \omega_c$, and the long-wavelength, ($k_x\ell_0(y) \ll 1$ if $\nu \leq 2$), conditions are satisfied. Then the current density induced by a wave can be calculated in the

Figure 1. 2DES, at $z = 0$, is embedded in heterostructure, $z < d$. We assume pined potential at $z = d$, to ensure that one may cover the surface by metallic-nonmagnetic film (gate) Ferromagnetic semi-infinite film ($y < 0$, $d < z < d(1 + \eta)$) induces inhomogeneity of magnetic field.

quasi-static approximation [3, 5] as

$$j_x(y) = \sigma_{xx}(y)E_x(y) - \sigma^0_{yx}(y)E_y(y), \quad j_y(y) = \sigma_{yy}(y)E_y(y) + \sigma^0_{yx}(y)E_x(y), \tag{1}$$

where we have suppressed the exponential factor $\exp[-i(\omega t - k_x x)]$ and common arguments ω, k_x in $j_\mu(y)$, $E_\mu(y)$. Further, we will neglect by a dissipation assuming a clean 2DES and sufficiently low temperatures T.

For the setup of Fig. 1 with the constant magnetic moment of the ferromagnetic semi-infinite film along the $z-$axis, $\mathbf{M}_0 = M_0\hat{\mathbf{z}}$, readily it follows [7] that $B(y) = B_{ext} - 2M_0\{\arctan(Y) - \arctan(Y/(1+\eta))\}$, where $Y = y/d$. Assuming $\max\{|B(y) - B_{ext}|\} \ll B_{ext}$ we can approximate $1/B(Y)$ as

$$\frac{1}{B(y)} = \frac{1}{B_{ext}} - \frac{1}{2B_0}\{\arctan(Y/(1+\eta)) - \arctan(Y)\}, \tag{2}$$

where $B_0 = B^2_{ext}/(4M_0)$. From Eq. (2) it follows that $B(\pm\infty) = B_{ext}$. We assume that $B_0 > 0$.

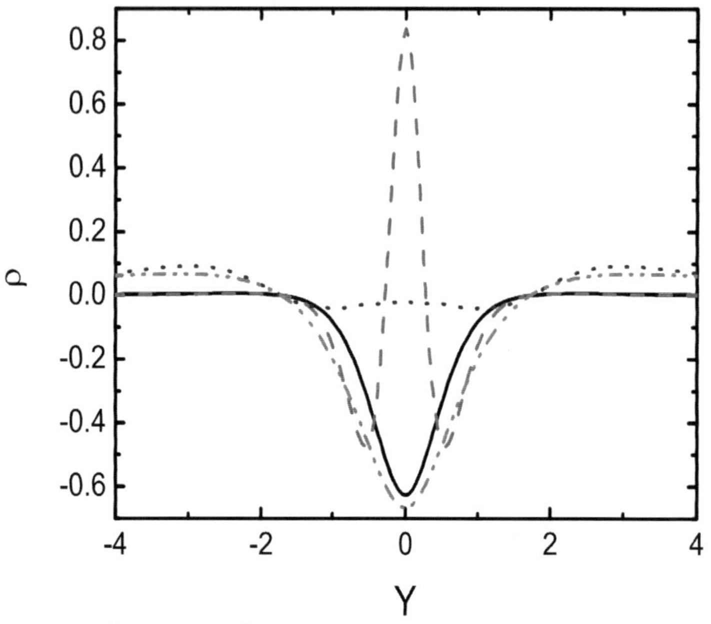

Figure 2. Spatial profile, at $\eta = 1$ and conditions of the table 1, for three fastest symmetric MEMPs W_{s1}, W_{s2}, and W_{s3} are plotted by the solid, the dashed, and the dotted curves. The dot-dot-dashed curve plots $g_0(Y)$.

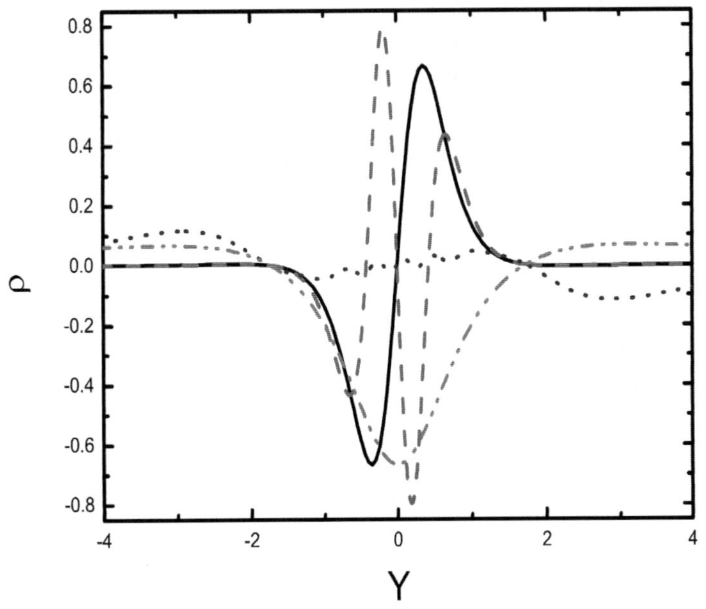

Figure 3. Spatial profile, at $\eta = 1$ and conditions of the table 1, for three fastest antisymmetric MEMPs W_{a1}, W_{a2}, and W_{a3} are plotted by the solid, the dashed, and the dotted curves. The dot-dot-dashed curve plots $g_0(Y)$.

Then using Eq. (1), the Poisson equation and the continuity equation we obtain for the wave charge density $\rho(\omega, k_x, Y)$ the integral equation

$$\frac{\omega}{k_x}\rho(\omega, k_x, Y) - \frac{2|e|cn_I}{\varepsilon B_0}g_0(Y)\int_{-\infty}^{\infty} dY'\rho(\omega, k_x, Y')R_g^{(0)}(|Y - Y'|; k_x d) = 0, \tag{3}$$

where

$$g_0(Y) = \{\frac{(1+\eta)^{-1}}{1 + Y^2/(1+\eta)^2} - \frac{1}{1+Y^2}\}, \tag{4}$$

$g_0(Y)$ is dimensionless gradient of magnetic field along y. In $R_g^{(0)}(|Y - Y'|; k_x d)$ we took into account screening of Coulomb potential by equipotential surface.

$$R_g^{(0)}(|Y - Y'|; k_x d) = K_0(|k_x d||Y - Y'|) - K_0(|k_x d|\sqrt{(Y - Y')^2 + 4}), \tag{5}$$

where $K_0(x)$ is the modified Bessel function.

From Eq. (3) it follows that its complete set of solutions can be presented as the set of symmetric solutions and the set of antisymmetric solutions. Further we will assume that $k_x d \ll 1$. Then from Eqs. (3)-(5) it follows readily that the magnetic edge waves have acoustic dispersion and their group and phase velocities have the same value. We use notation for the dimensionless wave velocity as $W = \frac{\omega}{k_x}\frac{\varepsilon B_0}{2|e|cn_I}$, given by an eigen value of the integral equation (3); its positive or negative sign corresponds to positive or negative chirality and the phase velocity. Equation (3) has infinite set of positive and negative eigen values, however, physically only a finite number of these modes, with largest $|W|$, meet assumed conditions as slower modes have the characteristic scale of the spatial structure along y of the order of magnetic length or smaller.

To solve equation (3) numerically we introduce a new variable $X = \frac{2}{\pi}\arctan(Y)$ that changes from -1 to 1 as Y changes from $-\infty$ to ∞. Then Eq. (4) as function of X we denote also as $g_0(X)$, to simplify notations. Then the symmetric solutions of Eq. (3) obtain the form

$$\rho_s(\omega, k_x; X) = g_0(X) \sum_{k=0}^{\infty} a_{ks}^{(1)}(\omega, k_x) \left[\cos(k\pi X) - \frac{1}{2}\delta_{k,0}\right], \tag{6}$$

and they are defined by the system of linear homogeneous equations

$$W a_{ns}^{(1)} - \sum_{k=0}^{\infty} r_{n,k}^{12} a_{k,s}^{(1)} = 0. \tag{7}$$

where $n = 0, 1, 2, ...,$ $\delta_{k,0}$ is the symbol Kroneckera, and the matrix elements

$$r_{n,k}^{12} = \frac{\pi}{2}\int_{-1}^{1} dX \cos(n\pi X) \int_{-1}^{1} dX' R_g^{(1)}(|\tan(\frac{\pi}{2}X) - \tan(\frac{\pi}{2}X')|, |k_x d|)$$
$$\times \left[\frac{1+\eta}{1+\eta(2+\eta)\cos^2(\pi X'/2)} - 1\right]\left[\cos(k\pi X') - \frac{1}{2}\delta_{k,0}\right]. \tag{8}$$

Point out, a similar treatment leads to pertinent equations and expressions for the antisymmetric modes density profile, $\rho_a(\omega, k_x; X)$, and their velocities. By keeping in the right hand side of Eq. (6) of the first n terms, $a_{0s}^{(1)}(\omega, k_x), ..., a_{n-1s}^{(1)}(\omega, k_x)$, and using for them the first n equations from Eqs. (7) we readily calculate the velocity of the first n MEMPs and their spatial structure. Our approach shows fast convergence.

In table 1 the velocity W for three fastest symmetric modes ($|W_{s1}| > |W_{s2}| > |W_{s3}|$) and three fastest antisymmetric modes ($|W_{a1}| > |W_{a2}| > |W_{a3}|$) are presented, within the model of $1/B(y)$ given by Eq. (2) for characteristic values of $\eta = 1$, 2

Table 1. Dimensionless wave velocity of three fastest symmetric MEMPs, W_s, and three fastest antisymmetric MEMPs, W_a.

η	W_{s1}	W_{s2}	W_{s3}	W_{a1}	W_{a2}	W_{a3}
1.0	-1.175	-0.2652	0.195 ± 0.005	$-$ 0.4190	-0.1893	0.184 ± 0.005
2.0	-1.672	-0.4028	0.247 ± 0.005	$-$ 0.6270	-0.2890	0.235 ± 0.005

Table 1 shows that the fastest mode is the fundamental symmetric MEMP which has negative chirality and velocity. The second most fast symmetric MEMP also have negative chirality and velocity for both $\eta = 1$ and 2. However, the third most fast symmetric MEMP has positive chirality (velocity) for these η. The first and the second most fast antisymmetric MEMPs have negative chirality and velocity as the third most fast antisymmetric MEMP has positive chirality and velocity for both $\eta = 1$, and 2. From table 1 it follows that the fastest symmetric MEMPs is much faster than the second fastest symmetric MEMP, $|W_{s1}|/|W_{s2}| > 4$, or the fastest antisymmetric MEMP, $|W_{s1}/|W_{a1}| \approx 2.75$. So among these modes the fastest symmetric MEMP, W_{s1}, should be manifested in experiment easier than any other MEMP as other modes are much slower and, respectively, must be much more damped.

In Figs. 2, 3 we plot charge density profiles of three fastest symmetric and antisymmetric modes for $\eta = 1$; in addition, $g_0(Y)$ is shown. It is seen that for the modes W_{s1}, W_{s2} and W_{a1},

W_{a2} with the negative chirality the wave charge density is strongly localized within the region of negative $g_0(Y)$, i.e, $-1.75 \le Y \le 1.75$. In addition, in Figs. 2, 3 the charge density of the MEMPs W_{s3} and W_{a3}, with the positive chirality, is more localized within the region of positive $g_0(Y)$, at $|Y| > 1.75$; notice that $\int_{-\infty}^{\infty} dY\, g_0(Y) = 0$.

We have assumed that $M_0 > 0$ and $B_{ext} > 0$. It is easy to see that the sign of chirality (velocity) of any MEMP will be changed if the sign of M_0 becomes opposite; however, it is independent of a sign of B_{ext}. If the ferromagnetic film has hysteresis then, in principle, it is possible to change a sign of M_0 while keeping the same B_{ext}. Then the chirality of a MEMP will be changed on opposite.

We have obtained that laterally inhomogeneous strong magnetic field applied to homogeneous 2DES allows "magnetic gradient" or special magnetic-edge magnetoplasmons. This mechanism is different from usual "density gradient" edge magnetoplasmons. Symmetric and antisymmetric families of magnetic-edge magnetoplasmons are obtained. The fundamental symmetric MEMP is much faster than any other MEMP and hence it should be less damped and easier to observe. The fastest MEMP is mainly localized within the region of largest (negative) magnetic gradient under the edge of magnetic film. MEMP of opposite chirality is mainly localized within the region of positive magnetic gradient that is close, but not immediately under the edge of magnetic film.

Acknowledgments

We are grateful to Lloyd W. Engel for many useful discussions on a possibility of MEMP, due to inhomogeneity of magnetic field, relevant experiments and experimental conditions. The research leading to these results has received funding from the European Union Seventh Framework Programme (FP7/2007-2013) under grant agreement n° PCOFUND-GA-2009-246542 and from the Foundation for Science and Technology of Portugal. We also acknowledge support by Brazilian FAPEAM grants: Universal Amazonas (Edital 021/2011) and PVS.

References

[1] Volkov V A and Mikhailov S A 1991 *Electrodynamics of Two-Dimensional Electron Systems in High Magnetic Fields*, in *Landau Level Spectroscopy, Modern Problems in Condensed Matter Sciences*, Ed. by G. Landwehr and E. I. Rashba (North-Holland, Amsterdam) vol 27.2 chapter 15 pp 855-907; Volkov V A and Mikhailov S A 1988 *Zh. Eksp. Teor. Fiz.* **94** 217 [1988 *Sov. Phys. JETP* **67** 1639]
[2] Kushwaha M S 2001 *Surface Science Reports* **41** pp 1-416
[3] Balev O G and Vasilopoulos P 1998 Phys. Rev. Lett. **81** 1481
[4] Khannanov M N, Fortunatov A A and Kukushkin I V 2009 *Pis'ma Zh. Eksp. Teor. Fiz.* **90** 740 [2009 *JETP Lett.* **90** 667
[5] Silva S and Balev O G 2010 *J. Appl. Phys.* **107** 104310
[6] Larkin I A and Sukhorukov E V, Phys. Rev. **49**, 5498 (1994).
[7] Jackson J D 1999 *Classical Electrodynamics* 3d ed. (John Wiley&Sons, New York)

Journal of Physics: Conference Series **456** (2013) 012024 doi:10.1088/1742-6596/456/1/012024

Visualization of wave function of quantum dot at Fermi-edge singularity regime

I A Larkin[1], Yu N Khanin[2], E E Vdovin[2], Sebastian Ujevic[3] and M Henini[4]

Abstract. We consider in this work many-body enhanced electron tunneling through an InAs quantum dot in a magnetic field applied perpendicular to the tunneling direction. We have examined in details the anisotropic behavior of the amplitude and shape of the resonant peaks.

[1] Department of Physics, Minho University, Braga 4710-057, Portugal
[2] Institute of Microelectronics Technology RAS, 142432 Chernogolovka, Russia
[3] Departamento de Ciências Exatas - EEIMVR, Universidade Federal Fluminense, 27255-125, Volta Redonda, RJ, Brazil
[4] School of Physics and Astronomy, Nottingham Nanotechnology and Nanoscience Centre, University of Nottingham, Nottingham NG7 2RD, UK

E-mail: vaniala2000@yahoo.co.uk

1. Introduction

Self-assembled quantum dots (QDs) have been thoroughly investigated in the recent fifteen years, using electronic and optoelectronic methods. The high interest in these nanoscopic systems stems from their potential application in laser devices or for quantum computing due to their controllable and discrete spectrum of energy levels. A special technique, which probes the wave function in k space, makes use of magneto-tunneling spectroscopy (MTS) [1, 2, 3]. Although the MTS technique allows us to observe clear quantization effects in the dot, a direct correspondence between the measured values of current or capacitance, and the probability density plots should be corrected by a number of factors. For example, when an electron tunnels from the emitter into the QD, it can be scattered by the local potential of impurities and on another electrons in the emitter. The later effect has many-body nature and increases with the decreasing of temperature. This enhancement is originated by the Fermi Edge Singularity (FES) phenomena that is relevant to the "Orthogonality Catastrophe" [4] phenomena: a state of a Fermi gas after a suddenly imposed localized potential is almost orthogonal to a state of an unperturbed Fermi gas. The theory proposed in Refs. [5, 6, 7] is based on techniques involving the summation of infinite perturbation series, what gives rise to a power-law divergence in the tunneling amplitude. Ohtaka and Tanabe [8] performed necessary calculations for finite temperatures. In our experimental situation, a sudden potential occurs as a result of charging a QD in the barrier.

The FES effect has been observed in the current-voltage characteristics $I(V)$ of several semiconductor heterostructure devices [9, 10, 11, 12, 13, 14, 15, 16, 17, 18]. Experiments [10, 11, 12, 16, 17, 18] involving additional perturbation on the electron system have revealed strong dependence of the FES on magnetic field $\mathbf{B}$ applied parallel or perpendicular to tunneling barrier. At low temperature, the shape of $I-V$ curve and amplitude of the peak is strongly

Journal of Physics: Conference Series **456** (2013) 012024 doi:10.1088/1742-6596/456/1/012024

field-dependent. The shape of this sharp current peak can be described by a steep ascent below resonant voltage V_0 and a more moderate decrease of the current toward higher voltages. We relate the rapid enhancement of the low voltage resonance in magnetic field at low temperature with effect of the Fermi-edge singularity associated with tunneling electron via a localized state [7]. In this paper we will investigated how the tunneling current through a QD at low temperature is influenced by the applied magnetic field parallel to the barrier or $\mathbf{B} \perp \mathbf{J}$.

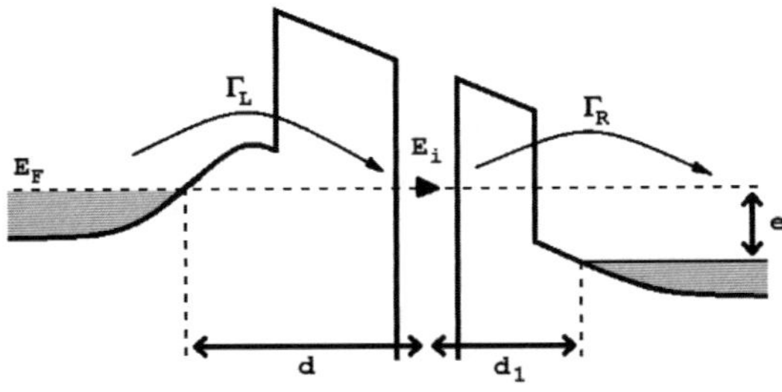

Figure 1. A sketch of the tunneling process through InAs quantum dots grown in between AlGaAs barriers with applied voltage V. d is the distance from the Fermi sea of emitter to the QD and d_1 is the distance from QD to collector. Γ_L and Γ_R are partial tunneling widths of the localized state $|E_i\rangle$ related to decay to emitter and to collector respectively. A detailed description of the device is available in Ref. [12].

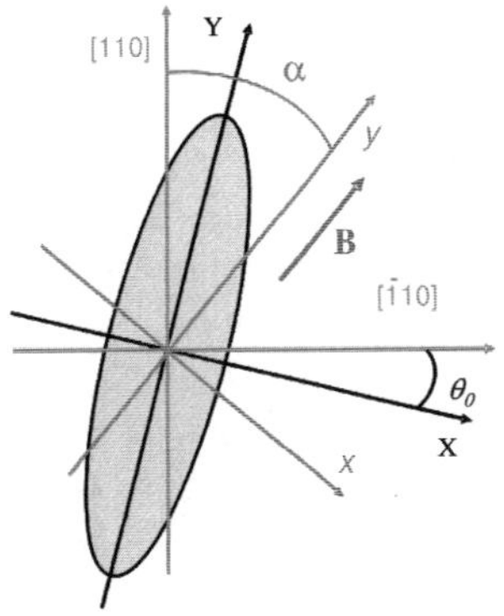

Figure 2. (Color online) Coordinate system relative to the crystallographic axes and the electron tunneling. The magnetic field is directed along the y axis. The green ellipse is the effective potential generated by the QD with main axes along X and Y. Electrons tunnel along the z direction.

2. Fermi edge singularity at magnetic field

The sketch of the conduction band profile of the central part of the hetero-structure is shown in Fig. 1. The InAs QDs create discrete zero-dimensional electronic states $|E_i\rangle$ in the tunnel barrier. An applied voltage V shifts the QD energy levels with respect to the Fermi energy (E_F) of the GaAs emitter layer. When a level of the QD coincides with E_F, the resonant tunneling of the electrons through the system leads to a sharp current increase. We define the deviation from the resonance as $\Delta \equiv E_i - E_F = e(V - V_0)l_f$, where V_0 is the resonant voltage at which the energy level of the localized states matches the Fermi level at the emitter and l_f is the leverage factor $l_f = d/(d + d_1)$, where d and d_1 are the distances from the QD to emitter and collector respectively. Figure 2 shows the cross section of the QD in the xy plane (electrons tunnel along z direction $[0, 0, 1]$). In general, the effective potential of the QD is anisotropic and can be approximated with an anisotropic harmonic oscillator potential with eigenfrequencies $\omega_X \neq \omega_Y$. The main axes of the oscillator (green ellipse on Fig.2) may be nonparallel to crystallographic axes. We denote by θ_0 the tilt of the main axes of the QD X and Y versus axes $[\bar{1}, 1, 0]$ and $[1, 1, 0]$ respectively. The magnetic field which is perpendicular to the current direction may also not be aligned with the crystallographic axes. We denote by α the angle between magnetic field and axes $[1, 1, 0]$.

In order to analyze the many-body effect on the tunneling through the system, we have investigated the shape of the resonant peak in the $I(V)$ curve. Combining the calculation methods of Refs. [7] and [8] we get [17, 18]

$$I(V) = \mathcal{A}\ \mathrm{Re}\left[i^{\gamma-1}\frac{\Gamma((1-\gamma)/2 - i\Omega)\Gamma(\gamma)}{\Gamma((\gamma+1)/2 - i\Omega)}\right], \tag{1}$$

where $\mathcal{A} = e\Gamma_L\hbar^{-1}(W_b/2\pi T)^\gamma$ is the peak amplitude, $\Gamma_L/\hbar$ is the tunneling rate from the emitter to the localized states, $W_b \sim E_F$ is the band width of the emitter, γ is a FES exponent and

$$\Omega = \frac{\Delta}{2\pi T} + i\frac{\Gamma_T}{4\pi T}, \tag{2}$$

where $\Gamma_T = \Gamma_R + \Gamma_L$ is the total level width, Γ_L is the partial tunneling width towards the emitter and Γ_R is the partial tunneling width that corresponds to decay towards the collector. Eq. (1) is valid for $\Gamma_R \gg \Gamma_L$. At $T \to 0$ and $\Gamma_T \to 0$, Eq. (1) simplifies to $I \propto (V - V_0)^{-\gamma}$, at $V > V_0$. The FES exponent γ and the tunneling rate $\Gamma_L/\hbar$ are very sensitive to the device geometry, the shape of wave function at localized state, the magnetic field, the spin polarization, and the spectrum of the Fermi electrons. Consistent analysis of experimental data may shed the light on both electrons in emitter and properties of the QD. In our previous study [11, 12], we have observed a similar behavior of $\gamma(B)$ at all angles α, namely, it increases substantially at magnetic fields around 6 T and drops rapidly at higher fields. According to theory [17], $\gamma(B)$ reaches its maximum at a magnetic field value of 6 T, in a good agreement with the experiment.

To explain the variation of the peak at high magnetic field, we take into account that the ground state function in the (k_X, k_Y) space that corresponds to the X-Y axes is given by

$$\psi(k_X, k_Y) \propto \exp\left[-\frac{\hbar}{2m}\left(\frac{k_X^2}{\omega_x} + \frac{k_Y^2}{\omega_y}\right)\right]. \tag{3}$$

It is visible from Fig. 3 that the current peak amplitude has a very strong angular dependence that reflects the anisotropy of the wave function in the QD [1]. In a magnetic field the wave vector of the electron is complemented by an additional term $p = eBd$. As the momentum of the electron at the edge of the Fermi sea is very small, the probability of a tunneling to the QD is determined by the squared absolute value of the wave function of Eq. (3). Therefore, for an anisotropic QD wave function, the level width Γ_L will have an strong magnetic field orientation dependence given by

$$\Gamma_L(\alpha) \propto \exp\left[-\frac{\hbar(eBd)^2}{m}\left(\frac{\cos^2(\alpha-\theta_0)}{\omega_x} + \frac{\sin^2(\alpha-\theta_0)}{\omega_y}\right)\right], \tag{4}$$

where θ_0 is the angle between the main axes of the QD and the crystallographic axes. This dependence is shown at Fig. 4 by red line. However, one must compare the amplitude of the currents flowing through individual QDs (open circles in Fig. 3). Moreover, the amplitude of the peak is $\mathcal{A}(\alpha) \propto \Gamma_L(\alpha)^{-1}(W_b/2\pi T)^{\gamma(\alpha)}$. The blue curve in Fig. 3 is a fit of the current amplitude with this function with a parameter $(W_b/2\pi T) \approx 30$. The function

$$\gamma(\alpha) = 0.67 - \frac{1}{6}\cos(2\alpha + \varphi_1) - \frac{1}{10}\cos(4\alpha + \varphi_2), \tag{5}$$

where $\varphi_1 = 22^o$ and $\varphi_2 = 15^o$ is obtained from independent fitting of $I - V$ curves for each α and shown as the blue curve in Fig. 4. Note, that only even-numbered harmonics exist in formula (5) what indicates that $\gamma(\alpha)$ independent of the sign of the magnetic field.

Notably, experimentally the strong variation of the FES exponent γ occurs at magnetic field values above 7 T. This variation is shown in Fig. 4 for $B = 8$ T. The white circles indicate that γ variates from 0.38 to 0.86 with a maximum value at $\gamma = 0.86$ for $\alpha = 60^o$ and 240°. These values have a visible departure from the ones obtained for the current maximums. These features are

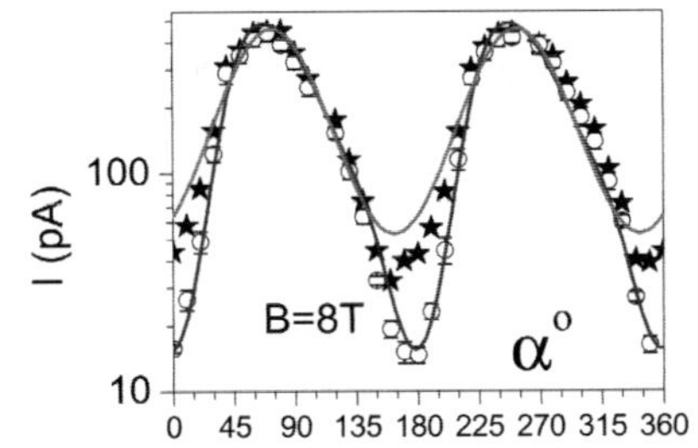

Figure 3. (Color online) Angular dependence of the peak current for $B = 8$ T in log scale at $T = 0.4$ K. Black stars are the maxima of the experimental current. White circles are the calculated current values at resonant bias. The red curve is a fit of the experimental data by Eq. 4. The blue curve is a fit of the peak amplitude $\mathcal{A}(\alpha)$, see text for details.

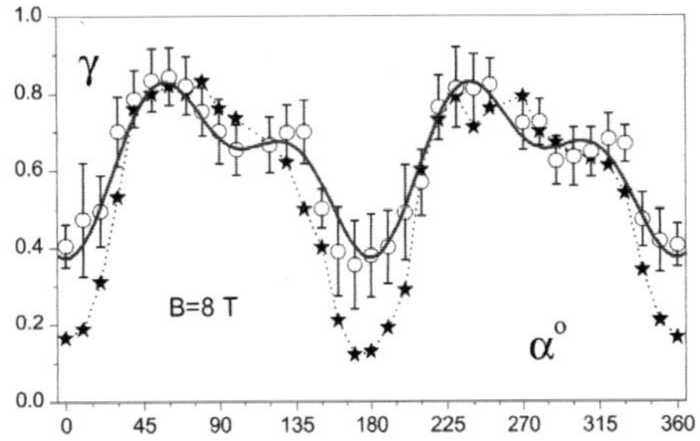

Figure 4. (Color online) Angular dependence of the FES exponent γ for $B = 8$ T at $T = 0.4$ K. Circles are results of the fitting using Eq. (1), error bars are 95% confidence intervals. The solid blue curve is the plot of Eq. (5). Black stars are the fitting results of the descent part of the experimental curves using the relation $I \propto (V - V_0)^{-\gamma}$.

repeatable and identical at opposite directions of the applied magnetic field. Overall, in Figs. 3 and 4 the white circles and black stars show that the resonant bias, the peak amplitude and the FES exponent has the periodicity 180°, *i.e.* they are independent of the magnetic field sign. The fitting procedure that takes into account the variation of $\gamma(\alpha)$ allows the extraction of the QD wave function parameters with better acuracy. From the data shown at Figs. 3 and 4, we may conclude that $\theta_0 \approx 11°$. The wave function size along the principal axes X and Y are $\sqrt{< X^2 >} \approx 3.5$ nm and $\sqrt{< Y^2 >} \approx 5$ nm with the assumption that $d \approx 25$ nm.

3. Conclusion

We have observed a strong enhancement of the tunnel current through the ground state of an InAs QD in the presence of a magnetic field. An unexpectedly strong angular anisotropy effect is observed when the magnetic field is rotated in the plane of the QD layer. A reason for the current *versus* the magnetic angular dependence is related to the strong anisotropy of the QD confinement potential in the lateral direction. Interplay of the lateral potential with the magnetic field causes a strong alteration of the tunneling rate with the magnetic field. The observed effect of a rapid increase of γ arises due to the effective localization of the electrons in the magnetic parabola. The electrons at the lowest Landau level interact more efficiently with the QD Coulomb potential, causing an enhancement in both Born amplitude and phase shift. These results are relevant to recent studies [10, 11, 12, 17] of many-body interactions effects on electrons tunneling through discrete quantum states. We examine in details the anisotropic behavior of the amplitude and shape of the resonant peaks of $I - V$ curves at FES regime and concluded that: (i) the magneto-tunneling spectroscopy at FES regime allows establishing position of resonant level in QD with high accuracy, (ii) the distinguishable shape of the FES peak allows extracting the amplitude with much better accuracy, and (iii) the FES exponent dependence on magnetic field gives additional information about potential distribution outside the QD.

The anisotropy of the tunneling amplitude is close to the major direction of the crystallographic axis, but we are confident that it differs about 10°. Overall, the effect of the renormalization of $\Gamma_L(B)$ as a function of the magnetic field for the current variation is comparable with the effect of the variation of $\gamma(B)$ due to magnetic field orientation.

Acknowledgments
The research leading to these results has received funding from the European Union Seventh Framework Programme (FP7/2007-2013) under grant agreement n° PCOFUND-GA-2009-246542 and from the Foundation for Science and Technology of Portugal. In Brazil these research were supported by FAPEAM. In Russia work was financially supported by the programs of the Presidium of the Russian Academy of Sciences and by the Russian Foundation for Basic Research (project numbers 12-02-00043-a).

References

[1] Patanè A *et al* 2002 *Phys. Rev.* B **65** 165308
[2] Wibbelhoff O S, Lorke A, Reuter D and Wieck A D 2005 *Appl. Phys. Lett.* **86** 092104
[3] Lei W, Notthoff C, Peng J, Reuter D, Wieck A, Bester G and Lorke A 2010 *Phys. Rev. Lett.* **105** 176804
[4] Anderson P W 1967 *Phys. Rev. Lett.* **18** 1049
[5] Mahan G D 1967 *Phys. Rev.* **163** 612
[6] Nozières P and De Dominicis C T 1969 *Phys. Rev.* **178** 1097
[7] Matveev K A and Larkin A I 1992 *Phys. Rev.* B **46** 15337
[8] Ohtaka K and Tanabe Y 1984 *Phys. Rev.* B **30** 4235
[9] Geim A K *et al* 1994 *Phys. Rev. Lett.* **72** 2061
[10] Hapke-Wurst I, Zeitler U, Frahm H, Jansen A G M, Haug R J and Pierz K 2000 *Phys. Rev.* B **62** 12621
[11] Khanin Yu N and Vdovin E E 2005 *JETP Lett.* **81** 267
[12] Vdovin E E, Khanin Yu N, Makarovsky O, Dubrovskii Yu V, Patanè A, Eaves L, Henini M, Mellor C J, Benedict K A and Airey R 2007 *Phys. Rev.* B **75** 115315
[13] Benedict K A, Thornton A S G, Ihn T, Main P C, Eaves L and Henini M 1998 *Physica* B **256** 519
[14] Gryglas M, Baj M, Chenaud B, Jouault B, Cavanna A and Faini G 2004 *Phys. Rev.* B **69** 165302
[15] Rüth M, Slobodskyy T, Gould C, Schmidt G and Molenkamp L W 2008 *Appl. Phys. Lett.* **93** 182104
[16] Frahm H, von Zobeltitz C, Maire N and Haug R J 2006 *Phys. Rev.* B **74** 035329
[17] Larkin I A, Vdovin E E, Khanin Yu N, Ujevic S and Henini M 2010 *Phys. Scripta* **82** 038106
[18] Ubbelohde N, Roszak K, Hohls F, Maire N, Haug R J and Novotný T *Scientific Reports* **2** 374 doi:10.1038/srep00374

Journal of Physics: Conference Series **456** (2013) 012025 doi:10.1088/1742-6596/456/1/012025

Shubnikov-de Haas effect in tilted magnetic fields in wide quantum well

I A Larkin[1], Sebastian Ujevic[2], S Wiedmann[3], N Mamani[4], G M Gusev[5], A K Bakarov[6] and J C Portal[7]

Abstract. We study the magneto-resistance of a two-dimensional electron systems in a wide quantum well. We have considered two occupied subbands subjected to the tilted magnetic field in the regime of the magneto-inter-subband oscillations. We report on both experimental and theoretical studies of such a phenomenon and deduced information on the inter-subband energy gap in the presence of the in-plane magnetic field.

[1] Department of Physics, Minho University, Braga 4710-057, Portugal
[2] Departamento de Ciências Exatas - EEIMVR, Universidade Federal Fluminense, 27255-125, Volta Redonda, RJ, Brazil
[3] Radboud University Nijmegen, Institute for Molecules and Materials, High Field Magnet Laboratory, Toernooiveld 7, 6525 ED Nijmegen, the Netherlands
[4] Instituto de Física de São Carlos, Universidade de São Paulo, 13560-970 São Carlos, SP, Brazil
[5] Instituto de Física da Universidade de São Paulo, CP 66318, São Paulo, SP, Brazil
[6] Institute of Semiconductor Physics, Novosibirsk 630090, Russia
[7] Laboratoire National des Champs Magnétiques Intenses, CNRS-UJF-UPS-INSA, 38042 Grenoble, France; INSA Toulouse, 31077 Toulouse Cedex 4, France; Institut Universitaire de France, 75005 Paris, France

1. Introduction

The electron transport of a two-dimensional (2D) electron gas subjected to a magnetic field exhibits different types of magneto-resistance oscillations. Among them, the Shubnikov-de Haas oscillation (SdHO) are periodic in inverse magnetic field $1/B$ and controlled by the filling factor $\nu = E_F/\hbar\omega_c$, where E_F is the Fermi energy, m is the electron effective mass and $\omega_c = eB/mc$ is the cyclotron frequency. SdH oscillations appear only at low temperature, and are strongly damped when thermal broadening of the Fermi distribution exceeds the cyclotron energy. In quantum wells with two occupied 2D subbands, the magneto-resistance exhibits another kind of oscillating behavior, the so-called magneto-inter-subband (MIS) oscillations [1]. These oscillations arise from the periodic modulation of the probability of transitions between the Landau levels belonging to different subbands. The MIS oscillation is periodic in $\Delta/\hbar\omega_c$, where Δ is the subband separation. Since the origin of the MIS oscillations is related to the alignment between the different Landau levels (LL) of the two subbands and not to the position of LL with respect to the Fermi energy, these oscillations survive at high temperatures when the SdH oscillations are suppressed. The experimental studies of MIS oscillations have been carried out in single quantum wells with two populated 2D subbands [2]. Recently, MIS oscillations with large amplitudes have been observed and investigated in two-subband systems based on double quantum wells [3], triple quantum wells [3] and wide quantum wells slit into layers electrostatically [4]. MIS oscillations are used to obtain information about electron-electron scattering and quantum life time at high temperatures when SdH oscillations are completely damped in the region of weak magnetic fields.

Another advantage of the MIS oscillations is the possibility of a determination of the inter-subband energy gap Δ. Recently it has been predicted that in high Landau levels the tunneling gap oscillates with the in-plane magnetic field [5], which has been confirmed experimentally in double quantum wells [6, 7] in the quantum Hall effect regime. At low magnetic field, the variation of the gap Δ has been interpreted as Aharonov-Bohm interference effect between cyclotron orbits in different layers [6, 8]. However, the theoretical model [9] describes the interlayer resistivity in the parallel magnetic field, while in the experiments [7] the longitudinal resistance parallel to the layers has been measured. Therefore the behavior of the SdH and MIS oscillations in the presence of the in-plane magnetic field requires further theoretical study and detailed comparison with the experiments.

In the present paper we study the magneto-resistance in the wide GaAs quantum well subjected to a tilted magnetic field. The charge distribution in a wide single quantum well is more subtle than the one in the double quantum well. Here the Coulomb repulsion of the electrons in the well leads to a soft barrier inside the well, which in turn results in a bilayer electron system [Fig. 1(b)]. We performed analytical and numerical calculations of the symmetric-antisymmetric level separation in the presence of the in-plane magnetic field and make detailed comparison with the behavior of the MIS oscillations in a wide quantum well.

2. Experimental results

We have studied wide GaAs quantum wells (w=45 nm) with an electron density of $n_s \simeq 9.2\times10^{11}$ cm^{-2} and a mobility of $\mu \simeq 1.9 \times 10^6$ cm^2/Vs at low temperatures. Samples in both Hall bar ($l \times w$= 250 μm $\times$ 50 μm) and van der Pauw (size 3 mm $\times$ 3 mm) geometries have been studied. The two lowest subbands are separated by the energy $\Delta_{12} = E_{AS} - E_S = 1.40$ meV, extracted from MIS oscillation periodicity. Measurements of the longitudinal resistance R_{xx} have been carried out in a perpendicular magnetic field B up to 15 T in a dilution refrigerator (base temperature 50 mK).

In Fig. 1(c) we present magneto-resistance for a wide quantum well at T=1.5 K. The figure shows MIS oscillations below 0.4 T. For temperature T=50 mK the MIS oscillations are also superimposed on low-field SdH oscillations. These oscillations are periodic in $1/B$ and have a maxima under the condition $\Delta_{SAS} = k\hbar\omega_c$. The peak at B=0.8 T corresponds to k=1. Numerous sweeps of the magnetic field sweeps have been taken for different tilt angles. Figure 1(a) shows the resulting phase diagrams, or the plots of the longitudinal resistance R_{xx}, in the $B_\perp - B_{||}$ plane at the balance point in magnetic field below B=1 T in the regime of the MIS oscillations. We recalculate $B_\perp$=$B\cos\theta$ and $B_{||}$ =$B\sin\theta$, where θ is a tilt angle between the normal to quantum well plane and magnetic fields. The phase diagram shown in Fig. 1(a) is similar to results obtained for double quantum well structures [7,8]. We may see that the vanishing of the resistance minima is correlated with evolution of the MIS oscillations maxima with tilt angle. When the tilt angle increases, even number k MIS peaks become more pronounced at filling factors $\nu = 4N + 1$ and $\nu = 4N + 3$, while MIS peaks with odd k number disappear.

3. Numerical and analytical calculations

Electrons subjected to a tilted magnetic field $(B_{||}, 0, B_\perp)$ in a quantum well potential are described by the Hamiltonian [10] in the Landau gauge.

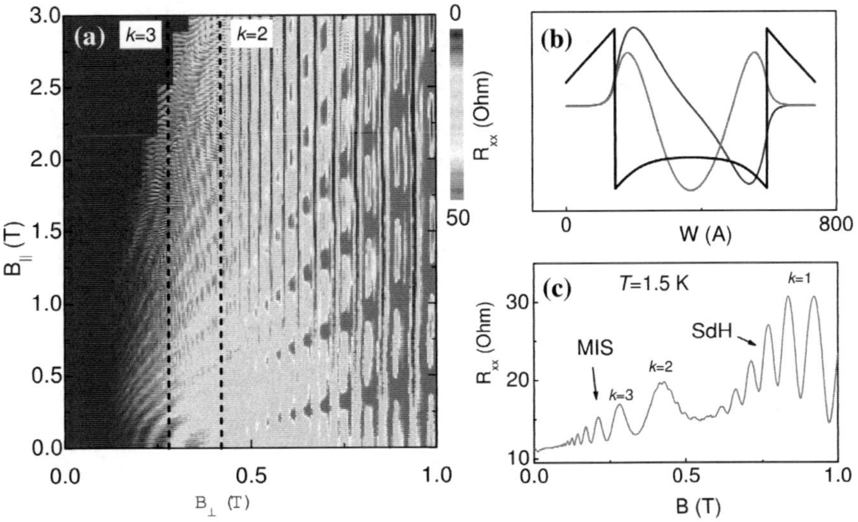

Figure 1. (Color online)(a) Experimental plot of the resistance in the $B_\perp - B_{||}$ plane for a wide GaAs well at T=50 mK. (b) Calculated confinement potential profile of our wide quantum wells and electron wave functions for the first two subbands. (c) The magneto-resistance R_{xx} as a function of the magnetic field at T=1.5 K. The magneto-resistance exhibits MIS oscillations together with SdH oscillations.

$$H = -\frac{\hbar^2}{2m}\frac{d^2}{dx^2} + \frac{q^2 B_\perp^2}{2m}x^2 - \frac{\hbar^2}{2m}\frac{d^2}{dz^2} + V(z) + \frac{q^2 B_{||}^2}{2m}z^2 - \frac{q^2 B_\perp B_{||}}{m}xz, \quad (1)$$

where q is the electron charge and m is the effective mass. The first four terms describe electrons in a perpendicular magnetic field, where energy levels are Landau Levels in symmetric (S) and anti-symmetric (AS) sub-bands. The small energy separation and the symmetry of the wave functions for the two lowest subbands show that the corresponding (symmetric and antisymmetric) states are formed as a result of a tunnel hybridization $t_0 = (E_{AS} - E_S)/2$ of the states in the two quantum wells near the interfaces. The last two terms stem from the in-plane magnetic field.

To diagonalize this Hamiltonian we use the method described in [11]. The matrix elements of H can be expressed by using the eigen-states in a perpendicular field (N,ξ) as a basis set, where $N = 0, 1, ...$ and $\xi_i = S_i, AS_i$ represent the Landau orbit and sub-band indexes respectively. In a perpendicular magnetic field, the eigen-energy of each basis is $E_\perp^{N,S(AS)} = \hbar\omega_{c\perp}(N + 1/2) + E_{S(AS)}$. When a finite parallel magnetic field is applied, the diagonal matrix elements undergo a diamagnetic shift given by

$$E_d(\xi) =< N, \xi|\frac{q^2 B_{||}^2}{2m}z^2|N, \xi >= \frac{\hbar\omega_{c||}}{2} < \xi|\frac{z^2}{l_{B_{||}}^2}|\xi > . \quad (2)$$

More importantly, off diagonal terms arise from the xz term, since x couples Landau Levels when the Landau indexes differ by one. As a result, the matrix element between (N, ξ_i) and $(N + 1, \xi_j)$ is

$$E_c(\xi_i,\xi_j) =< N,\xi_i|\frac{q^2 B_{||}B_{\perp}}{m}xz|N+1,\xi_j >= \hbar\omega_{c\perp}\tan(\theta)\sqrt{\frac{N+1}{2}} < \xi_i|\frac{z}{l_{B_\perp}}|\xi_j > . \tag{3}$$

For the calculations of this work we have considered 30 Landau Levels and two functions for each S and AS sub-band. The Hamiltonian built in this way produces two families of fan energy charts for the system, related to the symmetric and anti-symmetric solutions. We have checked that the accuracy of our calculations is about 1% by increasing the numbers of basis functions.

The magneto-resistance oscillation is then modeled through a function of the form

$$M(B) = C_1\cos[C_s/B+\phi_s(B)] + C_2\cos[C_{as}/B+\phi_{as}(B)], \tag{4}$$

where $C_{1,2}$ are constants, and $\phi_{s,as}(B)$ the symmetric and anti-symmetric corrections to the overall phase. The arguments $C_{s,as}$ are given by

$$C_{s,as} = \frac{2\pi mc(\mu - E_{S,AS})}{q\hbar}, \tag{5}$$

with a fixed chemical potential μ. Equation 4 synthesizes our numerical magneto-resistance oscillation model.

If in our Hamiltonian we leave only one S and one AS subband then energy spectrum reduces to [6],

$$E_{AS,S} = E_0 + \frac{md^2\omega_{c||}^2}{8} + \frac{\hbar^2}{2m}(k_x^2+k_y^2) \pm \sqrt{t_0^2 + \frac{\hbar^2 d^2\omega_{c||}^2 k_y^2}{4}}, \tag{6}$$

where d is the distance between the centers of the electrons wave functions at the different sides of the quantum well. The sign $+$ in the last term corresponds to the antisymmetric state and $-$ to the symmetric one. In quasi-classical approximation, the Bohr-Sommerfeld quantization rule reads

$$\oint dk_y\sqrt{\frac{2m}{\hbar^2}\left(E \pm \sqrt{t_0^2 + \frac{\hbar^2 d^2\omega_{c||}^2 k_y^2}{4}}\right) - k_y^2} = \frac{2\pi(n+1/2)qB_\perp}{\hbar}, \tag{7}$$

where n is the level number. Assuming that the Fermi energy of the 2D electron gas is much larger than the difference between the symmetric and antisymmetric energies, the integral can be reduced to

$$\pi k_F^2 \pm \frac{2m}{\hbar^2}t_0\,\mathbf{E}\left(-\frac{\hbar^2 d^2\omega_{c||}^2 k_F^2}{4t_0^2}\right), \tag{8}$$

where k_F is the Fermi wave vector of the 2DEG and $\mathbf{E}(...)$ is the full elliptic integral of second kind. At small magnetic field, the symmetric-antisymmetric level splitting Δ_{SAS} is

$$\Delta_{SAS} = 2t_0\left(1 + \frac{\hbar^2 d^2\omega_{c||}^2 k_F^2}{4t_0^2}\right), \tag{9}$$

and increases slightly at small $B_{||}$. At large $B_{||}$, when $\hbar^2 d^2\omega_{c||}^2 k_F^2 \gg t_0^2$, we obtain

$$\Delta_{SAS} = \frac{d\hbar k_F}{\pi m}qB_{||}. \tag{10}$$

As a result, the SdHOs experience a modulation that depends only on the ratio $B_{||}/B_{\perp}$ and the distance d. In this way, the longitudinal resistivity ρ_{xx} can then be written as

$$\rho_{xx} = \rho_0 \left(1 - 4D(T)\cos\left[k_F^2 \frac{\pi\hbar}{qB_{\perp}}\right]\cos\left[k_F d\frac{B_{||}}{B_{\perp}}\right]\exp[-\pi m/qB_{\perp}\tau]\right), \tag{11}$$

where τ is the relaxation time and $D(T)$ the temperature damping factor of the SdHOs. Equation (11) shows that the amplitude of the SdHOs has maximums at $k_F d \tan\theta_n = \pi n$, where n is a integer number. In experiment, the amplitude maximums were observed at θ=25°, 50°, 62°, 70°... at high magnetic field. This allows us to deduce a factor $k_F d \approx 4.6$ and distance $d \approx 39$ nm that agree well with the size of the well and with the thickness of the wave functions.

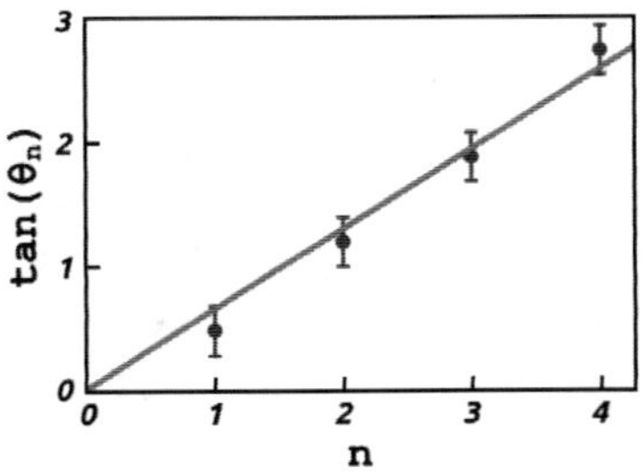

Figure 2. (Color on line) Blue points is the experimental data of $\tan\theta_n$ at which amplitude of SdHOs is maximal. Error bars correspond to half width of the maxima. Red line is a linear fit of the data points. The slope of the line is $\pi/k_F d$.

4. Conclusion

In conclusion, we have studied the magneto-resistance in a wide GaAs quantum well subjected to a tilted magnetic field. We have developed a very effective algorithm to calculate the spectrum of electrons in a quantum well in tilted magnetic field. We performed numerical and analytical calculations of the symmetric-antisymmetric level separation in the presence of a magnetic field, and compared it with the amplitude variation of the SdHO in a wide quantum well.

Acknowledgments

The research leading to these results has received funding from the European Union Seventh Framework Programme (FP7/2007-2013) under grant agreement n° PCOFUND-GA-2009-246542 and from the Foundation for Science and Technology of Portugal. In Brazil the research was supported by FAPESP, FAPEAM and CNPq.

References

[1] Polyanovsky V 1988 *Fiz. Tekh. Poluprovodn.* **22** 2230 (*Sov. Phys. Semicond.* **22** 1408)
[2] Leadley D R, Fletcher R, Nicholas R J, Tao F, Foxon C T and Harris J J 1992 *Phys. Rev.* B **46** 12439
[3] Mamani N C, Gusev G M, Lamas T E, Bakarov A K and Raichev O E 2008 *Phys. Rev.* B **77** 205327
[4] Wiedmann S, Mamani N C, Gusev G M, Raichev O E, Bakarov A K and Portal J C 2009 *Phys. Rev.* B **80** 245306
[5] Wiedmann S, Gusev G M, Raichev O E, Bakarov A K and Portal J C 2010 *Phys. Rev. Lett.* **105** 026804
[6] Hu J and MacDonald A H 1992 *Phys. Rev.* B **46** 12554
[7] Gusev G M, Bakarov A K, Lamas T E and Portal J C 2007 *Phys. Rev. Lett.* **99** 126804
[8] Gusev G M, Duarte C A, Lamas T E, Bakarov A K and Portal J C 2008 *Phys. Rev.* B **78** 155320
[9] Yakovenko V M and Cooper B K 2006 *Physica* E (Amsterdam) **34** 128
[10] Kumada N, Iwata K, Tagashira K, Shimoda Y, Muraki K, Hirayama Y and Sawada A 2008 *Phys. Rev.* B **77** 155324
[11] Medvedeva M V, Larkin I A, Ujevic S, Shchur L N and Ivlev B I 2008 *Phys. Rev.* B **78** 165325

Journal of Physics: Conference Series **456** (2013) 012026 doi:10.1088/1742-6596/456/1/012026

Magnetic field induced rearrangement of the vacuum charge in a graphene quantum dot with a mass gap

P. A. Maksym[1] **and H. Aoki**[2]

[1]Department of Physics and Astronomy, University of Leicester, Leicester LE1 7RH, UK
[2]Department of Physics, University of Tokyo, Hongo, Bunkyo-ku, Tokyo 113-0033, Japan

E-mail: map@le.ac.uk

Abstract. A graphene quantum dot provides a solid state analogue of the as yet unobserved *charged vacuum* predicted by quantum electrodynamics. The magnetic field affects the energy levels that contribute to the vacuum charge and either discharges a vacuum that is already charged or adds charge to it, depending on the position of the affected level relative to the Fermi level. These effects occur in external potentials of a few hundred meV and magnetic fields of a few T, an experimentally realisable regime.

1. Introduction

A charged vacuum is predicted to occur as the ground state of quantum electrodynamics (QED) in a strong electric field [1, 2, 3]. For some time, it has been thought that this situation can be realised in a collision between two heavy nuclei that briefly results in a nuclear charge, Z, that exceeds the threshold (~ 172) for charging the vacuum. The vacuum charging is accompanied by spontaneous positron emission, and in principle this gives a way of observing the charged vacuum in heavy ion collisions. But despite much experimental effort, the effect has not yet been observed, probably because the interaction time is not long enough to allow the charged vacuum to occur [4]. Recent work has shown that it may be possible to prepare graphene with a mass gap, while retaining its Dirac-line energy spectrum, and it has been suggested that a charged vacuum analogue may occur in graphene that has a gap and contains Coulomb impurities [5]. Experimental control of the impurity charge is difficult. However it has recently been shown that well-controlled charged vacuum states may be realised in a graphene quantum dot [6]. This work concerns the magnetic field dependence of this sold state, charged vacuum analogue.

The magnetic field affects the energy levels drastically and has a profound effect on both the atomic charged vacuum and its solid state analogue. At zero magnetic field the vacuum charges when the dot potential is strong enough to lower a bound state into the valence band or equivalently the negative energy continuum. At this critical strength the bound state becomes a resonance which stores charge if it is below the Fermi level. Simultaneously, holes or equivalently positrons are emitted so the total charge is conserved. The effective mass equation that gives the graphene states near the Dirac points at K and K' is exactly the 2D Dirac equation. The valley index acts as a pseudo-spin and the energy levels in a magnetic field show pseudo-spin splitting as well as splitting due to the real electron spin. In addition, they show a non-linear field dependence when the magnetic field is stronger. These effects cause energy levels to cross the Fermi level as the magnetic field increases. The vacuum *discharges* when an energy level rises above the Fermi level and *charges* when a level falls below it. In graphene with rest mass-energy $m_0c^2 \sim 100$ meV these effects occur at about 1 T but in an atomic system with $m_0c^2 \sim 0.5$ MeV the corresponding field is about 3×10^8 T. These ultra strong fields only occur in extreme astrophysical environments such as magnetised neutron stars [7] so graphene provides an accurate solid state model of atomic systems in extreme astrophysical fields as well as the charged vacuum. The effect of a magnetic field on the charged vacuum does not seem to have been investigated before.

Observation of the effects discussed here requires a material with a mass gap and a Dirac-like energy dispersion. This could be prepared by depositing single-layer graphene on a suitable substrate and mass gaps have been reported in ab-initio and experimental studies of graphene on BN ($m_0c^2 \sim$ 10-30 meV) [8, 9, 10] and experimental studies of graphene on Ru ($m_0c^2 \sim 100$ meV) [11] and SiC ($m_0c^2 \sim 100$ meV) [12, 13]. Although the SiC results appear to be controversial [14], it seems feasible to prepare a suitable material with a mass gap. In the present work, $m_0c^2 = 100$ meV is taken as a typical value and the consequences for vacuum charging are investigated.

The paper begins with a description of computational methods in section 2. Numerical results for the magnetic field dependence of the vacuum charge are presented in section 3 and experimental consequences are discussed in section 4.

2. Computational Model

Graphene is treated within the effective mass approximation. In existing models [5, 15], the mass is included by splitting the energies of the A and B sublattice sites in the graphene. Although the mechanism of mass generation is not yet fully understood, these models give a dispersion relation that is consistent with experimental data and ab-initio calculations. The 4×4 effective mass Hamiltonian consists of two 2×2 blocks,

$$H = \begin{pmatrix} H_K & 0 \\ 0 & H_{K'} \end{pmatrix} \equiv \begin{pmatrix} c\boldsymbol{\sigma} \cdot \boldsymbol{\pi} + m_0c^2\sigma_z & 0 \\ 0 & c\boldsymbol{\sigma}^* \cdot \boldsymbol{\pi} + m_0c^2\sigma_z \end{pmatrix} + V \begin{pmatrix} E & 0 \\ 0 & E \end{pmatrix}, \quad (1)$$

where $\boldsymbol{\sigma}$ and σ_z are Pauli matrices, V is the dot potential, E is the 2×2 unit matrix and $\boldsymbol{\pi} = \mathbf{p} + e\mathbf{A}$ with $\mathbf{p}$ the momentum and $\mathbf{A}$ the magnetic vector potential. The rest mass-energy is m_0c^2, $c = \gamma/\hbar$ and γ is taken to be 646 meV nm [16].

The dot is taken to be circularly symmetric and in a uniform magnetic field, B, perpendicular to the graphene sheet. The model potential is $V(r) = V_0 \exp(-(r/\lambda)^p/2)$ where r is the radial co-ordinate. The parameter λ sets the length scale, and is taken to be 50 nm, while p affects the potential shape. When $p = 2$ the potential is Gaussian and approximately parabolic at its minimum while potentials with larger p are flat-bottomed and rise rapidly at the edge. In the present work calculations are done for $p = 2$ and $p = 8$. The eigenstates are eigenstates of angular momentum and are labelled by the *total* angular momentum quantum number m. At both K and K' they have the form $\phi(\mathbf{r}) = (\chi_1(r)\exp(i(m-1)\theta), \chi_2(r)\exp(im\theta))^T$ where θ is the azimuthal angle. The radial functions χ_i are found numerically [6, 17].

The Landau levels, that is the energy levels of the system without an external potential, may be found analytically [18]. At both K and K' there are Landau levels of energy $\pm(m_0^2c^4 + 2eB\gamma^2N/\hbar)^{1/2}$ where $N \geq 1$ is the Landau quantum number. There is also one Landau level of energy $+m_0c^2$ at K and one of energy $-m_0c^2$ at K'. The energy spectrum is therefore symmetric about the centre of the mass gap and the 4×4 Hamiltonian has chiral symmetry. When $V = 0$, the chiral symmetry operator

$$S = \begin{pmatrix} 0 & i\sigma_y \\ i\sigma_y & 0 \end{pmatrix}, \quad (2)$$

Journal of Physics: Conference Series **456** (2013) 012026 doi:10.1088/1742-6596/456/1/012026

anticommutes with the Hamiltonian, so if ϕ is a state of energy E then $S\phi$ is a state of energy $-E$. Under S the state $\phi = (F_{AK}, F_{BK}, F_{AK'}, F_{BK'})^T$ becomes $S\phi = (F_{BK'}, -F_{AK'}, F_{BK}, -F_{AK})^T$. So the effect of S is to permute the components of ϕ and change their phase. This property is used to simplify the calculation of the vacuum charge.

The experimentally measurable vacuum charge is the additional charge stored in the system when the potential strength exceeds the charging threshold. The vacuum charge density may found from the induced charge density $\rho(V) - \rho(0)$ where $\rho(V)$ is the charge density when the external potential is V. It may also be found from one of the two QED charge density operators, $(-e/2)[\bar{\psi}, \gamma^0\psi]$ or $\hat{N}(-e\bar{\psi}\gamma^0\psi)$ where γ^0 is a Dirac matrix, ψ is the Dirac field operator, $\bar{\psi}$ is its adjoint and $\hat{N}$ denotes normal ordering. The two QED operators are identical [19] and their vacuum expectation value is identical to $\rho(V) - \rho(0)$. To show this, consider the vacuum expectation value of the commutator form, $\hat{\rho}(\mathbf{r}) = (-e/2)(\sum_{E_n<E_F,\alpha} |\phi_{n\alpha}(\mathbf{r})|^2 - \sum_{E_n>E_F,\alpha} |\phi_{n\alpha}(\mathbf{r})|^2)$, where α represents the component and valley indices and E_F is the Fermi level. $\rho(V) - \rho(0)$ may be written in a similar way:

$$
\begin{aligned}
\rho(V) - \rho(0) &= -e \sum_{E_n<E_F,\alpha} |\phi_{n\alpha}(\mathbf{r})|^2 - \rho(0) \\
&= (-e/2)\left(\sum_{E_n<E_F,\alpha} |\phi_{n\alpha}(\mathbf{r})|^2 - \sum_{E_n>E_F,\alpha} |\phi_{n\alpha}(\mathbf{r})|^2\right) \\
&+ \left[(-e/2)\left(\sum_{E_n<E_F,\alpha} |\phi_{n\alpha}(\mathbf{r})|^2 + \sum_{E_n>E_F,\alpha} |\phi_{n\alpha}(\mathbf{r})|^2\right) - \rho(0)\right]. \qquad (3)
\end{aligned}
$$

The first two terms in the square brackets may be simplified by replacing $|\phi_{n\alpha}(\mathbf{r})|^2$ with $\phi_{n\alpha}(\mathbf{r})^*\phi_{n\alpha}(\mathbf{r}')$ and using a completeness relation. Thus $\sum_{E_n<E_F,\alpha} \phi_{n\alpha}(\mathbf{r})^*\phi_{n\alpha}(\mathbf{r}') + \sum_{E_n>E_F,\alpha} \phi_{n\alpha}(\mathbf{r})^*\phi_{n\alpha}(\mathbf{r}') = \sum_{E_n,\alpha} \phi_{n\alpha}(\mathbf{r})^*\phi_{n\alpha}(\mathbf{r}') = \sum_\alpha \delta_{\alpha\alpha}\delta(\mathbf{r} - \mathbf{r}') = N_c\delta(\mathbf{r} - \mathbf{r}')$, where N_c is the number of components. Hence the first two terms become $(-eN_c/2)\delta(\mathbf{r} - \mathbf{r}')$. The $\rho(0)$ term may be replaced by $-e\sum_{E_n<E_F,\alpha} \phi_{n\alpha}(\mathbf{r}, 0)^*\phi_{n\alpha}(\mathbf{r}', 0)$, where the sum is over $V = 0$ states. The sum over states below E_F is not the full completeness sum however if the system has chiral symmetry, $S\phi$ is related to ϕ by a permutation and phase change of components. Hence if E_F is in the mass gap, $\sum_{E_n<E_F,\alpha} \phi_{n\alpha}(\mathbf{r}, 0)^*\phi_{n\alpha}(\mathbf{r}', 0) = \sum_{E_n>E_F,\alpha} \phi_{n\alpha}(\mathbf{r}, 0)^*\phi_{n\alpha}(\mathbf{r}', 0)$ therefore $\sum_{E_n<E_F,\alpha} \phi_{n\alpha}(\mathbf{r}, 0)^*\phi_{n\alpha}(\mathbf{r}', 0) = (1/2)\sum_{E_n,\alpha} \phi_{n\alpha}(\mathbf{r}, 0)^*\phi_{n\alpha}(\mathbf{r}', 0) = (N_c/2)\delta(\mathbf{r} - \mathbf{r}')$. So the $\rho(0)$ term becomes $(-eN_c/2)\delta(\mathbf{r} - \mathbf{r}')$ and cancels the first two terms for all $\mathbf{r}$ and $\mathbf{r}'$. Therefore $\rho(V) - \rho(0) = \hat{\rho}$. In the case of zero magnetic field, this can be proved directly by evaluating $\rho(0)$ with plane wave states but the present approach is valid for arbitrary field strengths. A similar approach can be used to show that $\rho(V) - \rho(0) = \hat{\rho}$ in the case of the 3D Dirac equation.

The total vacuum charge is found from $\hat{Q} = 2\pi \int_0^\infty \hat{\rho}(r) r dr = (-e/2)(\sum_{E_n<E_F,\alpha} - \sum_{E_n>E_F,\alpha})$. In QED the sums in $\hat{\rho}$ and $\hat{Q}$ are divergent and charge renormalization is needed to remove the divergence. However this problem does not occur in the case of graphene because the energy spectrum is bounded. The advantage of using $\hat{\rho}$ and $\hat{Q}$ for numerical calculations is that it is not necessary to compute the $V = 0$ states. Once the system is discretized, the analytic solutions for the Landau levels are no longer exact and if $\rho(V) - \rho(0)$ were used, it would be necessary to compute two sets of states to avoid numerical artifacts in the subtraction.

Physically, $\hat{\rho}$ gives the sum of the polarisation charge density that occurs in response to the external potential V and the real charge density stored in the vacuum. The effective mass model gives the real charge density accurately because the real charge is stored in a few states near K and K' where the effective mass approximation is accurate [6]. However the effective mass approximation loses accuracy away from K and K' so gives a poor approximation to the polarisation charge which results from all the states in the Brillouin zone. The real vacuum

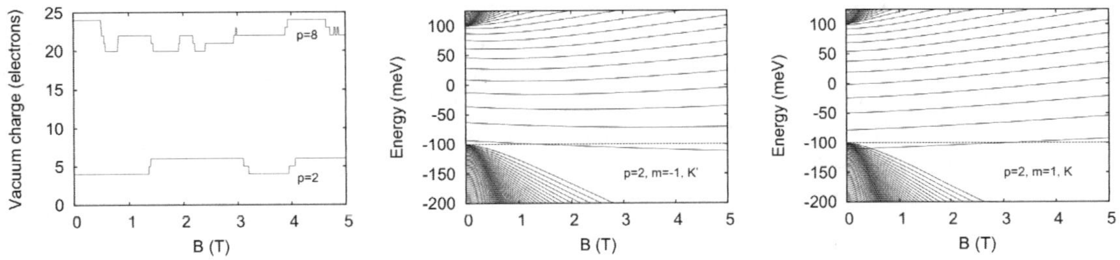

Figure 1. Total vacuum charge (left); $p = 2$ energies at K' (centre) and K (right).

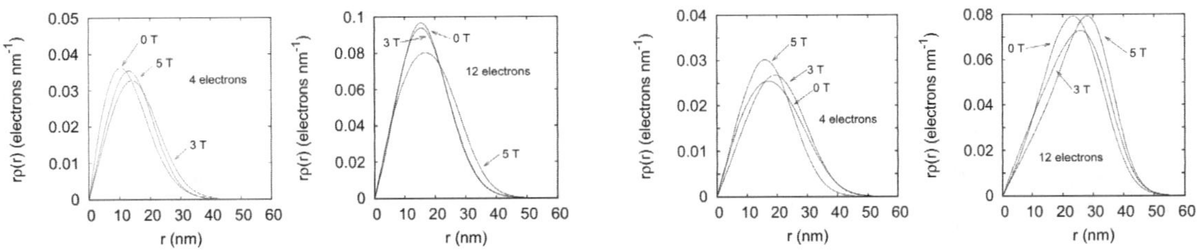

Figure 2. Vacuum charge density for $p = 2$ (left two frames) and $p = 8$ (right two frames).

charge is separated out by computing $\hat{\rho}(V_0 + \delta V_0) - \hat{\rho}(V_0)$, where the vacuum charge jumps abruptly between V_0 and $V_0 + \delta V_0$. This gives the real charge density to within a term of order δV_0 (taken to be -0.1 meV) that arises from the change in the polarisation charge density. This procedure is equivalent to the one described in ref. [3] and its consistency is shown in ref. [6].

The Fermi level is chosen so the graphene system provides an accurate model of the charged vacuum. Graphene has real spin splitting as well as pseudo-spin splitting, twice the number of degrees of freedom of an atomic system. When E_F is in the mass gap, a magnetic field cannot charge the vacuum in an atomic system. However in graphene, the real spin splitting of the Landau level of energy $-m_0c^2$ can cause it to rise above E_F and lead to a charged vacuum. To avoid this situation and model an atomic system, E_F is taken to be just above this Landau level.

3. Numerical Results

Figure 1 shows the total vacuum charge as a function of magnetic field, B, when the well depth is 230 meV i.e. above the threshold for charging the vacuum. Steps in the total charge are clearly visible and because of the real spin splitting each step is split into two sub-steps of height 1. The steps occur whenever an energy level crosses the Fermi level. This is shown in the figure for the case of $p = 2$ with the real spin splitting excluded for clarity. The bound states at K' are pseudo-spin down and their energy tends to decrease with magnetic field. Eventually an energy level crosses the Fermi level and the total vacuum charge *increases*. For example, the upward step at about 1.4 T occurs when the $m = -1$ level at K' (centre frame) crosses the Fermi level (dashed line). Similarly, the bound states at K are pseudo-spin up and their energy tends to increase with magnetic field. The total vacuum charge then *decreases* whenever a level crosses the Fermi level. For example, the downward step at about 3.2 T occurs when the $m = +1$ level crosses the Fermi level (right frame). Richer behaviour is associated with higher energy levels because they tend to go through a minimum as can be seen in the centre frame above about -50 meV. For $p = 8$, the behaviour is similar but the energy level spacing is smaller so the total vacuum charge is higher. In addition, there is strong hybridization between states localised in the dot and negative energy continuum states[6].

Figure 2 shows the magnetic field dependence of the vacuum charge density, $\rho(r)$, at fixed total vacuum charge. The vacuum charge is stored in a few continuum states, typically 1 per electron when $p = 2$ and 2-4 when $p = 8$. The quantum numbers of these states depend on the magnetic field and this results in a magnetic field dependent charge density. At zero magnetic field, the first states to enter the continuum are pseudo-spin degenerate and have *orbital* angular momentum $l = 0$. As the magnetic field increases, pseudo-spin splitting occurs and the pseudo-spin up levels rise out of the continuum and are replaced by pseudo-spin down levels of larger, negative l. These higher angular momentum states have a larger spatial extent and this leads to the charge density expansion shown in the figure. Eventually all the higher angular momentum states are exhausted and the charge density starts to contract because of the usual magnetic field induced compression of the states. This can be seen in the 4-electron case where the density at 5 T is clearly narrower than at 3 T. In contrast when the total vacuum charge is 12, more states are involved and the expansion persists up to 5 T.

4. Conclusion

A magnetic field can change the vacuum charge drastically because it is able to move energy levels through the Fermi level. While these effects have been examined in the context of a graphene quantum dot, they originate from the solutions of the Dirac equation and should also occur in any atomic system. Observation of these effects requires a potential strong enough to lower a bound state into the negative energy continuum. This requires a well whose depth exceeds the gap, that is between about 20 and 200 meV depending on the substrate used to induce the gap. The magnetic field dependent effects then occur between about 0 and 5 T. These values appear to be well inside the experimentally feasible regime. The experimental signature of the charged vacuum is hole emission corresponding to positron emission. In addition it may be possible to sense the charge stored in the vicinity of the dot directly.

Acknowledgments

We thank Seigo Tarucha, Michihisa Yamamoto and Robin Nicholas for helpful discussions. This work was supported by the UK Royal Society and the Japanese ministry of education.

References

[1] Gershtein S and Zeldovich Y B 1969 *Lett. Nuovo Cimento* **1** 835
[2] Pieper W and Greiner W 1969 *Z. Phys.* **218** 327
[3] Greiner W, Müller B and Rafelski J 1985 *Quantum Electrodynamics of Strong Fields* (New York:Springer)
[4] Ruffini R, Vereshchagin G and Xue She-Sheng 2010 *Phys. Repts.* **487** 1
[5] Pereira V M, Kotov V N and Castro Neto A H 2008 *Phys. Rev.* B **78** 085101
[6] Maksym P A and Aoki H, submitted.
[7] Lai D 2001 *Rev. Mod. Phys.* **73** 629
[8] Giovannetti G, Khomyakov P, Brocks G, Kelly P J and van den Brink J 2007 *Phys. Rev.* B **76** 073103
[9] Ci L, Song L, Jin C, Jariwala D, Wu D, Li, Y, Srivastava A, Wang Z F, Storr K, Balicas L, Liu F, and Ajayan P M 2010 *Nat. Mater.* **9** 430
[10] Yelgel C and Srivastava G P 2012 *Appl. Surf. Sci.* **258** 8342
[11] Enderlein C, Kim Y S, Bostwick A, Rotenberg E and Horn K 2010 *New J. Phys.* **12** 033014
[12] Zhou S Y, Gweon G -H, Fedorov A V, First P N, de Heer W A, Lee D -H, Guinea F, Castro Neto A H and Lanzara A 2007 *Nat. Mater.* **6** 770
[13] Vitali L, Riedl C, Ohmann R, Brihuega I, Starke U and Kern K 2008 *Surf. Sci.* **602** L127
[14] de Heer W A, Berger C, Wu X, Sprinkle M, HU Y, Ruan M, Stroscio J, First P N, Haddon R, Piot B, Faugeras C, Potemski M and Moon J 2010 *J. Phys. D: Appl. Phys.* **43** 374007
[15] Giavaras G and Nori Franco 2011 *Phys. Rev.* B **83** 165427
[16] Saito R, Dresselhaus G and Dresselhaus M S 1998 *Physical Properties of Carbon Nanotubes* (London:Imperial College Press)
[17] Giavaras G, Maksym P A and Roy M, in preparation
[18] Itzykson C and Zuber J 1980 *Quantum Field Theory* (New York: McGraw-Hill)
[19] Bjorken J D and Drell S D 1965 *Relativistic Quantum Fields* (New York:McGraw-Hill)

Temperature, electron density and in-plane magnetic field dependence of cyclotron relaxation time in the two-dimensional metallic phase

R. Masutomi[1], T. Chiba[1], K. Sasaki[1], I. Yasuda[1], A. Sekine[1], K. Sawano[2], Y. Shiraki[2] and T. Okamoto[1]

[1]Department of Physics, University of Tokyo, 7-3-1 Hongo, Bunkyo-ku, Tokyo 113-0033, Japan
[2]Research Center for Silicon Nano-Science, Advanced Research Laboratories, Tokyo City University, 8-15-1 Todoroki, Setagaya-ku, Tokyo 158-0082, Japan

E-mail: masutomi@phys.s.u-tokyo.ac.jp

Abstract. We have performed cyclotron resonance measurements on two-dimensional electrons in the metallic phase. The temperature, electron density and in-plane magnetic field dependence of cyclotron and transport scattering time are reported for a high-mobility Si/SiGe quantum well.

1. Introduction

The metallic temperature (T) dependence of resistivity (ρ) has been observed in various two-dimensional (2D) systems [1, 2, 3]. A common feature of these systems is a strong electron-electron (e-e) interaction, which is characterized by the Wigner-Seitz radius $r_s \equiv (\pi N_s)^{-1/2}/a_B = \pi^{1/2}(e/h)^2(m^*/\kappa\epsilon_0)N_s^{-1/2}$. Here N_s is the electron density, a_B is the effective Bohr radius, m^* is the effective mass and κ is the dielectric constant. According to the scaling theory of localization, which neglects e-e interactions, there can be no metallic state in 2D at a zero magnetic field and the system should become an insulator in the zero-temperature limit [4]. Thus the observation of the 2D metallic behavior and the metal-insulator transition (MIT) has caused much controversy and is still a widely debated subject. It is also suggested that the spin and/or valley degeneracies play an important role in the appearance of the metallic behavior [5, 6].

According to the Drude formula, the $B = 0$ dc resistivity is related to the transport scattering time τ_t as $\rho^{-1} = e^2 N_s \tau_t / m^*$. Since τ_t is not sensitive to small-angle scattering events, it is not identical to the single-particle relaxation time τ_s [7]. If $W_{k,k'}$ is proportional to the probability of scattering from state k to k' and θ is the scattering angle, τ_t and τ_s are given by

$$\frac{1}{\tau_t} = \int W_{k,k'}(1-\cos\theta)dk', \tag{1}$$

$$\frac{1}{\tau_s} = \int W_{k,k'}dk'. \tag{2}$$

It is believed that the ratio is large ($\tau_t/\tau_s \gg 1$) for long-range scattering potentials and small ($\tau_t/\tau_s \approx 1$) for short-range scattering potentials. Experimentally, the single-particle relaxation time have been deduced from the magnitude of the Shubnikov-de Haas oscillations ($\tau_{\rm SdH}$) [8, 9] and from the cyclotron resonance ($\tau_{\rm CR}$) [10, 11]. The ratio $\tau_t/\tau_{\rm SdH}$ was measured for various semiconductor 2D systems in order to identify the main low-temperature scattering mechanisms. However, it has been pointed out that $\tau_{\rm SdH}$ is severely affected by density inhomogeneities especially in high mobility samples [11]. In contrast, $\tau_{\rm CR}$ is thought to be insensitive to density inhomogeneities. Furthermore, $\tau_{\rm CR}$ can be measured even at high temperatures where the Shubnikov-de Haas oscillations disappear. In this paper, we report the temperature, electron density [12] and in-plane magnetic field dependence [13] of $\tau_{\rm CR}$ and τ_t in the 2D metallic phase.

2. Experiment

We used a n-Si/SiGe heterostructure with a 20-nm-thick strained Si channel sandwiched between relaxed $\rm Si_{0.8}Ge_{0.2}$ layers [14]. The electrons are provided by a Sb-δ-doped layer 20 nm above the channel. We can control N_s by varying bias voltage $V_{\rm BG}$ of a p-type Si(001) substrate 2.1 μm below the channel at 20 K. The 2D electron system (2DES) has a high mobility of 49 $\rm m^2/V$ s at $N_s = 2.36 \times 10^{15}$ $\rm m^{-2}$ ($V_{\rm BG} = 0$ V) and $T = 0.4$ K. The use of the extremely high-mobility sample allows us to probe low N_s region, where the Coulomb interaction between electrons is sufficiently larger than the Fermi energy.

Although the sample was mounted in a pumped ^{3}He refrigerator, a thermal insulation system enabled us to vary the sample temperature up to 10 K on the condition that the base temperature of the carbon bolometer was kept constant at 0.35 K. The schematic drawing of the equipment is described elsewhere [15, 16]. The cyclotron resonance (CR) measurements were performed using 100 GHz millimeter-wave radiation. The radiation power was kept low enough so that electron heating effects can be neglected. In order to measure ρ and N_s, the sample was fabricated in Hall bar geometry and Ohmic contacts were made outside the irradiation area. For the measurement of in-plane magnetic field $B_{||}$ dependence, we took electron heating effect in the 2DES under excitation at 100 GHz [13]. A two-axis vector magnet system was used to apply an in-plane magnetic field $B_{||}$ and a perpendicular magnetic field $B_\perp$ for CR measurements.

3. Results and discussion

Fig. 1 shows T-dependence of ρ at 0 T (solid line) for different N_s without millimeter-wave radiation. Clear metallic T-dependence of ρ can be seen for $N_s > 0.5 \times 10^{15}$ $\rm m^{-2}$, while it disappears for $N_s < 0.5 \times 10^{15}$ $\rm m^{-2}$. In order to study the effect of the magnetic field on the transport property, we also plotted the longitudinal resistivity ρ_{xx} at 0.7 T ($\sim B_{\rm CR}$) in Fig. 1 (dash line), where $B_{\rm CR}$ is the resonance magnetic field. The deviation of $\rho_{xx}(T)$ at 0.7 T from $\rho(T)$ at 0 T is small. This indicates that the magnetic field applied for the CR measurements is small enough not to affect the transport property significantly.

Fig. 2 shows T-dependence of $\tau_{\rm CR}$ for different N_s. From the half-width at half maximum ΔB of a CR curve, $\tau_{\rm CR}$ is obtained as $\tau_{\rm CR} = B_{\rm CR}/(\omega \Delta B)$. Here ω is the microwave frequency ($\omega/2\pi = 100$ GHz). At all N_s, $\tau_{\rm CR}$ is metallic, i.e., it increases with decreasing T. This result indicates that the metallic T-dependence of the scattering time is shown over a wide frequency range from dc to 100 GHz. At 100 GHz, a characteristic length $l_\omega = v_F/\omega$ becomes much shorter ($l_\omega = 50 \sim 70$ nm) than the electron mean free path $\lambda = 120 \sim 1900$ nm, where v_F is the Fermi velocity. Furthermore, the photon energy $\hbar\omega = 4.8$ K exceeds the typical temperatures where the metallic behavior is observed clearly. At this stage, it is unclear why the metallic T-dependence of the scattering time occurs under very different conditions. However, we believe that our observations will provide a strong constraint on theoretical models.

In Fig. 3, $\tau_{\rm CR}$ and τ_t at 0.4 K are plotted as a function of N_s. In order to discuss the 2D metallic behavior further, it is important to identify the type of disorder potential. For high N_s between 1 and 2 $\times 10^{15}$ $\rm m^{-2}$, $\tau_{\rm CR}$ is smaller than but comparable to τ_t. This suggests that short-range potential fluctuations which lead to large-angle scattering play a dominant role in

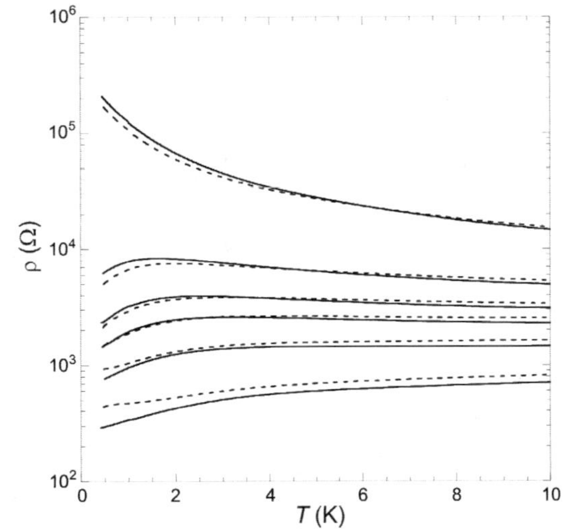

Figure 1. Temperature dependence of ρ at $B = 0$ T (solid line) and 0.7 T (dash line). The electron density is 0.44, 0.55, 0.66, 0.74, 0.90, and 1.22 $\times 10^{15}$ m^{-2} (from top to bottom).

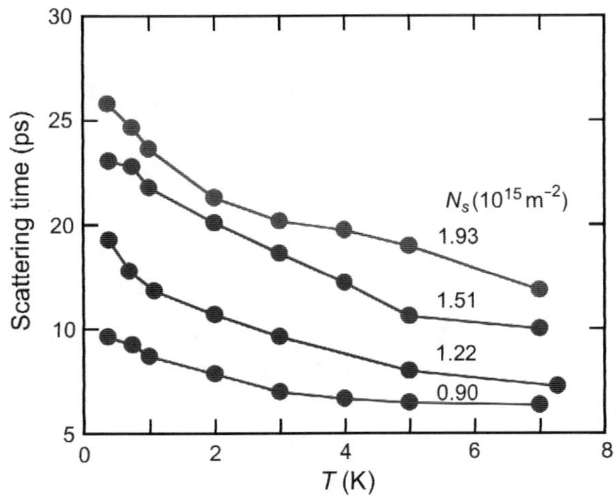

Figure 2. Temperature dependence of $\tau_{\rm CR}$ for $N_s = 1.93, 1.51, 1.22, 0.90 \times 10^{15}$ m^{-2} .

the metallic regime, although the Si/SiGe 2DES studied is a high-mobility modulation-doped-heterostructure. This is consistent with the recent work of Clarke *et al.* who compared the strength of the metallic behavior in GaAs 2D hole systems dominated by long- and short-range disorder [17].

As N_s approaches the critical density for the MIT, τ_t rapidly decreases, where τ_t is determined from $\tau_t = m^*/e^2 N_s \rho$ with the effective mass $m^* = 0.19 m_e$. In contrast, the N_s-dependence of $\tau_{\rm CR}$ seems to be gradual. Moreover, the ratio $\tau_{\rm CR}/\tau_t$ exceeds unity near the MIT. This cannot be understood in a single-particle picture on which Eqs. (1) and (2) is based. Our results for low N_s imply that electron correlation effects are important even in the metallic region at low

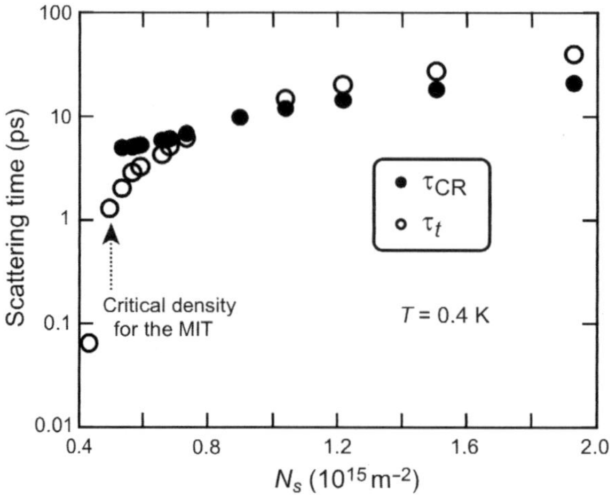

Figure 3. Electron density dependence of $\tau_{\rm CR}$ and τ_t at 0.4 K.

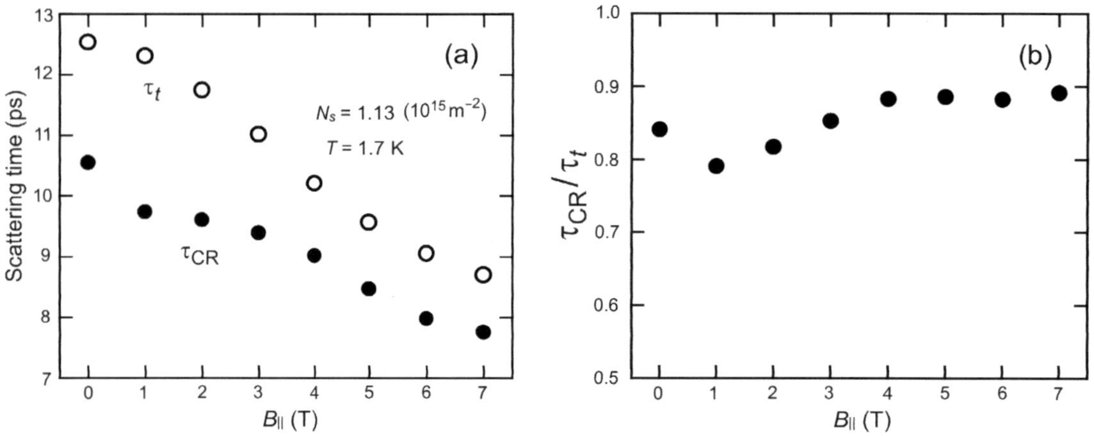

Figure 4. In-plane magnetic field dependence of (a) $\tau_{\rm CR}$, τ_t and (b) $\tau_{\rm CR}/\tau_t$.

temperatures.

Fig. 4(a) shows $B_{||}$-dependence of $\tau_{\rm CR}$ and τ_t at $T = 1.7$ K and $N_s = 1.13 \times 10^{15}$ m^{-2}. The critical magnetic field for the full spin polarization at $T = 0$ K is estimated to be 5.0 T at this density. The T-dependence of ρ exhibits apparent metallic behavior below $B_{||} = 7$ T [18, 19]. The cyclotron relaxation time $\tau_{\rm CR}$ have a negative $B_{||}$ dependence, which is similar to that of τ_t. Fig. 4(b) shows $\tau_{\rm CR}/\tau_t$ as a function of $B_{||}$. The ratio $\tau_{\rm CR}/\tau_t$ is close to unity, which suggests that short-range potential fluctuations which lead to large-angle scattering play a dominant role even in in-plane magnetic fields. In our measurements, $B_{\rm CR}$ shows an unexpected negative shift (~ 10 %) with increasing $B_{||}$ up to 7 T. An in-plane magnetic field can modify the wave function in the confinement direction [20]. However this effect increases $B_{\rm CR}$. Further measurements are needed to clarify this behavior.

Acknowledgments

This work has partly supported by Grant-in-Aid for Young Scientists (B) (No. 23740233) and Grant-in-Aid for Scientific Research (A) (No. 21244047) from MEXT, Japan.

[1] E. Abrahams, S. V. Kravchrenko, and M. P. Sarachik, Rev. Mod. Phys. **73**, 251 (2001).
[2] S. Das Sarma, and E. H. Hwang, Solid State Commun. **135**, 579 (2005).
[3] B. Spivak, S. V. Kravchrenko, S. A. Kivelson, X. P. A. Gao, Rev. Mod. Phys. **82**, 1743 (2010).
[4] E. Abrahams, P. W. Anderson, D. C. Licciardello, and T. V. Ramakrishnan, Phys. Rev. Lett. **42**, 673 (1979).
[5] T. Okamoto, K. Hosoya, S. Kawaji, and A. Yagi, Phys. Rev. Lett. **82**, 3875 (1999).
[6] O. Gunawan, T. Gokmen, K. Vakili, M. Padmanabhan, E. P. DE Poortere, and M. Shayegan, Nat. Phys. **3**, 388 (2007).
[7] S. Das Sarma and F. Stern, Phys. Rev. B **32**, 8442 (1985).
[8] J. P. Harrang, R. J. Higgins, R. K. Goodall, P. R. Jay, M. Laviron, and P. Delescluse, Phys. Rev. B **32**, 8126 (1985).
[9] P. T. Coleridge, Phys. Rev. B **44**, 3793 (1991).
[10] H. Linke, P. Omling, P. Ramvall, B. K. Meyer, M. Drechsler, C. Wetzel, R. Rudeloff, and F. Scholz, J. Appl. Phys. **73**, 7533 (1993).
[11] S. Syed, M. J. Manfra, Y. J. Wang, R. J. Molnar, and H. L. Stormer, Appl. Phys. Lett. **84**, 1507 (2004).
[12] R. Masutomi, K. Sasaki, I. Yasuda, A. Sekine, K. Sawano, Y. Shiraki, and T. Okamoto, Phys. Rev. Lett. **106**, 196404 (2011).
[13] T. Chiba, R. Masutomi, K. Sawano, Y. Shiraki, and T. Okamoto, Phys. Rev. B **86**, 045310 (2012).
[14] A. Yutani and Y. Shiraki, Semicond. Sci. Technol. **11**, 1009 (1996); J. Cryst. Growth **175/176**, 504 (1997).
[15] R. Masutomi, A. Sekine, K. Sasaki, K. Sawano, Y. Shiraki, and T. Okamoto, Physica E **42**, 1184 (2010).
[16] R. Masutomi, K. Sasaki, I. Yasuda, A. Sekine, K. Sawano, Y. Shiraki, and T. Okamoto, Journal of Physics: Conference Series **334**, 012057 (2011).
[17] W. R. Clarke, C. E. Yasin, A. R. Hamilton, A. P. Micolich, M. Y. Simmons, K. Muraki, Y. Hirayama, M. Pepper, and D. A. Ritchie, Nat. Phys. **4**, 55 (2008).
[18] T. Okamoto, K. Hosoya, S. Kawaji, A. Yagi, A. Yutani, and Y. Shiraki, Physica **6E**, 260 (2000).
[19] T. Okamoto, M. Ooya, K. Hosoya, and S. Kawaji, Phys. Rev. B **69**, 041202(R) (2004).
[20] L. Smrŏcka and T. Jungwirth, J. Phys.: Condens. Matter **6**, 55 (1994); **7**, 3721 (1995).

Journal of Physics: Conference Series **456** (2013) 012028 doi:10.1088/1742-6596/456/1/012028

Theory for optical Hall conductivity in the trilayer graphene in the quantum Hall regime

Takahiro Morimoto
Condensed Matter Theory Laboratory, RIKEN, Saitama, 351-0198, Japan

Mikito Koshino
Department of Physics, Tohoku University, Sendai, 980-8578, Japan

Hideo Aoki
Department of Physics, University of Tokyo, Hongo, Tokyo 113-0033, Japan

E-mail: tmorimoto@riken.jp

Abstract. We theoretically study optical Hall conductivities for ABA and ABC trilayer graphenes. In ABA-stacked trilayer, the resonance spectrum is shown to be a superposition of effective monolayer and bilayer contributions with band gaps, while ABC trilayer exhibits a distinct spectrum peculiar to the cubic-dispersed bands with a strong trigonal warping, where we found the signals associated with low-energy Dirac cones are conspicuous enough to be directly observable owing to a large Lifshitz-transition energy ($\sim$ 10 meV). We have also revealed how the presence or absence of the inversion symmetry affects the optical Hall conductivities.

1. Introduction

While the observation of the anomalous quantization of the static quantum Hall effect in graphene has established the existence of massless Dirac-quasiparticle in graphene [1, 2], physics of the optical (ac) Hall effect has only begun. The *optical Hall* conductivity $\sigma_{xy}(\omega)$ is an ac-extension of the static Hall conductivity, which has theoretically been proposed to be an interesting dynamical property to look at in the quantum Hall regime [3]. Experimentally, the optical Hall conductivity is measurable through Faraday rotation, since Faraday rotation angle is proportional to $\sigma_{xy}(\omega)$, where the angle is of the order of the fine-structure constant in the quantum Hall regime, as subsequently detected in a two-dimensional electron gas [4]. For graphene, giant cyclotron resonances in Faraday rotation is observed in the quantum Hall regime [5].

In the graphene physics, on the other hand, there are growing interests toward multilayer graphene, whose electronic structures are distinct from that of monolayer graphene. Especially, trilayer graphenes show various electronic structures according to the stacking order, where an ABA stacked trilayer has a monolayer-like linearly dispersed band plus a bilayer-like pair of parabolic bands, while an ABC stacking has a pair of cubic-dispersed bands. While the optical longitudinal conductivity $\sigma_{xx}(\omega)$ is mainly discussed so far, the the optical Hall conductivity

$\sigma_{xy}(\omega)$ has not been fully studied for multi-layer graphene in the quantum Hall regime. We have previously studied the optical Hall conductivities in trilayer graphene [6].

In the present paper, we discuss how the presence or absence of the inversion symmetry affects the optical Hall conductivities. It is argued that, when both the inversion symmetry and the time reversal symmetry are respected, the gaplessness of the bands are topologically protected in the graphene systems [7].

For ABA trilayer, we show the optical responses basically comes from Dirac contribution and bilayer contribution, which are both massive due to the lack of the inversion symmetry. The inversion symmetry is retained in ABC trilayer, so that the bands are gapless and zero energy LLs appear consequently.[8] We also show for ABC trilayer that the Lifshitz transition due to the trigonal warping affects the optical responses, where even away from the Lifshitz transition the trigonal warping effect manifest itself as satellite resonances.

2. ABA-stacked trilayer

For ABA stacked trilayer, the effective Hamiltonian around K_+/K_- points is given by a 6×6 matrix (the dimension being 2 sublattices $\times$ 3 layers) as [9, 10, 11, 12]

$$H_{\mathrm{ABA}} = \begin{pmatrix} 0 & v\pi^\dagger & 0 & v_3\pi & \gamma_2/2 & 0 \\ v\pi & \Delta' & \gamma_1 & 0 & 0 & \gamma_5/2 \\ 0 & \gamma_1 & \Delta' & v\pi^\dagger & 0 & \gamma_1 \\ v_3\pi^\dagger & 0 & v\pi & 0 & v_3\pi^\dagger & 0 \\ \gamma_2/2 & 0 & 0 & v_3\pi & 0 & v\pi^\dagger \\ 0 & \gamma_5/2 & \gamma_1 & 0 & v\pi & \Delta' \end{pmatrix}, \tag{1}$$

where $\pi = \xi\pi_x + i\pi_y$, $\pi^\dagger = \xi\pi_x - i\pi_y$, $\pi = \mathbf{p} + e\mathbf{A}$, with $\mathbf{A}$ being the vector potential arising from the applied magnetic field, and the valley index $\xi = \pm 1$ for $K_\pm$ points. The coefficient $v = \sqrt{3}a\gamma_0/(2\hbar)$ is the band velocity for monolayer graphene, $a \approx 0.246$ nm the distance between the nearest A sites, and $v_3 = \sqrt{3}a\gamma_3/(2\hbar)$ a velocity related to γ_3 that causes the trigonal warping. Δ' is the on-site energy difference between the atoms with and without vertical bond γ_1, and $\gamma_2(\gamma_5)$ are the next-nearest interlayer hoppings between $A1$ and $A3$ ($B1$ and $B3$). We adopt the values for bulk graphite: γ_0=3.2eV, γ_1=0.39eV, γ_3=0.32eV, γ_2=-0.020eV, γ_5=0.038eV, and Δ'=0.050eV. [13, 14] There is another interlayer hopping parameters, γ_4, connecting A1 (B1) and A2 (B2), which introduces a small electron-hole asymmetry to the band structure and we do not consider in this paper.

With a unitary transformation [12, 15, 16, 17, 18], this Hamiltonian is decomposed into two blocks corresponding to a 2×2 massive Dirac Hamiltonian with a shift in Fermi energy and a 4×4 gapped bilayer Hamiltonian with another energy shift. So the low-energy physics of ABA trilayer amounts to a superposition of gapped monolayer and gapped bilayer band contributions as seen in the low-energy band structure (Fig.1(a)).

In Fig.1(b), LLs are plotted against magnetic field B, which are labeled with M (B) for monolayer (bilayer) blocks, Landau index n, and band index $\pm$ (for $n \geq 1$). For moderate magnetic fields $B \sim 1$ T, the Landau level spacing $\propto \sqrt{B}$ for monolayer is much larger than that for bilayer $\propto B$. ABA trilayer lacks an inversion center and bands are gapped. As a consequence, zero-energy LLs of monolayer and bilayer appear at the bottom (top) of the conduction (valence) bands for K_+ (K_-) valley.

Now let us move on to the optical Hall conductivity ($\sigma_{xy}(\omega)$), which is evaluated from the Kubo formula as

$$\sigma_{\alpha\beta}(\omega) = \frac{\hbar}{iL^2}\sum_{ab} j_\alpha^{ab} j_\beta^{ba} \frac{f(\epsilon_b)-f(\epsilon_a)}{\epsilon_b-\epsilon_a}\frac{1}{\epsilon_b-\epsilon_a-\hbar\omega-i\eta}, \tag{2}$$

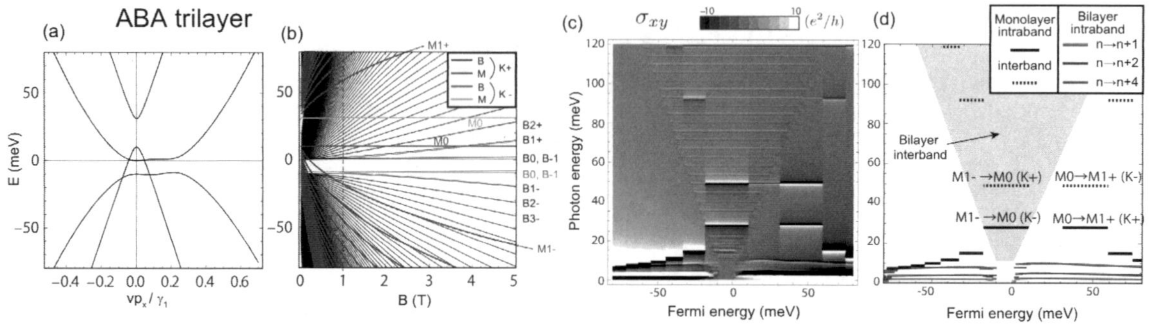

Figure 1. (a) Band structure and (b) Landau levels against magnetic field for ABA-stacked trilayer graphene. (c) Optical Hall $\sigma_{xy}(\epsilon_F, \omega)$ grey-scale plotted against the Fermi energy ϵ_F and the frequency ω for a magnetic field $B = 1T$ (a dashed line in (b)). (d) A diagram indicating allowed resonances in σ_{xy}.

where $f(\varepsilon)$ is the Fermi distribution, ϵ_a the energy of the eigenstate $|a\rangle$, $\mathbf{j}^{ab} = \langle a|\mathbf{j}|b\rangle$ the matrix element of the current operator $\mathbf{j} = \partial H/\partial \mathbf{A}$, and η a small energy cutoff for a stability of the calculation[3]. In order to show the resonance structures in optical responses clear, we set to $\eta = 0.15$ meV, which is a small value compared to realisitic situations. Larger η makes the resonances more broadened, while it does not alter their qualitative behaviors.

The result for the optical Hall conductivity $\sigma_{xy}(\epsilon_F, \omega)$ is plotted against the Fermi energy ϵ_F and frequency ω for ABA stacked trilayer graphene QHE system in Fig.1(c). We can discern contributions from monolayer-like Dirac LLs and those from bilayer LLs, both of which exhibit intra-band and inter-band transitions. Since Dirac cone is massive due to a breaking of an inversion symmetry and M0 LL is situated at the bottom of conduction band for K_+ valley and the top of valence band for K_-, $M0 \to M1+$ resonance occurs at a lower energy than $M1- \to M0$ for K_+, and vice versa for K_- valley. A cancellation of resonances in σ_{xy}, due to opposite signs in current matrices, occurs between $M1- \to M0$ for K_+ and $M0 \to M1+$ for K_- for a region of Fermi energy between $M0(K_+)$ and $M0(K_-)$, while this is not the case with σ_{xx}. For bilayer contributions satellites appear due to the trigonal warping from the transitions $(n, s) \leftrightarrow (n+1+3m, s')$ and $(n, s) \leftrightarrow (n+2+3m, s')$, since the trigonal warping due to γ_3 term mixes (n, s) and $(n + 3m, s')$.

The resonance frequency for intra-band transition within the conduction band is larger than those within the valence band, which is a consequence of an electron-hole asymmetry in the bilayer bands. Different cyclotron masses for electron and hole bands prevent a complete cancellation between $(n, -) \to (n + 1, +)$ and $(n + 1, -) \to (n, +)$ transitions, which results in small interband transitions in a wide region of Fermi energy.

3. ABC-stacked trilayer

If we turn to ABC stacked bilayer graphene, the effective Hamiltonian around K point is a 6×6 matrix as [9, 19, 20]

$$H_{\rm ABC} = \begin{pmatrix} 0 & v\pi^\dagger & 0 & v_3\pi & 0 & \gamma_2/2 \\ v\pi & 0 & \gamma_1 & 0 & 0 & 0 \\ 0 & \gamma_1 & 0 & v\pi^\dagger & 0 & v_3\pi \\ v_3\pi^\dagger & 0 & v\pi & 0 & \gamma_1 & 0 \\ 0 & 0 & 0 & \gamma_1 & 0 & v\pi^\dagger \\ \gamma_2/2 & 0 & v_3\pi^\dagger & 0 & v\pi & 0 \end{pmatrix}. \quad (3)$$

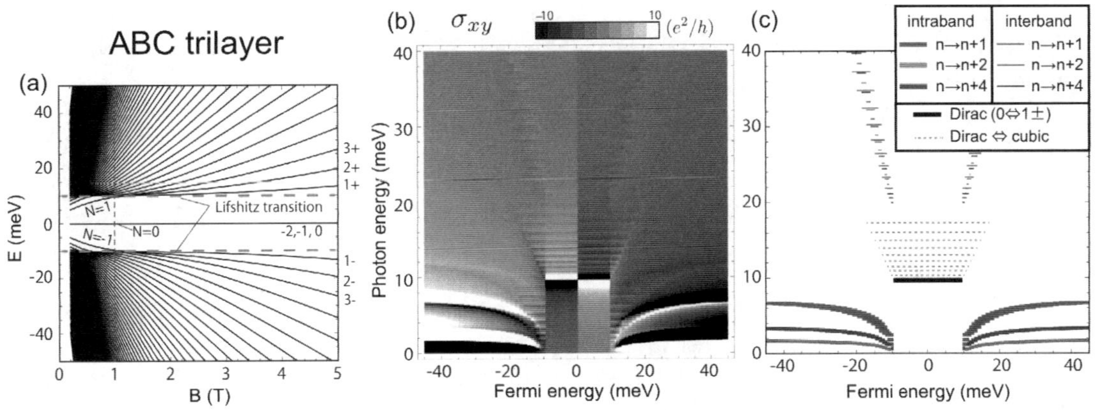

Figure 2. (a) Landau levels against magnetic field for ABC-stacked trilayer graphene. (b) Optical Hall $\sigma_{xy}(\epsilon_F, \omega)$ grey-scale plotted against the Fermi energy ϵ_F and the frequency ω for a magnetic field $B = 1T$ (a dashed line in (a)). (c) A diagram indicating allowed resonances in σ_{xy}.

We can derive a low-energy effective Hamiltonian as a 2×2 matrix with basis for A1 and B3, where we eliminate the states coupled by γ_1. As in the case of bilayer, a perturbation in ε/γ_1 gives the effective Hamiltonian for ABC trilayer graphene as [20],

$$H_{\rm ABC}^{\rm (eff)} = \frac{v^3}{\gamma_1^2}\begin{pmatrix} 0 & (\pi^\dagger)^3 \\ \pi^3 & 0 \end{pmatrix} + \left(\frac{\gamma_2}{2} - \frac{2vv_3\pi^2}{\gamma_1}\right)\begin{pmatrix} 0 & 1 \\ 1 & 0 \end{pmatrix}, \tag{4}$$

where $\pi^2 = (\pi^\dagger\pi + \pi\pi^\dagger)/2$. In zero magnetic field, the first term gives a pair of cubic-dispersed bands touching at zero energy, while the second term involving γ_2 and v_3 causes a trigonal warping in the band dispersion. In a low-energy region, the cubic bands are reformed into three Dirac cones at off-center momenta located in 120° symmetry around $K_\pm$ point. The Lifshitz transition occurs at $E_{\rm Lifshitz} \approx \frac{\gamma_2}{2} \sim 10\,{\rm meV}$, which is an order of magnitude *greater* than in bilayer's.

If we first neglect γ_2 and v_3 to consider the cubic part alone in Eqn. 4, LLs are[9, 20]

$$\varepsilon_{n,s} = s\hbar\omega_{\rm ABC}\sqrt{n(n+1)(n+2)}, \tag{5}$$

where $n \geq -2$ is the Landau index, $s = \pm$ (only for $n \geq 1$) the band index, and $\hbar\omega_{\rm ABC} = \frac{v^3}{\gamma_1^2}(2\hbar eB)^{\frac{3}{2}}$. A peculiar magnetic field dependence of cyclotron energy $\propto B^{\frac{3}{2}}$ for ABC trilayer graphene implies a smaller LL spacing as compared with the monolayer's LL $\propto B^{\frac{1}{2}}$ and bilayer's LL $\propto B$ for weak magnetic fields.

The trigonal warping due to γ_2 and v_3 hybridizes $\psi_{n,s}$ with $\psi_{n+3m,s'}$. In the low-energy region $|E| < E_{\rm Lifshitz}$, the spectrum is reconstructed into the monolayer-like Landau levels from small Dirac cones as a series $\varepsilon_N = {\rm sgn}(N)\sqrt{N}\hbar\omega_{\rm trig}$, with $\hbar\omega_{\rm trig} \simeq 10\sqrt{\frac{B}{1{\rm T}}}$ meV.

Now we go back to the original 6×6 Hamiltonian (Eqn.3) to discuss the optical Hall conductivity calculated with diagonalization and Kubo formula. In the LL spectrum in Fig.2(a), a Lifshitz transition is clearly identified at $E_{\rm Lifshitz} = 10$meV, which separates Dirac LLs and ABC cubic LLs. Since the inversion symmetry is *preserved* in ABC trilayer so that the band cannot be gapped, 3 zero-energy LLs appear which correspond to 3 gapless Dirac cones within the Lifshitz transition energy.

Figure 2(b) shows the optical Hall conductivities $\sigma_{xy}(\epsilon_F, \omega)$ plotted against the Fermi energy ϵ_F and frequency ω for ABC stacked trilayer graphene QHE system with $B = 1$T. At this magnetic field, there are three Landau levels $n = -1, 0, 1$ in each Dirac cone, where the $n = \pm 1$ levels are close to the Lifshitz transision. The higher levels are outside the Dirac cones, and can be regarded to belong to the the cubic dispersion. Accordingly we see cyclotron resonances between $1- \to 0$ and $0 \to 1+$ in $E_F < E_{\rm Lifshitz}$, while outside we see transitions between cubic LLs $\propto B^{\frac{3}{2}}$ with much smaller level spacings than for Dirac LLs $\propto \sqrt{B}$. In addition to resonances $n \leftrightarrow n+1$, the trigonal warping again gives rise to satellite transitions, $n \leftrightarrow n+4$ and $n \leftrightarrow n+2$, with opposite resonance weights. There also emerge small resonances between Dirac $n = 0$ LL and cubic LLs across the Lifshitz transition energy. Due to the ABC trilayer LLs arising from the cubic dispersion (Eqn.5) the intra-band transition energies show behaviors $\propto n^{1/2}$ against the Landau index n, which is different from monolayer ($\propto n^{-1/2}$) or bilayer (constant), while the inter-band transition energies are qualitatively similar to $\sim 2\epsilon_F$.

4. Summary

We have studied the optical Hall conductivities for ABA and ABC trilayer graphenes. The ac optical responses in the ABA-stacked trilayer graphene accommodates a mixture of contributions from an effective massive monolayer and from an effective gapped bilayer with the trigonal warping effect. In the case of ABC trilayer graphene the trigonal warping effect is an order of magnitude enhanced than in bilayer graphene, so it will be experimentally more feasible to access the effects induced by the trigonal warping. We have discussed how the differences between ABA and ABC trilayers are related to the absence or presence of the inversion symmetry.

HA acknowledges the financial support by Grants-in-Aid for Scientific Research No.23340112 from JSPS, while MK the Grants-in-Aid for Scientific Research No.24740193 from JSPS, and by JST-EPSRC Japan-UK Cooperative Programme Grant No. EP/H025804/1.

References

[1] Novoselov K, Geim A, Morozov S, Jiang D, Katsnelson M, Grigorieva I, Dubonos S and Firsov A 2005 *Nature* **438** 197–200
[2] Zhang Y, Tan Y W, Stormer H L and Kim P 2005 **438** 201
[3] Morimoto T, Hatsugai Y and Aoki H 2009 *Phys. Rev. Lett.* **103** 116803
[4] Ikebe Y, Morimoto T, Masutomi R, Okamoto T, Aoki H and Shimano R 2010 *Phys. Rev. Lett.* **104** 256802
[5] Crassee I, Levallois J, Walter A, Ostler M, Bostwick A, Rotenberg E, Seyller T, Van Der Marel D and Kuzmenko A 2011 *Nature Physics* **7** 48–51
[6] Morimoto T, Koshino M and Aoki H 2012 *Phys. Rev. B* **86**(15) 155426
[7] Mañes J L, Guinea F and Vozmediano M A H 2007 *Phys. Rev. B* **75**(15) 155424
[8] Koshino M and McCann E 2010 *Phys. Rev. B* **81**(11) 115315
[9] Guinea F, Castro Neto A H and Peres N M R 2006 *Phys. Rev. B* **73**(24) 245426
[10] Partoens B and Peeters F 2006 *Physical Review B* **74** 075404
[11] Lu C, Chang C, Huang Y, Chen R and Lin M 2006 *Physical Review B* **73** 144427
[12] Koshino M and Ando T 2007 *Phys. Rev. B* **76**(8) 085425
[13] Charlier J, Gonze X and Michenaud J 1991 *Phys. Rev. B* **43** 4579
[14] Dresselhaus M and Dresselhaus G 2002 *Advances in Physics* **51** 1–186
[15] Koshino M and Ando T 2008 *Phys. Rev. B* **77**(11) 115313
[16] Koshino M and McCann E 2009 *Phys. Rev. B* **79** 125443
[17] Koshino M and Ando T 2009 *Solid State Communications* **149** 1123–1127
[18] Koshino M and McCann E 2011 *Phys. Rev. B* **83**(16) 165443
[19] Lu C L, Lin H C, Hwang C C, Wang J, Lin M F and Chang C P 2006 *Appl. Phys. Lett.* **89** 221910
[20] Koshino M and McCann E 2009 *Phys. Rev. B* **80** 165409

Giant tunnel magnetoresistance in double quantum well structures under an in-plane magnetic field

E N Morozova[1], V A Volkov[1] and J – C Portal[2]

[1] V.A. Kotelnikov Institute of Radio-engineering and Electronics of RAS, Moscow, Russia
[2] IUF-INSA, Toulouse and CNRS-LNCMI, Grenoble, France

E-mail: elena.morozova@gmail.com

Abstract. The vertical Ohmic magnetotransport in the structures with low but wide barrier $Al_{0.1}Ga_{0.9}As$ between two quantum wells GaAs, in temperature range T=3K – 45K under in-plane magnetic fields B= 0 – 9 T is investigated. At zero-bias voltage, two mechanisms that determine the transitions well-barrier-well observed. The low-temperature regime (3K – 20K) is corresponded to the tunnel transport. The above-barrier ballistic transport with activation energy 50 meV is realized in high-temperature regime (35K – 45K). Magnetoresistance R(B)/R(0) in the high-temperature regime doesn't depend on T and is increased two times at B=0 – 5 T. At low temperatures R(B)/R(0) is increased 10 times at B=0-5 T and 400 times at B=0 – 9 T. The effect is explained by suppression by in-plane magnetic field of resonant tunneling processes conserving the in-plane momentum.

1. Introduction.

Properties of the 2D tunnel heterostructures have been investigated for a long time. The basic attention is paid to research of 2D electron systems separating by thin barriers. Influence of an in-plane magnetic field ***B*** on the Ohmic resistance was studied [1-3]. The field ***B*** turns the momentum of tunnelling electron [2] and changes resistance. The resulting magnetoresistance is small, as a rule. In this work three-barrier tunnel heterostructure with wide barrier separating two narrow quantum wells was examined, see figure 1. The vertical Ohmic magnetotransport in this structure is investigated. It is shown that the large width of the central barrier, which is a feature of the investigated system, leads to the giant magnetoresistance in the Ohmic regime. Tunneling electron experiences the action of the Lorentz force in magnetic field ***B***= (0, *B,0*). It leads to the 2D momentum change of the tunneling electron. This variation is proportional to the magnetic field and tunnel length [4], [5]:

$$\Delta p_x = eBd\ .$$

It is expected this influence would be increased with the barrier width increasing. However, the in-plane magnetic field ***B*** leads to an increase in effective barrier height and a strong suppression of the tunneling current [6], [7]. So in this work structures with wide but low barrier $Al_{0.1}Ga_{0.9}As$ was used.

2. Structures. Heterostructures were grown on the high doped substrate n^+-GaAs by MBE.

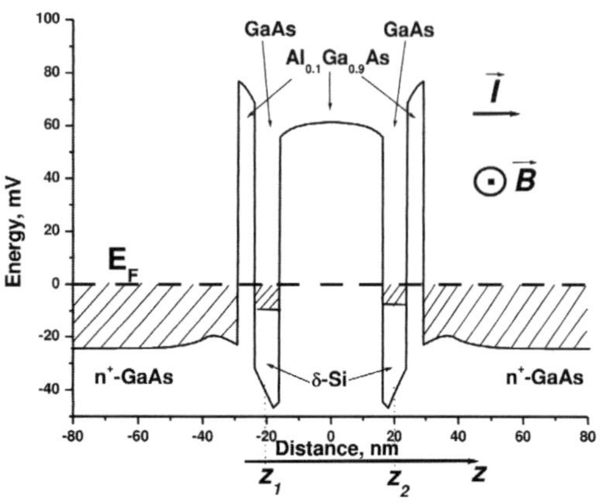

Figure 1. Energy band diagram of the three-barrier structure tunnel structure. Current *I* flows along the growth direction OZ and is perpendicular to an in-plane magnetic field *B*.

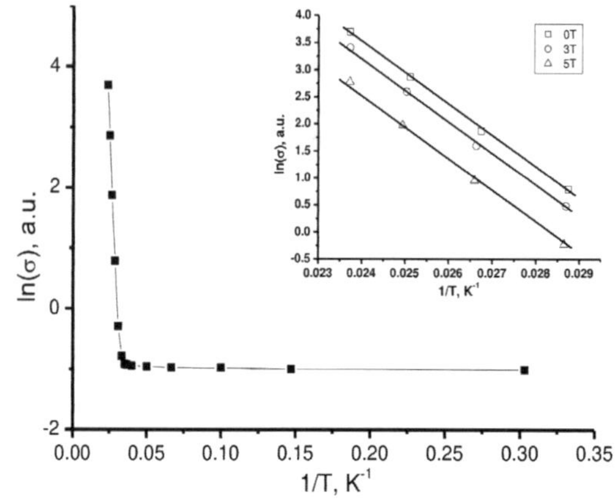

Figure 2. The temperature dependence of the Ohmic conductivity of the three-barrier structure.

The thickness of the external barriers of symmetric heterostructures is 5,1 nm, the central barrier is 33,6 nm, the each quantum well is 8,4 nm. A relatively small value of the barrier height (about 10 meV) is provided by a low Al content (10%). Two-dimensional electron layers with an impurity concentration of $2.8*10^{11}$ cm^{-2} were formed in each quantum well by using the silicon delta-doping. Undoped GaAs layers with thickness of 10,8 nm of each separate the external barriers from the heavily doped n^+-GaAs layers (a concentration of $3*10^{17}$ cm^{-3} and a thickness of 60 nm). This was followed by a transition layer to the substrate with a concentration of 10^{18} cm^{-3}, on the one hand, and a contact layer of 0.3 μm n^+-GaAs with the same concentration, on the other.

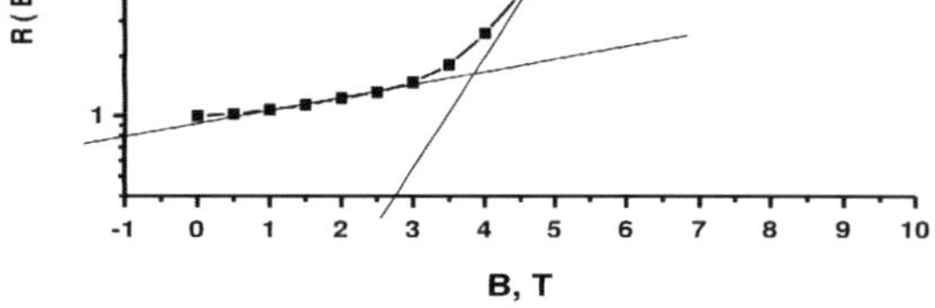

Figure 3. The dependence of the resistivity R (B) / R (0) on the magnetic field B at T = 3 K. The inset - this same dependence for different temperatures.

The measured current flows from the heavily doped GaAs-emitter to the GaAs-collector. The electrons tunnel through the narrow outer barrier quickly and are accumulated in the well, and the bias voltage drops on the wide barrier. Measurements were carried out under the in-plane magnetic field ***B*** from 0 to 9 T at the temperatures from 3 to 45 K.

3. Results and discussions.

The differential tunnelling conductance near zero-bias voltage has a broad peak associated with the 2D-2D interwell tunnel resonance. The zero-bias anomaly of the dip-type is observed in the centre of the peak at low temperatures (it is not discussed in the paper). By increasing the temperature to 30K, the anomaly and the peak disappear. In this work the conductance σ (or resistance

Journal of Physics: Conference Series **456** (2013) 012029 doi:10.1088/1742-6596/456/1/012029

$R = 1/\sigma$) are investigated at zero-bias voltage.

Depending on the ohmic conductivity of the inverse temperature, see figure 2, there are revealed two qualitatively different temperature regimes: tunnel behaviour at low temperatures and an activation regime at high T (associated with the above-barrier activation). In tunnel regime (T=3-20K) the conductivity at B=0 is increased slightly (10%) with increasing temperature. In the activation mode (T=30-45K) the conductance is increased exponentially by almost two orders of magnitude with the activation energy of 50 meV. Magnitude of the B-independent activation energy is correlated with the calculated value of the barrier height U (approximately 80 meV). Magnetic field dependence of the differential resistance demonstrates the same two temperature regimes, low-temperature and high-temperature ones. In each of these modes, the magnetoresistance R(B)/R(0) does not depend on temperature.

In the both temperature regimes, a giant magnetoresistance is revealed. In activation regime the magnetoresistance increases in two times in the range from 0 to 5 T. In low-temperature regime R(B)/R(0) increases two times in the same magnetic fields range and 400 times in the range B =0-9T. Let us consider each of these modes in more detail.

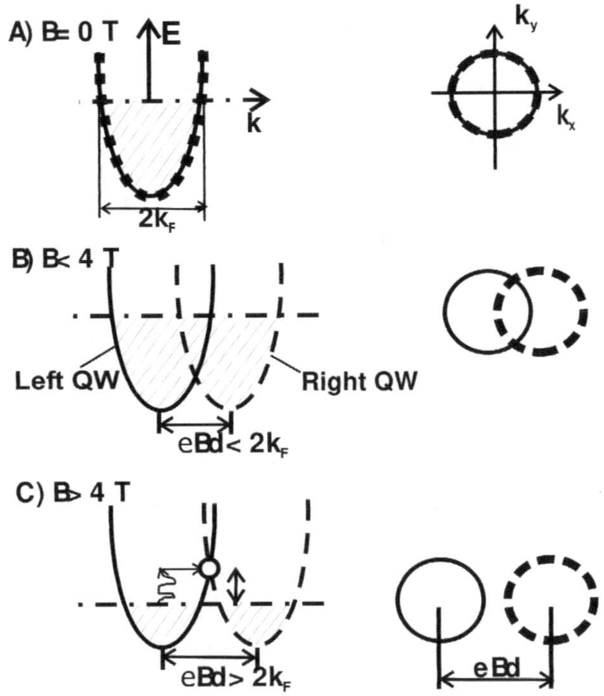

Figure 4. Combined dispersion of electrons in the left and right wells for B = 0 (figure 4A), 0 <B <4 T (figure 4B), B> 4T (figure 4C).

In the activation mode, see figure 3, resistance remains practically unchanged at B<4T, while it begins to increase at B>4T. The electrons in this regime fill a shallow well, formed by a top of the wide barrier and the narrow lateral barriers (height of about 10 meV). Calculations show that the Larmor diameter is compared at B = 4 T with the width of this well. Therefore, for B> 2T the electrons are localized in this well and the resistance is increased.

In tunnel mode, an increase in resistance with in-plane magnetic field is apparently due to the influence of the Lorentz force on the momentum of the tunnelling electrons. There is a crossover of the mechanisms of formation of R (B) near B = 4 T, see figure 4. It can be attributed to the *B*-induced change of momentum of tunnelling electrons [1].

The combined dispersion of electrons i.e. the superposition of two dispersion parabolas for electron in the left and right wells (before and after the tunneling event) is shown on fig. 4. Parabolas are coincided at *B*=0, all electrons are included in the resonance tunneling with conserving of the 2D momentum. The number of these electrons decreases with increasing *B* and disappears at e*B*d= 2 k_F (d is width of the central barrier, k_F is the Fermi momentum). With a further increase of the magnetic field *B* tunneling resonance is possible only through the circle region, that unreal at T=0. As a result, the magnetoresistance is increased dramatically with magnetic field increase from 5 to 9 T.

4. Conclusion.

The zero-bias resistance and conductivity of a wide, but low barrier $Al_{0.1}Ga_{0.9}As$, separating two narrow quantum wells GaAs, were investigated in a temperature range 3- 45 K under in-plane magnetic fields B = 0 -9 T. At low temperatures (T = 3 -20 K) the conductivity does not depend on temperature practically (tunneling regime). At high temperatures (T = 35 -45 K) the conductivity has an activation character with an activation energy of 50 meV that is independent of B (over-barrier regime). In both temperature regimes, giant magnetoresistance was revealed. In the low-temperature regime R (B)/R (0) increases by 10 times at B = 0 -5 T and 400 time as B increases from 0 to 9 T. This is associated with suppression (B = 0 -4 T) and disappearance (B > 4 T) of a channel of resonant tunneling between the QWs conserving the 2D electron momentum.

Acknowledgments
This work is supported by RFBR (Russia) and PICS (France).

References:
[1] Eisenstein J P, Gramila T J, Pfeiffer L N, West K W 1991 *Phys. Rev. B.* V.**44**. 6511
[2] Smoliner J, Demmerle W, Berthold G, Gornik E, Weimann G and Schlapp W 1989 *Phys. Rev. Lett.* **63**. 2116
[3] Berk Y, Kamenev A, Palevski A, Pfeiffer L N, West K W 1995 *Phys. Rev. B.* **51**. 2604
[4] Snell et al 1987 *Phys Rev Lett.* **59**, 2806
[5] Hayden, Maude, Eaves L et al 1991 *Phys. Rev. Lett.* **66**, 1749
[6] L. Eaves, K. W. H. Stevens, and F. W. Sheard, in Physics and Fabrication of Microstrustures and Devices, Springer Proceedings in Physics, vol.13 (Springer-Verlag, Berlin, 1986), p.343
[7] Gueret P, Baratoff A and Marclay E 1987 *Europhys. Lett* **3**, 367

Journal of Physics: Conference Series **456** (2013) 012030 doi:10.1088/1742-6596/456/1/012030

Probing Quantum Hall State Homogeneity with Surface Waves

M E Msall,[1,2] W Dietsche,[2] S Schmult[2,3] and K von Klitzing[2]

[1] Dept. of Physics & Astronomy, Bowdoin College, Brunswick, ME 04011 USA
[2]Max-Planck-Institute for Solid State Research, Heisenbergstraße 1, D-70569 Stuttgart
[3]Technical University Dresden, Noethnitzer Str. 64, D-01187 Dresden
mmsall@bowdoin.edu

Abstract. We study single quantum wells and matched density bilayer samples. Simultaneous measurements of the Hall voltages using low frequency lock-in techniques and of the changes in the 232 MHz (12.6 μm) SAW propagation measured with a vector network-analyzer allow comparison of the complex bulk and edge conductivities. The $\nu_{total} = 1$ bilayer state is seen directly in the SAW measurement only when the conductivity is below ~ 6 x 10^{-7} Siemens and is destroyed at moderate SAW powers by localized heating. The simultaneous reduction of the conductivity minima extracted from Hall data and from SAW data conclusively demonstrates that the $\nu_{total} = 1$ state disappears simultaneously throughout the bulk and not by the formation of competing domains or conducting filaments.

1. Introduction

Past use of SAW probes has provided insight into the composite fermion state and the domain structure in quantum hall ferromagnets. [1, 2] Flexible techniques using proximity effects of the 2-DEG on SAW travelling on adjacent $LiNbO_3$ surfaces have allowed the extraction of the complex conductivity, whereby the ratio of imaginary to real components is a measure of the fraction of localized charge carriers. [3] SAW can also provide momenta to probe the energy dispersion of neutral excitations through microwave spectroscopy, e.g., the roton minima at filling factor 9/2. [4] The general benefit is that SAW wavelengths can be tailored to probe structure at particular length scales and provide information about bulk states without edge effects. This work presents the first SAW measurements of a bilayer quantum well system along with calibrated measurements of the complex conductivity of a single 2-DEG derived directly from SAW measurements. We find strong SAW sensitivity to the $\nu_{total} = 1$ state, where an excitonic condensate forms due to coupling between electrons and holes in the upper and lower layer which individually have filling factor 1/2 of the lowest Landau level.

2. Simultaneous SAW and Resistivity Measurements

In our experiments a standard 300 x 840 μm^2 Hall Bar is flanked by a set of interdigital 12.6 μm (232 MHz) interdigital Al transducers (IDT). SAW pulses are generated and detected with an Agilent 5070 Network Analyzer (NA). Electrical voltages are monitored with Stanford 830 DSP Lock-in Averagers. The IDT in our single quantum well experiments has a width of 50 μm and can be aligned with the center or the edge of the Hall bar as seen if Fig. 1. For the bilayer measurements the IDT is centered on and matched in width (300 μm) to the Hall bar.

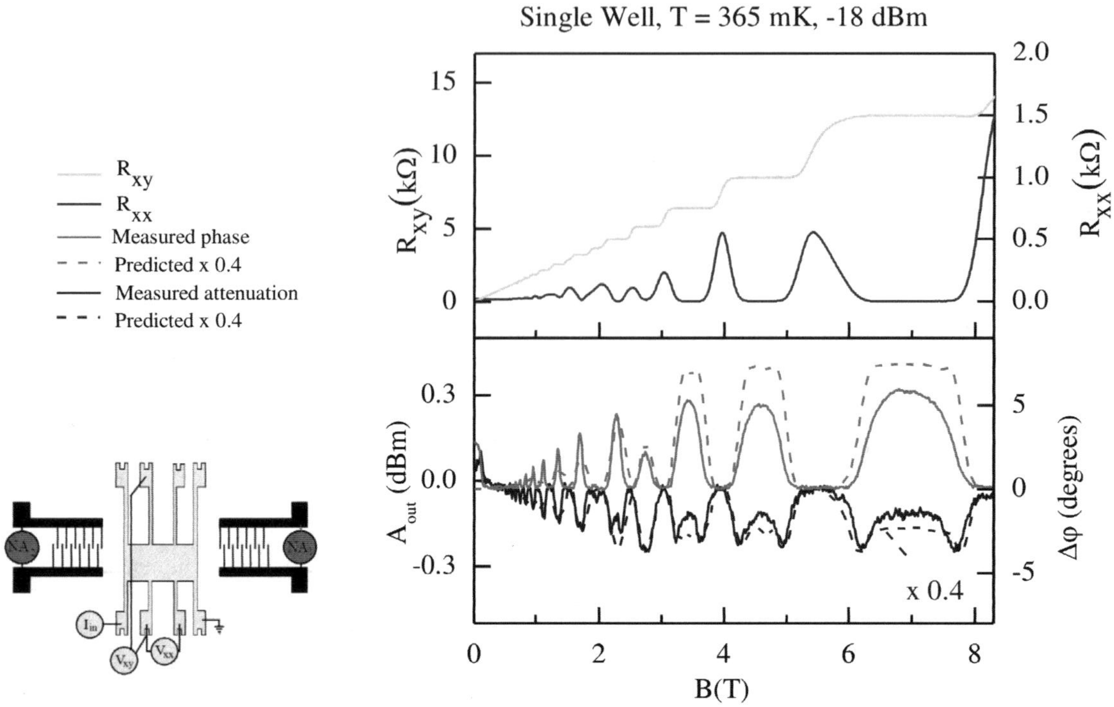

Figure 1. (left) Top view of the Hall bar contact pattern and interdigital SAW transducers. The two transducers are controlled with an Agilent Network Analyzer (NA). The surface wave launched by NA_1 is detected after transport through the Hall bar at NA_2. Echo signals can be detected for up to 5 passes through the Hall bar. **(right)** Changes in SAW phase and attenuation measured in a time window centered on the first SAW arrivals show the coincidence of SAW changes with resistance and conductivity minima in the Hall bar. Single Quantum Well sample has $\rho = 3.4 \times 10^{11}$ cm^{-2}, $\mu = 7.5 \times 10^{5}$ cm^{2}/ Vs.

2.1. Predicting SAW Phase Change and Attenuation from Conductivity Data

In order to calibrate the SAW sensitivity to changes in the 2-DEG we first investigate the effect of a single quantum well sample on the SAW signal. Simultaneous DC resistance and SAW measurements (Fig. 1) show the clear correspondence between longitudinal resistance (and conductance) minima and peak SAW phase shifts. The SAW magnitude is minimal (maximum attenuation) at magnetic field values corresponding to the edges of Hall plateaus, where the longitudinal resistance begins to rise. A simple relaxation model relates the expected attenuation per meter, and relative phase shift to the conductivity:

$$\Gamma = \frac{\alpha k_{ref}\left(\sigma_{xx}/\sigma_m\right)}{\left(1+\sigma_{xx}'/\sigma_m\right)^2+\left(\sigma_{xx}/\sigma_m\right)^2} \qquad \frac{\Delta\phi}{\phi_{ref}} = \frac{-\Delta v}{v_{ref}} = \frac{-\alpha\left(1+\sigma_{xx}'/\sigma_m\right)}{\left(1+\sigma_{xx}'/\sigma_m\right)^2+\left(\sigma_{xx}/\sigma_m\right)^2} \tag{1}$$

where v is the SAW wavespeed, k the wavenumber and α is the piezoelectric coupling constant (~6.4 x 10^{-4}). σ_{xx} and σ_{xx}' are the real and imaginary parts of the complex conductivity of the 2-DEG and σ_m is the GaAs conductivity (~3.7 x 10^{-7}). The subscript *ref* refers to expected values for the SAW on GaAs in the absence of the 2-DEG. For our low frequency (lock-in at 8.3 Hz) electrical measure-

Journal of Physics: Conference Series **456** (2013) 012030 doi:10.1088/1742-6596/456/1/012030

ments, the conductivity is assumed to be purely real. The predicted SAW phase maximum from this mode, shown in Fig. 1, is in good agreement with the measurement. The higher than predicted values of attenuation are attributed to parallel conduction. The distance between the free surface and the buried 2-DEG (50-70 nm) is not likely to affect attenuation for these samples.

The interdigital transducers are excited with a bandwidth of 10-20 MHz. Thus, a range of excited wavelengths inversely proportional to the frequency are present in the SAW pulse. The phase shift with frequency can be used to estimate the maximum expected phase shift from Eqn. 1, $\boldsymbol{\Delta\phi_{max}} = \boldsymbol{\alpha\phi_{ref}}$. For a typical measurement after transit over a surface distance of 1.9 mm including an 0.84 mm Hall bar, the phase difference measured at 231 and 232 Hz was 2.5 radians at zero field, yielding an estimated maximum phase shift in the Hall bar of:

$$\Delta\phi_{max} = \alpha\,\phi_{ref} \approx \alpha\frac{f\Delta\phi}{\Delta f}\frac{\mathrm{d}}{\mathrm{x}} = \mathbf{0.16}\ \boldsymbol{radians}\ (\mathbf{9.4}°) \quad (2)$$

where x and d represent the total surface distance and the distance in the 2-DEG traversed by the wave. The maximum measured phase shift of 6.8^{0} and maximum predicted phase shift of 7.4^{0} shown in Fig. 1 are both low, suggesting a lower bound to the measurable conductivity.

2.2. Predicting SAW Phase Change and Attenuation from SAW Data

The Network Analyzer measures the phase shift and attenuation of the surface acoustic wave (SAW) with outputs in the form:

$$\mathbf{A}_{out}(\textbf{in dBm}) = \mathbf{10\ log}\left(\frac{P_{transmitted}}{1\,mW}\right) = -\mathbf{10}\,\boldsymbol{\Gamma}\,\boldsymbol{d} + \boldsymbol{C_A} \quad (3)$$
$$\boldsymbol{\varphi_{out}} = \boldsymbol{\Delta\varphi} + \boldsymbol{C_\varphi} = \boldsymbol{d}\left(\boldsymbol{k} - \boldsymbol{k_{ref}}\right) + \boldsymbol{C_\varphi}$$

The Analyzer phase output is in degrees modulo 360. The constants $\boldsymbol{C_A}$ and $\boldsymbol{C_\varphi}$ represent the accumulated attenuation and phase in the connecting wires and in the region of SAW travel unaffected by the 2-DEG. These are removed by a simple baseline subtraction from the output data.

The SAW conductivity peaks shown in Fig. 2 are more symmetric with field than the Hall conductivity peaks, because they are less affected by edge states. Similar conductivity values are derived from 1.162 GHz SAW data and from data taken with SAW positioned at the Hall Bar center. These measurements on single quantum wells prove that the accumulated phase and attenuation changes, which essentially average over a 50 μm wide propagation strip, are dominated by the bulk conductivity and do not show size effects at these filling factors.

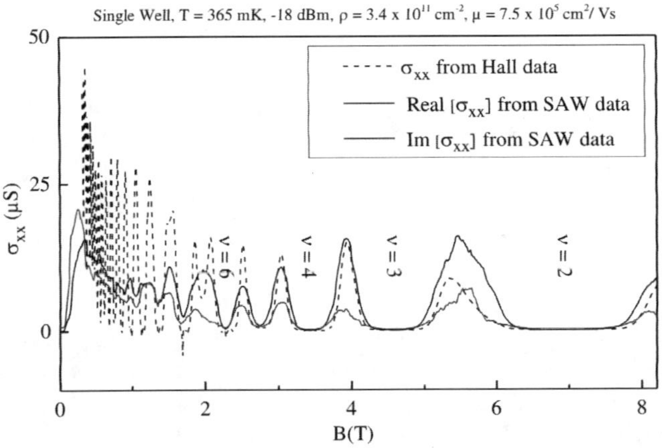

Figure 2. The SAW data cannot be used to calculate conductivity at low B fields because of large offsets in the output of the superconducting Al transducers (which disappear when higher B fields). The SAW conductivity peaks are symmetric with field, because they are less affected by edge channels.

2.3. Bilayer SAW response

Our bilayer samples consist of two 19 nm GaAs QWs, separated by a 10 nm GaAs/AlAs superlattice barrier. Two Si-doping layers are embedded 430 nm away in an $Al_{0.24}Ga_{0.76}As$ matrix. [5] At low temperature (T = 20 mK), the bilayer samples show very clear Hall plateaus at 1.3 and 2.6 Tesla, corresponding to filling factor 1 (ν_{total} = 1 + 1 = 2) and filling factor ½ (ν_{total} = ½ + ½ = 1) in the individual bilayers. The bilayer SAW signals are detected at 232 MHz with a 300 μm wide transducer centered on the Hall bar. Higher frequency SAW were not detected in the bilayer experiments, possibly due to scattering from the high mesa formed in etching the Hall bar.

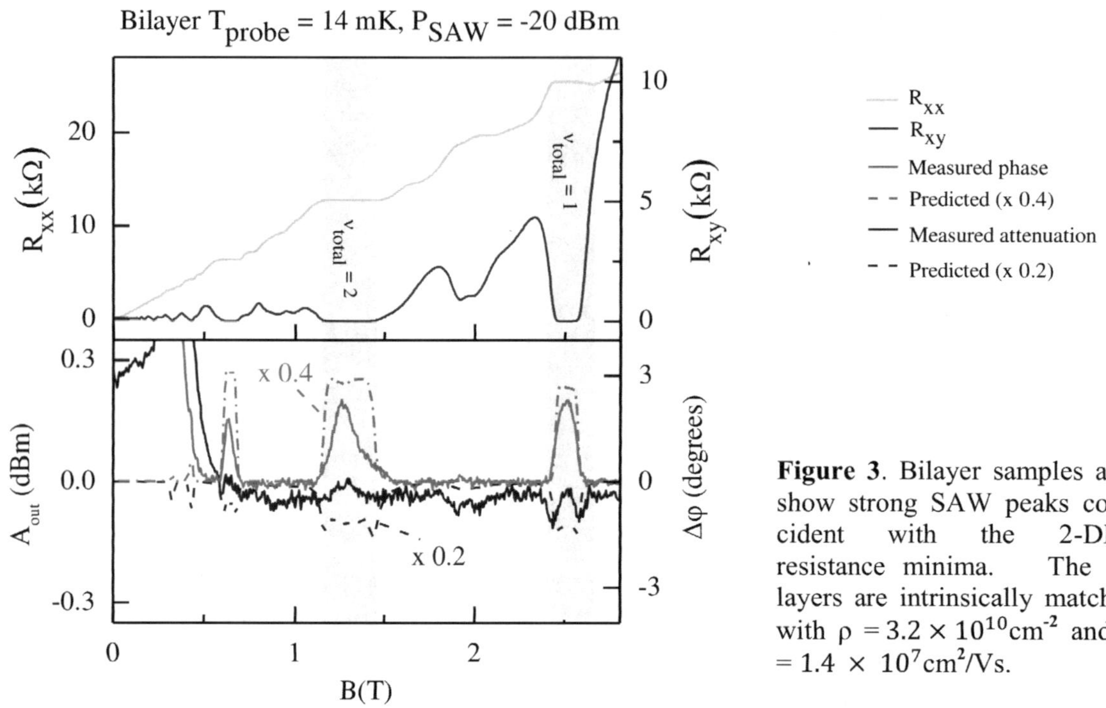

Figure 3. Bilayer samples also show strong SAW peaks coincident with the 2-DEG resistance minima. The bilayers are intrinsically matched with $\rho = 3.2 \times 10^{10}cm^{-2}$ and $\mu = 1.4 \times 10^{7}cm^{2}/Vs$.

In comparison to the single quantum well data of Fig. 1, the field dependent SAW changes in the bilayer system shown in Fig. 3 are a factor of 3 smaller. This factor is consistent over multiple samples, and must be attributed to differences in the growth structure. The expected maximum phase shift, based upon the measured phase change with frequency at zero field (Eqn. 2) for this sample is very close to the single well value (10°). The overall signal level in the bilayer sample is 10 dB higher than that measured in the single well, most likely due to decreased scattering at lower temperature. The sample does not have top or back gates, but the field dependent SAW signal could be affected by the upper Si doping layer.

The ν_{total} = 1 state is sensitive to temperature. The longitudinal resistance minimum and the associated conductivity minimum is shallower at higher probe temperatures and at higher SAW power, which raises the temperature locally. Fig. 4 shows the correlated changes in the peak height and conductivity extracted from the Hall data with increasing temperature (right) and increasing SAW power (left). The reported temperatures here are those recorded with a low temperature thermistor mounted on the probe stage of the dilution refrigerator. Judging the effect of heating with SAW by the effect on peak height, it appears that P_{SAW} = 0 dBm corresponds to T_{probe} ~ 200 mK. At this level, the SAW

Journal of Physics: Conference Series **456** (2013) 012030 doi:10.1088/1742-6596/456/1/012030

peak at $\nu_{total} = 1$ is completely suppressed, although a shallow longitudinal minima persists. In contrast, the peak height and conductivity minimum at $\nu_{total} = 2$ remain constant up to 10 dBm.

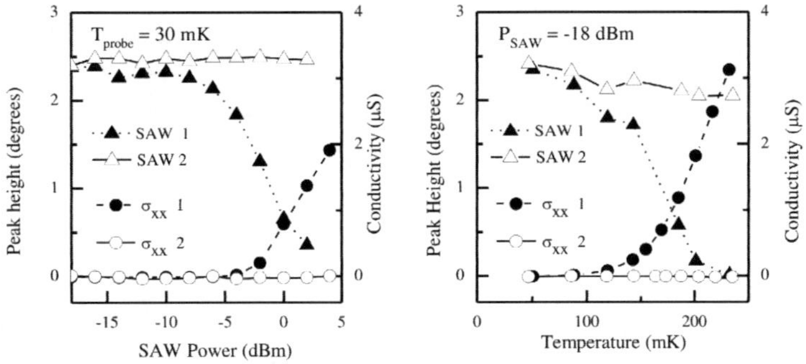

Figure 4. (left) SAW peak heights and conductivity changes show the disappearance of the $\nu_{total} = 1$ state at 0 dBm SAW power **(right)** Similar changes in conductivity are seen at higher probe temperatures. A shallow Rxx minima persists and σ_{xx} (extracted from the Hall data) continues to increase up to the 10 dBm.

3. Conclusions

We have studied single quantum wells along with matched density bilayer samples. Our simultaneous measurements of the SAW propagation and Hall voltages allow direct comparison of the complex bulk and real (~DC) edge conductivities. The $\nu_{total} = 1$ state is seen directly in the SAW measurement only when the conductivity is below ~ 6 x 10^{-7} Siemens and is destroyed at moderate SAW powers by localized heating. Heating does not diminish the SAW peak at $\nu_{total} = 2$, though there are small changes in the field range of this state. The strong correlation between the disappearance of the conductivity minima and the SAW signal at $\nu_{total} = 1$ demonstrates that the homogeneous excitonic state disappears across the entire sample simultaneously. In contrast, Quantum Hall Breakdown in a single quantum well shows persistent SAW signal well beyond the breakdown current (where conductivity minima are no longer measureable). [6] This conclusively demonstrates that the $\nu_{total} = 1$ excitonic state is uniformly distributed in the bulk and not restricted to small domains or filaments.

References

[1] R.L. Willett, R.R. Ruel, K. W. West, and L.N. Pfeiffer, PRL **71**, 3846 (1993).
[2] O. Stern, D. Dini, N. Freytag, W. Dietsche, K. von Klitzing and W. Wegscheider, phys. stat. sol. (b) **245**, 428-438 (2008).
[3] I.L. Drichko, I. Yu. Smirnov, A.V. Suslov, and D.R. Leadley, PRB **83**, 235318 (2011).
[4] I.V. Kukushkin, V. Umanskz, K. von Klitzing and J.H. Smet, PRL **106**, 206804 (2011).
[5] S. Schmult, L. Tiemann, W. Dietsche, and K. von Klitzing, J. Vac. Sci. Technol. B **28** C3C1 (2010).
[6] M.E. Msall, W. Dietsche and K. von Klitzing, in preparation.

Journal of Physics: Conference Series **456** (2013) 012031 doi:10.1088/1742-6596/456/1/012031

THz Magneto-photoresponse of an InAs-based quantum point contact in the region of cyclotron resonance

M Pakmehr[1], V R Whiteside[1], N Bhandari[2], R Newrock[3], M Cahay[2] and B D McCombe[1]

[1]Department of Physics, University at Buffalo, the State University of New York, Buffalo, NY 14260, USA
[2]Department of Electrical engineering, University of Cincinnati, Cincinnati, OH 45221, USA
[3]Department of Physics, University of Cincinnati, Cincinnati, OH 45221, USA

Email: mpakmehr@buffalo.edu

Abstract. We have studied the THz magneto-photoresponse of a 2DEG in an InAs quantum well with an embedded Quantum Point Contact in the frequency/field region where electron cyclotron resonance (CR) dominates the response. The photoresponse near CR is manifested as an envelope of the amplitude of the Shubnikov-de Haas oscillations of the 2DEG with a peak near the CR field. Clear spin-splitting of the quantum oscillations is observed for $B > 4$ T. Data were simulated by a model of resonant carrier heating, and from the simulations the carrier density, the CR effective mass, scattering times and the g-factor were obtained. We find a significantly enhanced g-factor apparently due to exchange interaction.

1. Introduction

For more than a decade there has been strong worldwide interest in spin effects and the spin-orbit (SO) interaction in semiconductors. Materials with large inherent SO interaction, and narrow gaps are of particular interest because of the possibility of manipulating spins with electric fields. There have been predictions of extremely large SO effects in InAs-based structures lacking inversion symmetry (the Rashba effect) including the possibility of exciting electron spin resonance via the electric field of an EM wave (Electric Dipole Spin Resonance) and the concomitant promise of efficient spin manipulation [1]. Our previous photoresponse (PR) results [2], which show complex multiple-line structure in the field region of the expected THz ESR that we have interpreted in terms of Rashba effective fields in quantum Hall edge channels, have led us to explore the effects of the Rashba SO interaction, the g-factor of electrons, and the possibility of observing and studying electron spin resonance in quasi-1D structures. The present experiments focus on understanding in detail the PR itself, and effects of a quantum point contact embedded in a 2DEG in an InAs quantum well. We explore PR in the region of cyclotron resonance at two THz laser frequencies, and we examine how the voltages on two lateral gates of the QPC affect the PR signals. We have also determined the electron g-factor from model fits at high fields.

2. Experimental Details

The samples for these studies contained a 3.5 nm InAs QW sandwiched between 10 nm $In_{0.53}Ga_{0.47}As$ neighboring barriers and $In_{0.52}Al_{0.48}As$ outer barriers. These structures were symmetrized vertically by

Journal of Physics: Conference Series **456** (2013) 012031 doi:10.1088/1742-6596/456/1/012031

doping with donors in the $In_{0.52}Al_{0.48}As$ barriers "below" the quantum well. The 2DEG has low temperature mobility ≈ 50,000 cm^2/V-sec, and carrier density $n_e \approx 10^{12}$ cm^{-2}. The samples investigated were lithographically fabricated rectangular Hall bars. A Quantum Point Contact (QPC) fabricated by e-beam lithography and chemical etching of isolation trenches was embedded in the 2DEG in the center of the Hall bar (see inset to figure 1c). The minimum physical separation between the trenches was ≈ 280 nm. Contacts to the 2DEG outside the trenches (G in the inset) provide lateral gates to change the electrostatic width of the constriction in a symmetric or asymmetric manner and to pinch off the channel [3].

We have measured both magnetotransport and PR of several samples including a reference Hall bar without the QPC. An Edinburgh Instruments FIR L100 optically pumped THz laser was used as a source of strong (tens to 100 mW), monochromatic CW radiation for the PR measurements. Samples were placed in the Faraday geometry at pumped liquid helium temperature in an Oxford Instruments optical-access, 10 T superconducting magnetic system with a variable temperature insert. The THz beam was focused on the sample by an-off axis parabola. Measurements of the longitudinal (R_{xx}) and transverse (R_{xy}) resistances were made with both ac current and lock-in detection at low audio frequencies (I = 50 nA to 1 μA), or with standard dc methods (I ≈ 1 μA). The photoresponse, ΔR_{xx}, was detected synchronously with the chopped laser beam either with an ac current and a double modulation technique (multiple lock-in amplifiers), or with a dc current and a single lock-in amplifier

3. Results and Discussion

The PR signal at two laser frequencies (1.4 and 2.52 THz) was measured for several gate voltages and two polarizations (parallel or perpendicular to the current shown in the inset to figure 1c)) and

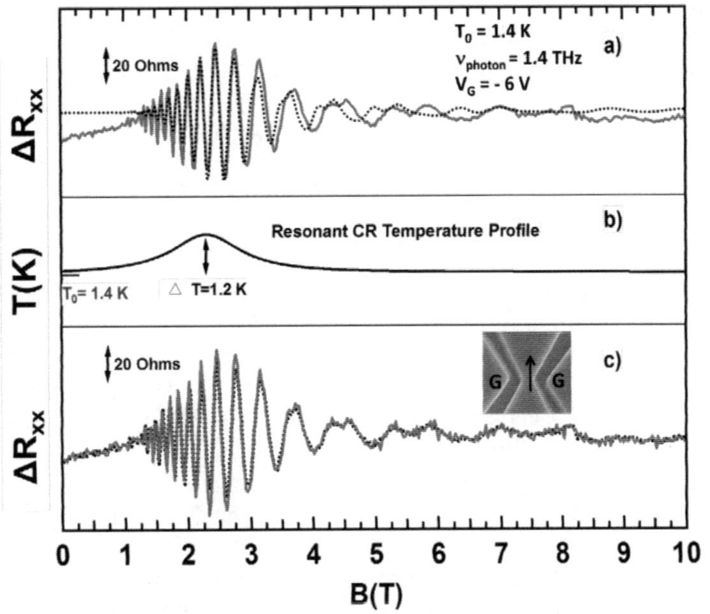

Figure 1. Photoresponse (ΔR_{xx}) and assumed carrier temperature vs magnetic field. a) Experiment and simulation. Solid line – experiment; dashed line – model. b) Carrier temperature profile. c) PR at V_G = -6 V (solid line) and 0 (dashed line). Inset: QPC; arrow - current direction; G - lateral gates.

modeled by assuming that the 2DEG is resonantly heated by cyclotron resonance absorption of the laser light, which creates a Lorentzian temperature profile as shown in figure 1b). The PR signal is the difference in R_{xx} with laser-on and -off. The oscillatory longitudinal magnetoresistance is given by [4]:

$$R_{xx} \propto \sum_{n=1}^{\infty} \frac{(-1)^n}{n^{\frac{1}{2}}} \frac{\left(n\frac{2\pi^2 k_B T}{\hbar\omega_c}\right)}{\sinh\left(n\frac{2\pi^2 k_B T}{\hbar\omega_c}\right)} e^{-n\frac{\pi}{\omega_c \tau_{sdh}}} \cos\left(n\frac{\pi g m^*}{2m_0}\right) \cos\left(n\frac{2\pi\, h n_e}{eB} + \frac{\pi}{4}\right); \quad (1)$$

here, n is the harmonic index, ω_c is the cyclotron angular frequency, τ_{SdH} is a scattering time that describes the Landau level broadening, m* is the effective mass, m_0 is the free electron mass, g is the effective g-factor, n_e is the 2DEG density, and the first cosine term describes the spin splitting of Landau levels. This expression, which does not include the monotonic magnetoresistance, was used to generate ΔR_{xx} by taking the difference at each magnetic field between $R_{xx}(B,T)$ at temperature T with the laser on (e.g., the profile of figure 1b)) and $R_{xx}(B,T_0)$, where T_0 is the bath temperature (laser off).

Figure 1a) shows the photoresponse signal at V_G = -6 V (an asymmetric configuration in which one gate is connected to the drain contact, and the other is biased negatively), and a fit to these data with the model described above. In figure 1c) the PR data at V_g = 0 is compared with the -6 V data; except for small differences in amplitude at the lower fields, the data are in excellent agreement, indicating that over this range of bias voltage the PR is dominated by the 2DEG and the lateral gate bias has little effect. The fit to the – 6V data is generally very good except for the amplitude at low fields and the differences in field position of the peaks above about 4 T. The periodicity of the oscillations determines the sheet carrier density ($n_e = 1.12 \times 10^{12}\ cm^{-2}$), which matches very well the density determined from SdH (R_{xx}) oscillations, as expected. The field at which the maximum of the envelope of the oscillations occurs is determined by the effective mass ($m^* = 0.046m_0$) through the field position of the peak in CR absorption. The scattering time, τ_{SdH}, strongly affects the onset field for the oscillations and also the resolution of the spin split Landau levels at high B; a value of approximately 10^{-13} sec provides a reasonable fit to both. The spin-splitting of the Landau levels is determined by the product $gm^*/2m_0$. To model the spin-split peaks at high fields it was necessary to increase the electron density (see below for further discussion). The fit combined with the measured m* yields $g = -14 \pm 0.5$for the spin-split peaks at about 4.5 T, much larger than expected from single particle non-parabolicity (> -8). Such large g-factor enhancement in InAs-based 2DEGs has recently been predicted theoretically [5] from exchange effects. The present results are consistent with these predictions.

As mentioned above, although the present resonant carrier-heating model fits the data at 1.4 THz reasonably well at fields up to about 4 T, the field positions of the spin-split peaks deviate from the model at higher fields. We have found that the model with a constant density is incapable of fitting the oscillatory data over the entire range of fields. It appears that above 4 T, the low-temperature carrier density increases continuously by approximately 10% up to 10 T. This is presently not understood. The origin of the poorer fit between 1 and 2 T is also uncertain.

Figure 2 shows a comparison of the PR data with the model at a laser frequency of 2.52 THz (note that the laser power $P_{2.52\ THz} \approx 2P_{1.4THz}$). In this case the fit to the PR signal is very poor except for the oscillatory period, which fits well with the same value of n_e obtained at 1.4 THz. In this case there are two distinct regions where the fit deviates greatly from the data: 1) fields between 1 and 3 T, where the CR heating model predicts very small amplitude, and 2) the region between 3 and 4.5 T, where the CR absorption has its maximum. At low fields the experimental oscillations begin to be observable for B slightly above 1 T, and rapidly increase in amplitude with field. The model with only CR absorption included cannot reproduce this behavior. We have also determined that the large amplitude low-field oscillations with onset just above 1 T cannot be reproduced by any reasonable constant background absorption, which fails to yield a significant increase in onset and greatly increases the oscillatory amplitude at high fields, in contrast to the data. We have also attempted to model this behavior by introducing an additional strong resonant absorption around 1 T; this increases the amplitude significantly above 1.5 T, but does not reproduce the observed onset, which seems to require a

Journal of Physics: Conference Series **456** (2013) 012031 doi:10.1088/1742-6596/456/1/012031

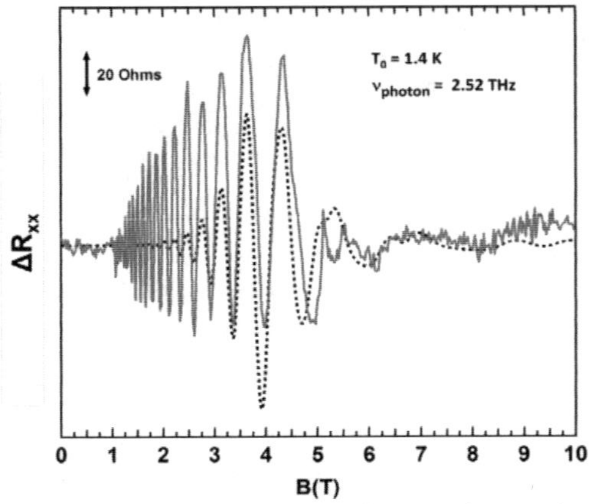

Figure 2. Comparison of experimental PR (solid line) with the model (dashed line) for ν_{Laser}=2.52 THz and V_G = 0 V.

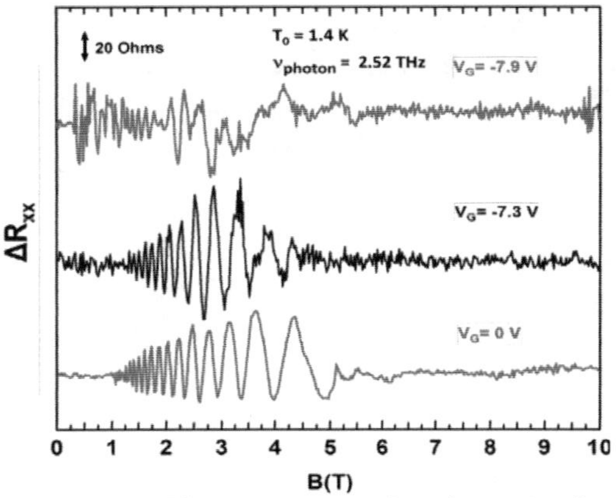

Figure 3. Photoresponse signal vs. B for ν_{photon} = 2.52 THz at three gate voltages; the upper two curves are very close to pinch-off of the QPC channel for asymmetric gate voltages (see text).

significantly longer scattering time. In the field region around CR the amplitude of the oscillations in the data is very asymmetric (the maxima of the oscillations are about a factor of two larger than the minima relative to the reference background signal at 1 T). We have adjusted the resonant CR temperature profile in the model to reproduce the peak-to-peak amplitude of the oscillations at the maximum of the envelope ($\Delta R_{xx}^{pk-pk} \approx 90\ \Omega$) and the variation of the positive-going peaks with magnetic field; this yields an effective mass, m* = $0.042m_0$, with considerable systematic uncertainty. We have not attempted to account for resonant changes in the non-oscillatory part of R_{xx}, which we believe is the origin of the large asymmetry, as it could yield a background that peaks at the CR field, but has been ignored in the model.

To explore the effect of the QPC on the PR signal we have examined the PR near pinch-off of the channel by applying appropriate asymmetric gate voltages. When the channel is completely pinched off the resistance becomes very large and no PR signal is expected. The PR signals at two gate bias voltages slightly before pinch-off are compared with the PR signal with no gate bias in figure 3. At –7.3 V the PR signal is clearly affected with more noise appearing and a change in the period of the oscillation. At -7.9 V the PR signal is dramatically changed with some residual evidence of 2DEG oscillations but apparent strong beating and other behavior between 3 and 5 T. We are unable to interpret these data at present, but there are clearly strong effects of the QPC constriction near pinch-off.

4. Summary and Conclusions

We have investigated the THz magneto-photoresponse of InAs-based quasi-2DEGs with an embedded QPC, focusing on the region of frequency and fields corresponding to CR. The observed PR signal in the vicinity of CR has the form of quantum oscillations, the amplitude envelope of which is peaked slightly above the CR field. We have modeled this behavior assuming that the CR power absorbed leads to a corresponding resonant increase in the carrier temperature and find good agreement for a photon frequency of 1.4 THz. We have shown that the PR signal, which is effectively a differential of $R_{xx}(B,T)$ with temperature enhances the low field oscillations and is very sensitive to the spin splitting of the LLs at high fields. We have used detailed fits to determine the carrier density, the cyclotron effective mass and the g-factor (at fields above about 5T). We find g ≈ - 14 at 4.5 T, much larger in magnitude than is predicted by single particle band theory at the Fermi energy. This is apparently a result of many body exchange corrections, which have recently been shown to have a large effect in InAs [5]. Over a range of QPC gate voltages (0 to -6 V), the PR is dominated by the response of the

2DEG; however near pinch-off the behavior changes dramatically, but as yet, we have not been able to interpret details of these data.

These studies have shown that the magneto-PR method is promising for studying the g-factor and effects of many electrons, and that with a suitable enhancement of the EM wave coupling via antennas, it should be possible to study the Rashba effect and electron spin resonance in quasi-1D channels.

5. Acknowledgements

Work at UB was supported by NSF MWN 1008138 and the Office of the Provost at the University at Buffalo; work at the University of Cincinnati was supported by NSF ECCE 1028483.

6. References

[1] E.I. Rashba and Al.L. Efros, *Phys. Rev. Lett.* **91**, 126405 (2003); Appl. Phys. Lett. **83**, 5295 – 5297 (2003)

[2] A.V. Stier, C.J. Meining, V.R. Whiteside, B.D. McCombe, E.I. Rashba, P. Grabs and L.W. Molenkamp, *Bull. Am. Phys. Soc.* **54,** #1 (2009) MAR.Z22.2

[3] P. Debray, S.M.S Rahman, J. Wan, R.S. Newrock, M. Cahay, A.T. Ngo, S.E. Ulloa, S.T. Herbert, M. Muhammad and M. Johnson, *Nature Nanotechn*ology **4,** 759-764 (2009)

[4] L.M. Roth and P.N. Argyres, *Semiconductors and Semimetals*, **Vol 1** Physics of III-V Compounds, Ed. by R.K. Willardson and A.C. Beer, Academic Press (New York and London), 1966, pp. 159-202

[5] S.S. Krishtopenko, V.I. Gavrilenko, M. Goiran, *J.Phys.:Condens Matter* **23**, 385601 (2011)

Journal of Physics: Conference Series **456** (2013) 012032 doi:10.1088/1742-6596/456/1/012032

Disorder and de-coherence in graphene probed by low-temperature magneto-transport: weak localization and weak antilocalization

S.Pezzini[1,2], **C.Cobaleda**[2], **E.Diez**[2] **and V.Bellani**[1]

[1] Dipartimento di Fisica and CNISM, Università degli studi di Pavia, I-27100 Pavia, Italy
[2] Laboratorio de Bajas Temperaturas, Universidad de Salamanca, E-37008 Salamanca, Spain

E-mail: sergio.pezzini@unipv.it

Abstract. We studied weak localization (WL) and and weak antilocalization (WAL) in a eight-contacts Hall bar made of exfoliated monolayer graphene on Si-SiO_2, by means of magneto-transport experiments, at temperatures between 0.3 K and 15 K. At low carrier density ($n \simeq 7 \times 10^{11}$ cm^{-2}) we observed a transition from WL to WAL driven by the increasing of the magnetic field, while at high carrier density ($n \simeq 2 \times 10^{12}$ cm^{-2}) only WL was observable. We analyzed the magnetic field driven WL-WAL transition and we evaluated the temperature dependence of the de-coherence parameters using an alternative method compared to previous studies. The values we obtained were corroborated by a root-mean-square analysis of the amplitude of highly-reproducible universal conductance fluctuations.

1. Introduction

The two-dimensional material graphene represents one of the most intriguing systems in present-day condensed-matter physics [1, 2]. Among its numerous outstanding properties, the relativistic behavior of chiral charges in graphene can be considered the most prominent [3, 4, 5]. The charge carriers are indeed governed by the Dirac-like Hamiltonian $H_{\mathbf{K}} = v_f \boldsymbol{\sigma} \cdot \mathbf{p}$ (where $v_f \simeq 10^6$ $\frac{m}{s}$ is the Fermi velocity of electrons in graphene, $\mathbf{p}$ is the momentum operator and $\boldsymbol{\sigma}$ the *pseudospin* operator) in the proximity of the so-called Dirac points **K** and **K'**, two classes of special points in the reciprocal space, placed at the vertexes of the hexagonal first Brillouin zone. The eigenvalues $E_{\pm} = \pm v_f \hbar |\mathbf{q}|$ describe a linear energy-momentum dispersion, with degeneracy at the Dirac points. The eigenstates of $H_{\mathbf{K}}$ are two-components spinors and simultaneous eigenstates of the helicity operator $h = \boldsymbol{\sigma} \cdot \mathbf{p}/|\mathbf{p}|$,with positive or negative eigenvalues depending on the relative orientation of momentum and pseudospin: +1 for electrons in **K** valley (holes in **K'** valley), −1 for electrons at **K'** (holes at **K**) (see Fig. 1(a)). In addition, as a consequence of the topology of the energy bands, a non-trivial Berry phase $\gamma = \pi$, is associated with those states (i.e. they change their sign when covering a closed circuit in reciprocal space including **K** or **K'**).

Chirality and Berry phase are especially crucial when coherent backscattering events in graphene are considered. In case of *intravalley* events (black arrow in Fig. 1(a)), the conservation of the helicity eigenvalue forbids backscattering, namely the chiral charges are expected to perform the so-called *Klein tunneling* paradox [6, 7]. On the other hand, backscattering is

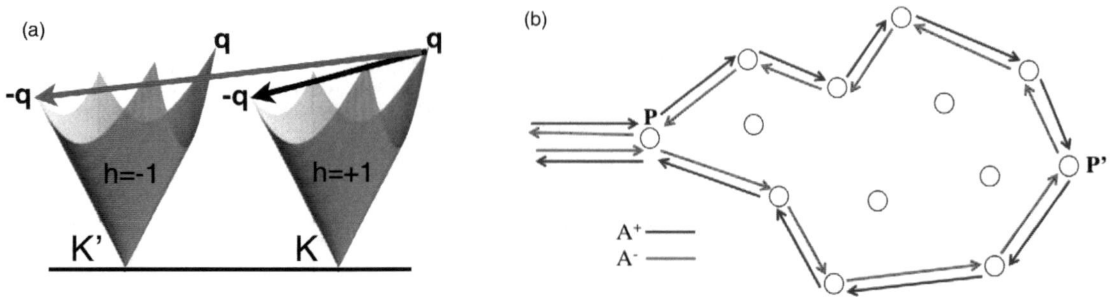

Figure 1: (a) Pictorial view of graphene's conduction band at K and K', with sketched intravalley (black arrow) and intervalley (red arrow) backscattering events. (b) Schematic picture of two closed PP'P electronic paths. A probability amplitude $A^+ = ae^{i\varphi^+}$ ($A^- = ae^{i\varphi^-}$) is associated to the clockwise (counterclockwise) trajectory, giving a backscattering probability $p = |A^+ + A^-|^2 = 2a^2(1 + \cos(\varphi^+ - \varphi^-))$.

allowed for *intervalley* events (black arrow in Fig. 1(a)), as the change of sign of the momentum is accompanied by a change of sign in helicity.

From a complementary point of view, the quantum interference between time-reversed counter-propagating closed electronic paths (for instance the two trajectories sketched in Fig. 1(b)) has to be taken into account. Depending on the phase difference acquired by the charge carrier while covering the two paths, constructive or destructive interference can occur. In the first case the backscattering probability is enhanced, so that the electrical conductivity is lowered with respect to the standard Drude's σ_0 [8], i.e. weak localization (WL) takes place [9, 10]. In the second case the backscattering probability is lowered, then the electrical conductivity is increased with respect to σ_0, i.e. one observes weak antilocalization (WAL). The π Berry phase implies destructive interference (and therefore WAL) for intravalley events, while produces constructive interference (and then WL) for intervalley ones. As a consequence, both WL and WAL can occur in graphene, their grade and presence depending on different types of elastic scattering mechanisms. Long-range potentials (such as ripples, dislocations and charged impurities) are responsible for intravalley events, while short-range potentials (adatoms, vacancies) induces intervalley ones.

The study of the WL-WAL corrections to conductivity in a graphene sample can therefore provide significant information about the disorder affecting the sample itself. A mean scattering time τ_s can be associated to intravalley events, while τ_i is the analogous for intervalley ones. Nevertheless, another contribution to intravalley scattering τ_w has to be taken into account, for the trigonal warping of the energy bands far from the Dirac points restores some backscattering probability as helicity ceases to be a good quantum number. Accordingly, the WL-WAL behavior experimentally observable depends on the relative incidence of these parameters.

Quantum interference corrections are suppressed by the breaking of time-reversal symmetry and the loss of phase coherence of charge carriers. Time-reversal can be broken by applying a non-zero magnetic field perpendicular to the graphene plane, which sets an upper time limit $\tau_B \propto B^{-1}$ to the the scatterings contributing to WL and WAL. Accordingly, WL manifests as a conductivity increasing with the magnetic field, while WAL corresponds to a decreasing magneto-conductivity. Decoherence due to inelastic scatterings is taken into account by introducing a characteristic time-scale (length-scale) parameter τ_ϕ (l_ϕ), that competes with the above mentioned scattering times in determining the quantum interference corrections to conductivity.

A complete expression for the magneto-conductivity in graphene, which embodies all the described effects, has been derived by E. McCan *et al.* in 2006 [11]; the analysis of the experimental data is usually based on the use of such function [12, 13, 14]. Nevertheless, we will try to follow a more intuitive procedure hereafter. We will not perform a fit of our data, we will rather make use of approximated relations in order to relate the WL-WAL magneto-conductivity to the mean elastic and inelastic scattering rates, at different carrier densities and temperatures.

2. Measurements and data analysis

We studied the electrical transport properties of an eight-contacts Hall bar made of mechanically exfoliated graphene, placed on a standard Si/SiO_2 (300 nm thick) substrate, with dimension $W \times L = 5.1\ \mu m \times 10.6\ \mu m$ (corresponding to a geometrical factor $W/L \simeq 0.5$). The longitudinal and Hall measurements were performed with standard low-frequency (10 Hz $< f <$ 15 Hz) ac lock-in technique, forcing a current of intensity 1 nA $< I <$ 10 nA to flow across the graphene. The sample was placed in a cryogen-free ^{3}He refrigerator with base temperature $\simeq$ 0.3 K, located in a cryogen-free superconducting magnet.

As a preliminary step, we performed measurements of longitudinal resistivity as a function of back-gate voltage at zero magnetic field and $T = 0.3$ K, reconstructing the so-called Dirac peak of the sample (Fig. 2(a)). Following the fit procedure developed by Kim *et al.* [15], we estimated the charge neutrality point at $V_D \simeq 4.66$ V, with a residual charge density $n_{res} \simeq 2 \times 10^{11}$ cm^{-2}. A chearged impurities' concentration at SiO_2-graphene interface $n_{imp} \simeq 10^{12}\ cm^{-2}$ was estimated making use of the relation $n_{res} \simeq 0.2 \times n_{imp}$ [16]. The average separation between such impurities resulted therefore $l_{imp} \simeq 10\ nm$, two order of magnitude larger than the nearest-neighbor distance $a = 1.42\ \AA$, thus satisfying the condition $l_{imp} \gg a$. The slowly-varying long-range potential associated with the charges randomly-located below the graphene plane can result in a finite electrostatic gradient inside the unit cell (i.e. an asymmetry between the two sublattices), which has been shown to produce an effective breaking of time-reversal symmetry (and therefore sets a temporal cutoff to the interference events) [17].

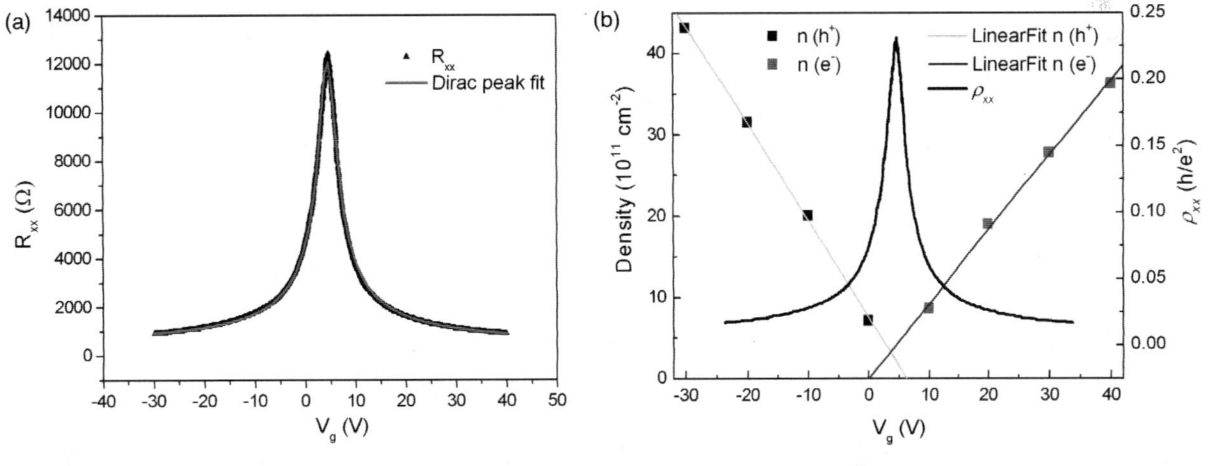

Figure 2: (a) Field effect in graphene: Dirac peak of the sample, fitted with the function of Kim *et al.* [15]. (b) Two dimensional density of carriers and resistivity as a function of gate voltage.

The two dimensional density of charge carriers (n, see Fig. 2(b)) was estimated both in the hole and electron regime at $T = 0.3$ K, from the Hall resistance (R_{xy}) measurements, making use of the relation n=$1/(es)$ (where $s = dR_{xy}/dB$ was obtained form a linear fit of $R_{xy}(B)$). The Hall mobility $\mu = 1/(ne\rho^0_{xx})$ (where ρ^0_{xx} is the longitudinal resistivity at $B = 0$ T) was also evaluated and resulted $\mu \sim 4 \times 10^3\ cm^2V^{-1}s^{-1}$ at carrier density $n \simeq 7 \times 10^{11}\ cm^{-2}$. A slight decrease of the mobility was found as a consequence of the increase of the carrier density (e.g. $\mu \sim 3.6 \times 10^3$ $cm^2V^{-1}s^{-1}$ at $n \simeq 2 \times 10^{12}$ cm^{-2})).

We successively performed measurements of the magneto-conductivity at different gate voltages (in the hole regime) and temperatures (between 0.3 K and 15 K), in order to determine the evolution of the quantum-interference corrections. The most significant data are shown in Fig. 3.

At very low field all the curves present a deep minimum at $B = 0$ T and a subsequent increasing magneto-conductivity, indicating WL behavior. At $V_g = 0$ V (corresponding to $n \simeq 7 \times 10^{11}$ cm^{-2}, see Fig. 3(a)) the magneto-conductivity curves reach a maximum at $B \simeq 0.06$ T (which is independent from temperature) and decrease at higher field: a B-driven transition from WL to WAL occurs. Such transition is expected to take place at a characteristic magnetic field B_i at which $\tau_B \approx \tau_i$ [11]. However, in order to take into account the field dependence of the WL correction (given by the function presented in Ref. 11), such value must be renormalized by a factor $4e^{\gamma}$ (where γ is the Euler-Mascheroni constant). From the renormalized value of B_i, we estimated for our sample a mean intervalley scattering rate $\tau_i^{-1} \simeq 0.5 \times 10^{12}$ s^{-1} at low carrier density. The fact that WAL is observable for $B > B_i$ (i.e. WAL "survives" at much smaller τ_B) indicates that intravalley scatterings take place at lower time-scale with respect to intervalley events; we can therefore argue that $\tau_s^{-1} \gg \tau_i^{-1}$, namely that the intravalley long-range scattering centers (such as charged impurities) dominate over short-range intervalley ones in our sample.

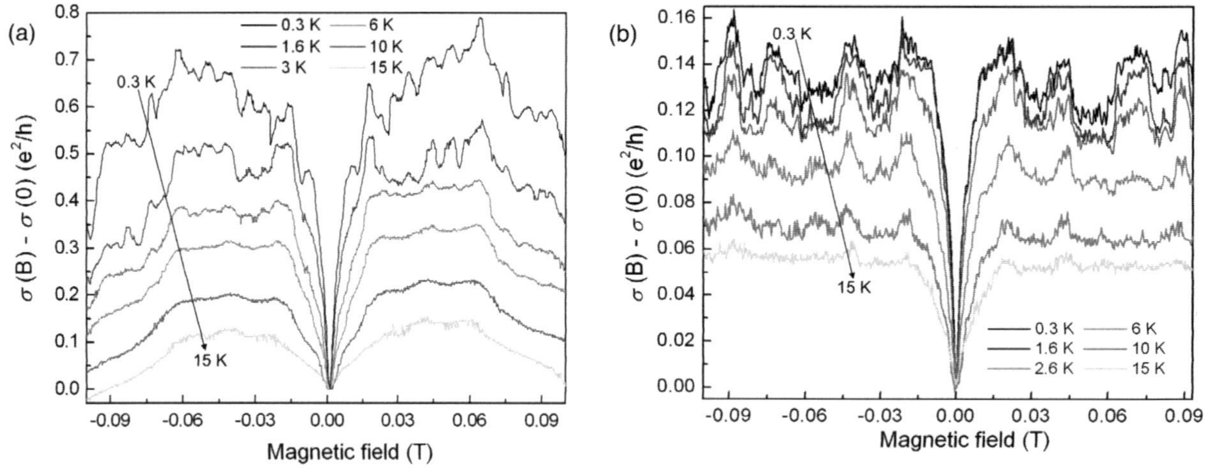

Figure 3: Evolution with temperature of magneto-conductivity, for $V_g = 0$ V (a) and $V_g = -10$ V (b).

The data collected at $V_g = -10$ V ($n \simeq 2 \times 10^{12}$ cm^{-2}, see Fig. 3(b)), on the other hand, does not show evidence of WAL. The applied gate potential sets the Fermi level far from the Dirac points, where the trigonal warping suppresses the chirality conservation and, accordingly, the WAL effect. Therefore, passing from $V_g = 0$ V to $V_g = -10$ V the sample is switched from a regime where $\tau_s^{-1} > \tau_w^{-1}$ to a trigonal warping-dominated regime, where $\tau_s^{-1} < \tau_w^{-1}$. Also for these data, a characteristic magnetic field B_i (at which $\tau_B \approx \tau_i$) marks the suppression of WL. We estimated $B_i \simeq 0.02$ T (for $B > B_i$ only fluctuations are observable), which gives, after proper renormalization with the factor $4e^{\gamma}$, $\tau_i^{-1} \simeq 0.3 \times 10^{12}$ s^{-1}. We believe that the intervalley scattering rate is lowered as a consequence of the increasing of Coulomb screening on the short-range scatterers for $n = 2 \times 10^{12}$ cm^{-2}.

After evaluating the relative strength of the different elastic mechanisms, we centered on the de-coherence of charge carriers. Considering that in graphene only intervalley scattering processes contribute to WL, we approximated of the function of Ref. 11 in order to obtain a simple analytical expression relating the amplitude of the WL correction to the de-coherence time

$$\delta\sigma^{WL}(\frac{e^2}{h}) = \frac{1}{\pi}\ln(1+2\frac{\tau_\phi}{\tau_i}) \qquad (1)$$

Making use of Eq. 1 we could easily estimate the de-coherence parameters from the peak-to-valley value of the WL conductivity dips of Fig. 3 and the mean intervalley scattering time. In Fig. 4(a) we can see that the phase-breaking rate τ_ϕ^{-1} grows linearly with temperature and that in the high density regime it has larger values (and increases more rapidly) then in the low density regime. The linear dependence of τ_ϕ^{-1} indicates that the carrier-carrier Coulomb interaction is the dominant de-phasing mechanism, in accordance with previous studies [12].

A finite offset is clearly observable in the de-phasing rate at high carrier density, indicating the presence of some other mechanisms competing with e-e de-coherence in determining the temporal cutoff to the interference processes at low temperature. A possible explanation can be found taking into account the effective breaking of time-reversal symmetry induced by the charged impurities at the graphene-SiO_2 interface [17]. This effect is expected to be relevant in case of decoupling of the **K** and **K'** valleys, namely for low intervalley scattering rate, which is the case of our high-density regime at $V_g = -10$ V. This mechanism can be mapped into the presence of an effective magnetic field, resulting into a suppression of the WL amplitude at low temperature: from the saturation value $\tau_\phi^{-1} \simeq 0.75 \times 10^{12}\ s^{-1}$ (obtained by a linear fit, see Fig. 4(a)), we estimated $B_{eff} \simeq 8$ mT.

A root-mean-square analysis of the magnitude of highly-reproducible universal conductance fluctuations (UCF, see Fig. 4(b)), based on the methodology of of P. A. Lee *et al.* [18], gave us a confirmation of the reliability of the values of Fig. 4(a). The analysis was performed considering the magneto-conductivity in the range 0.03 T $< B <$ 0.06 T for the V_g= 0 V case and the range 0.03 T $< B <$ 0.09 T for $V_g = -10$ V (i.e. the ranges where the UCF are well resolved). The extracted de-phasing rates at $T = 0.3$ K, $\tau_\phi^{-1}(V_g\text{=0 V}) \simeq 0.16 \times 10^{12}\ s^{-1}$ (orange triangle in Fig. 4(a)) and $\tau_\phi^{-1}(V_g\text{=-10 V}) \simeq 0.97 \times 10^{12}\ s^{-1}$ (orange square), are in a very satisfying agreement with the values obtained from Eq. 1.

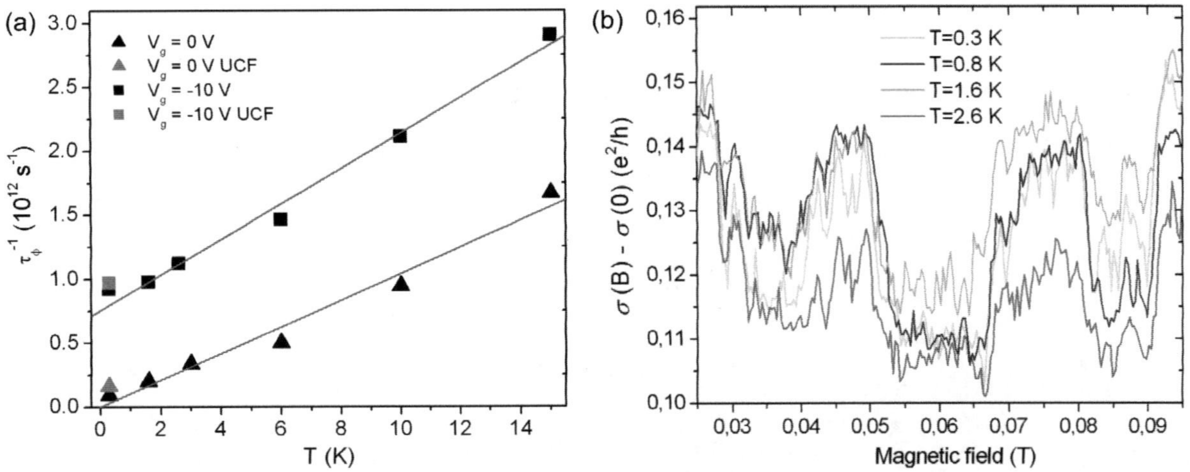

Figure 4: (a) De-coherence length and rate as a function of temperature both for $V_g = 0$ V and $V_g = -10$ V. (b) Universal conductance fluctuations measured at $T \leq 3$ K for $V_g = -10$ V.

3. Conclusions

In conclusion, we studied the quantum interference corrections to magneto-conductivity in mechanically exfoliated monolayer graphene at different charge carrier densities in the hole region. We found evidence of both WL or WL combined with WAL depending on the relative strength of the various elastic scattering mechanisms. The temperature dependence of de-coherence rate has been estimated using a simplified method with respect to previous studies, and indicates that the carrier-carrier Coulomb interaction is the main de-phasing mechanism in graphene at our densities. The low-temperature saturation of the de-coherence rate in the high carrier density regime can be related to an effective breaking of time-reversal symmetry due to the presence of long-range slowly-varying potentials, that have been shown to be the dominant disorder. The extreme sensitivity of WL-WAL to various elastic scattering mechanisms hints the use of magneto-transport measurements as a tool of characterization of both unintentional and intentional disorder in graphene.

Acknowledgements

We are thankful to David López and Mario Amado and to the ISOM for their help during the sample process. This work has been supported by the following projects: MICINN (Spain) FIS2009-07880 and PCT310000-2009-3 and Junta de Castilla y León SA049A10.

References

[1] K. S. Novoselov, A. K. Geim, S. V. Morozov, D. Jiang, Y. Zhang, S. V. Dubonos, I. V. Grigorieva and A. A. Firsov, Science **306**, 666 (2004).
[2] K. S. Novoselov, A. K. Geim, S. V. Morozov, D. Jiang, M. I. Katsnelson, I. V. Grigorieva, S. V. Dubonos, and A. A. Firsov, Nature **438**, 197 (2005).
[3] K. S. Novoselov, S. V. Morozov, T. M. G. Mohiuddin, L. A. Ponomarenko, D. C. Elias, R. Yang, I. I. Barbolina, P. Blake, T. J. Booth, D. Jiang, J. Giesbers, E. W. Hill, and A. K. Geim, Phys. Stat. Sol. (b) **244**, 4106 (2007).
[4] A. H. Castro Neto, F. Guinea, N. M. R. Peres, K. S. Novoselov, and A. K. Geim, Rev. Mod. Phys. **81**, 109 (2009).
[5] R. N. M. Peres, Rev. Mod. Phys. **82**, 2673 (2010).
[6] O. Klein, Z. Phys. **53**,157 (1929).
[7] M. I. Katsnelson, K. S. Novoselov, and A. K. Geim, Nature Physics **2**, 620 (2006).
[8] N. W. Ashcroft and N. D. Mermin, Solid State Physics, Holt, Rinehart and Winston, New York (1976).
[9] C. W. J. Beenakker and H. van Houten, Solid State Phys. **44**, 1 (1991).
[10] G. Bergmann, Phys. Rev. B **28**, 2914 (1983).
[11] E. McCann, K. Kechedzhi, V. I. Fal'ko, H. Suzuura, T. Ando, B. L. Altshuler, Phys. Rev. Lett. **97**, 146805 (2006).
[12] S. V. Morozov, K. S. Novoselov, M. I. Katsnelson, F. Schedin, L. A. Ponomarenko, D. Jiang, and A. K. Geim, Phys. Rev. Lett. **97**, 016801 (2006).
[13] F. V. Tikhonenko, D. W. Horsell, R. V. Gorbachev, and A. K. Savchenko, Phys. Rev. Lett. **100**, 056802 (2008).
[14] F. V. Tikhonenko, A. A. Kozikov, A. K. Savchenko, and R. V. Gorbachev, Phys. Rev. Lett. **103**, 226801 (2009).
[15] S. Kim, J. Na, I. Jo, D. Shahrjerdi, L. Colombo, Z. Yao, E. Tutuc, and S. K. Banerjee, Appl. Phys. Lett. **94**, 062107 (2009).
[16] S. Adam, E. H. Hwang, V. M. Galitski, and S. Das Sarma, Proc. Natl. Acad. Sci. U.S.A. **104**, 18392 (2007).
[17] A. F. Morpurgo and F. Guinea, Phys. Rev. Lett. **97**, 196804 (2006).
[18] P. A. Lee, A. D. Stone and H. Fukuyama, Phys. Rev. B **35**, 1039 (1987).

Journal of Physics: Conference Series **456** (2013) 012033 doi:10.1088/1742-6596/456/1/012033

Magnetic-field interaction of spatially confined quantum-well exciton-polaritons

A Rahimi-Iman, C Schneider, J Fischer, S Holzinger, M Amthor, L Worschech, S Reitzenstein*, S Höfling, A Forchel and M Kamp

Technische Physik, Physikalisches Institut, and Wilhelm Conrad Röntgen Research Center for Complex Material Systems, Universität Würzburg, Am Hubland, D-97074

E-mail: arash.rahimi-iman@physik.uni-wuerzburg.de

Abstract. We report on pronounced magneto-optical effects of trapped polariton modes in a single InGaAs quantum well microresonator with a lithographically defined modulation of the cavity length. In our optical polariton traps with diameters ranging from 1 to 10 μm, a confinement potential of 7.5 meV is achieved. In magnetic-field dependent experiments, a diamagnetic shift and a Zeeman splitting of the trap-modes are observed which confirms that the polaritonic nature of the quantized emission modes are preserved even for traps as small as 1 μm. Furthermore, focusing on theoretical estimates using a simple model, we have identified a clear correlation between the polaritons' magnetic response and their excitonic fraction, corresponding to the Hopfield coefficients of the composite particles.

1. Introduction

Exciton-polaritons are the result of strong light–matter coupling in quantum-well (QW) microcavities [1]. These bosonic quasiparticles which combine properties of photons and excitons in a linear superposition have proven to be a model system for the demonstration of dynamical Bose-Einstein condensation (BEC) in semiconductors [2-5]. Since the realization of BEC is not possible in a homogeneous and infinite two-dimensional (2D) system [6], experimentally the condensation of polaritons in a planar system relies on local potential variations. This explains the strong interest in a lateral trapping of the exciton-polaritons.

Various technological approaches have been pursued to structurally tailor the confinement of polaritons [4,7-10]. The low polariton effective mass hereby enables the observation of three-dimensional (3D) quantization effects already for trap diameters on the order of a few microns. With lateral confinement of polaritons, new opportunities for the realization of future quantum light emitters arise that have confined polaritons as the source of indistinguishable single photons and can be utilized for quantum information processing [11]. According to theoretical considerations on the single-polariton level, a photonic dot with high confinement potential for polaritons can act as a single-photon source on the basis of polariton quantum blockade [12].

Our magneto-optical studies of discrete optical modes originating from spatially trapped, zero-dimensional (0D) polaritons, presented in this work, reveal a diamagnetic shift and a Zeeman splitting for magnetic fields up to 5 T. These effects were observed in *k*-space resolved measurements for different trap diameters, ranging from 1 to 10 μm, confirming that the polaritonic properties of the emission modes are preserved for various trap sizes, even for the smallest traps studied. By the use of an external magnetic field, we probe light–matter coupling in our systems via direct manipulation of the excitonic component of the formed quasiparticles, with both the spectral shift and the splitting being attributed to the excitonic fraction of the trapped composite particles [13,10]. We present good agreement between experiment and theory using a simple Hopfield coefficient model to describe both effects. Our results hereby indicate that magneto-optical studies present a reliable tool to identify strong coupling for various exciton-polariton systems.

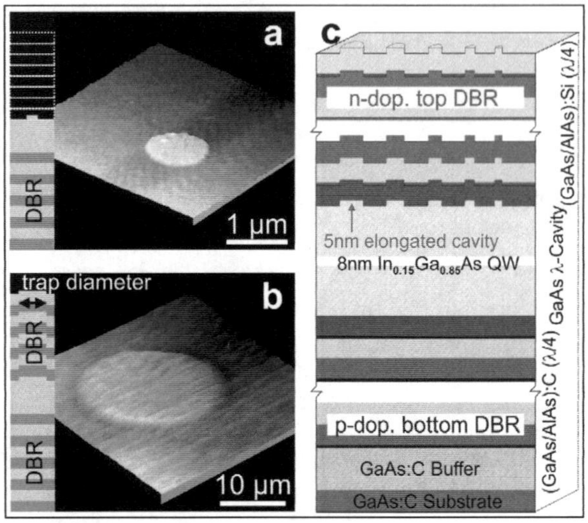

Figure 1: Atomic force microscope image of an etched cavity spacer with (a) a 1-µm diameter before and (b) a 10-µm diameter after top DBR overgrowth. Inset: Schematic layout of the structure. (c) Schematic drawing of a planar microcavity design with a single $In_{0.15}Ga_{0.85}As$ QW [10].

Trapping of polaritons is achieved by a gentle modulation of the optical confinement in a one-λ thick GaAs cavity spacer which is sandwiched between AlAs/GaAs distributed Bragg reflectors (DBRs) with 23/27 top/bottom mirror pairs in our microcavity system [10]. Circular trap diameters d were systematically varied from 0.5 to 10 µm. The active layer of our high-quality planar microresonator, grown by molecular beam epitaxy, consists of a single $In_{0.15}Ga_{0.85}As$ QW. The traps were defined by electron-beam lithography and the structure was patterned by etching the cavity spacer before overgrowth [cf. figures 1a and 1b]. The pattern is preserved during overgrowth. In our system, the unetched area introduces a trap potential V of about −7.5 meV for the locally 5-nm-thicker cavity layer. In such a structure, the unmodified QW provides free 2D excitons, which then couple strongly to 3D localized photon modes and result in 0D polaritonic traps [7,14]. A schematic drawing of our microcavity system is shown in figure 1c.

2. Theoretical Background

The strong coupling of cavity photons and QW excitons in our structure results in the lower and upper polariton energy-momentum dispersion branch (LP/UP), the two eigenstates of the coupled system [1]. In general, the photonic ($|C|^2$) and excitonic ($|X|^2$) contents — referred to as Hopfield coefficients [15] — of the composite particle sum up to unity for each branch at given in-plane wavevector $k_{||}$. For a polariton system at a given detuning $\Delta(k_{||}) = E_C(k_{||}) - E_X(k_{||})$ between the bare exciton and the cavity modes one can calculate both coefficients, $|C(k_{||})|^2$ and $|X(k_{||})|^2$, at corresponding $k_{||}$ [16]. Strong coupling manifests itself by the anticrossing behavior and the characteristic Rabi splitting of $E_{\text{Rabi}} = 2\hbar\Omega_{\text{Rabi}}$ of the polaritonic modes, with the Rabi frequency Ω_{Rabi}. In polaritonic systems, mode tuning can typically be achieved for instance by changing the temperature, and by other means such as an external magnetic field, exploiting the diamagnetic shift as a tuning mechanism.

Magneto-optical effects in polaritonic systems were particularly studied with respect to an increase in the oscillator strength and Rabi splitting with increasing magnetic field [17-21] and to the Zeeman splitting of the exciton-polariton system [13,20]. Furthermore, magneto-optical studies have been recently extended to exciton-polariton condensates [22,23] and quantum-dot polaritonic systems for which a decrease of the oscillator strength has been observed [24,25]. In the presented work, we utilized magneto-optical effects to identify polariton emission from traps.

While the uncoupled photon mode of the QW cavity system does not interact with an external magnetic field, the unperturbed exciton of a QW shifts in energy due to a diamagnetic response

$$\delta E_{\text{dia},X} = \kappa_X B^2 \quad (1)$$

according to Refs. 26 and 27, where κ_X denotes the effective diamagnetic factor of the QW exciton. At the same time, the magnetic field couples to the total angular momentum of the excitonic spin states which causes Zeeman splitting proportional to the field density, given by [28]

$$\Delta E_{ZS,X} = g_X \mu_B B \qquad (2)$$

where g_X denotes the effective excitonic *g*-factor and μ_B the Bohr magneton. By the circular polarization of the emitted light, the two Zeeman-split exciton states can be analyzed and distinguished [13,20]. The diamagnetic shift of a polaritonic system depends on the detuning situation and can be estimated by the associated exciton fraction in terms of the magnetic field–dependent Hopfield coefficient $|X|^2(B)$, i.e. by weighting Eqs. (1) and (2) with $|X|^2(B)$, respectively. The effective mode detuning at present field determines $|X|^2(B)$, taking into account the diamagnetic shift of the QW exciton mode $E_X(B)$ and an increase of $E_{Rabi}(B)$. In fact, if calculating the system's energy eigenstates [16] with both magnetic-field dependent parameters $E_X(B)$ and $E_{Rabi}(B)$, the polaritons' diamagnetic shift, which is particularly the result of a detuning change with increasing *B*, can be well estimated for given *B*. Taking into account spin-split $E_X(B)$, the same can be done for the Zeeman splitting.

3. Experiment

Utilizing angular resolution of emission in Fourier-space (FF: far-field) spectroscopy (cf. Ref. 5), trap polaritons were investigated for an accessible range of *k*-vectors and angular mode numbers in micro-photoluminescence (μPL) under high-above-band continuous wave excitation by a diode laser with a 4 μm wide pump spot.

The emission pattern of discrete trap modes reflects the planar energy-momentum dispersion when trap diameters are on the length scale of the particle's wavefunction, which is on the order of a few µm for cavity polaritons [10]. Representative energy-momentum dispersions from polariton modes are shown in figure 2 in false-color intensity profiles collected from a 10-, 2-, and 1-μm trap at 10 K [figures 2a–2c, respectively]. Discrete trap modes are indicated by colored boxes or dotted lines in the graphs. Additionally, background emission from the planar unpatterned system of the surrounding area is observed with the LP and UP dispersion (labeled 'Planar LP/UP'). While a polaritonic-like dispersion is formed by a number of discrete modes which appear for traps with a diameter of 10 μm [figure 2a], the 2- and 1-μm trap signal consist of few discrete modes with respective confinement blue shift in the spectral domain and delocalized features in *k*-space [figures 2b and 2c], representing the transition from shallow (10 μm) to strong (1 μm) confinement.

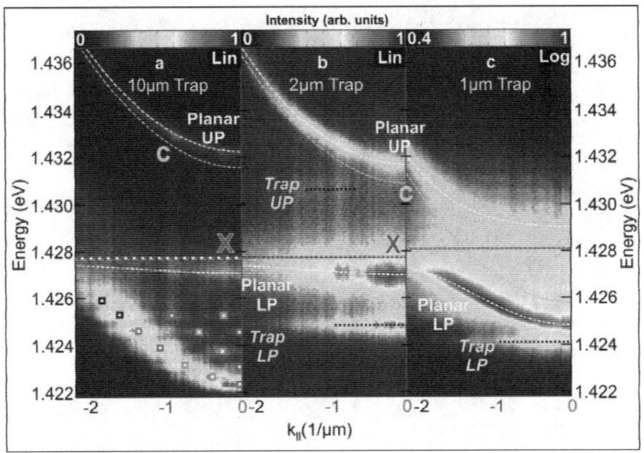

Figure 2: *k*-space resolved µPL from a (a) 10-μm, (b) 2-μm, and (c) 1-μm polariton trap [10]. Trap modes show quantization effects and are marked by (a) colored boxes and (b) and (c) dotted lines, while both planar polariton branches are detected as background emission. Calculated planar mode dispersions are plotted for the polariton branches (LP/UP, white), as well as for the uncoupled planar photon (C, green) and QW exciton (X, red) mode in dashed lines. The PL intensity in (c) is displayed in logarithmic scale for clarity.

The spectra in figures 2a and 2b were acquired for the same planar detuning $\Delta_{planar}(k_{||} = 0) = \Delta_{planar} = 3.7$ meV and are displayed in linear scale, while in figure 2c the detuning is $\Delta_{planar} = -2.5$ meV and the intensity is plotted with a logarithmic scale for clarity. In our system, the presented trap modes were brought close to resonance with the exciton mode at 1.4277 eV and the effective trap-mode detunings amounting to about −5 to 0 meV (10 μm) and about −2.5 meV (2 and 1 μm). More details about photoluminescence features of the sample can be found in Ref. 10.

4. Results and Discussion

The Zeeman splitting of polaritons has previously been studied in QW microcavities and its dependence on the cavity-exciton detuning identified [13,20]. It was demonstrated that Zeeman-split polaritons result from the independent coupling of left or right circular polarized photon modes with the spin-resolved exciton. Moreover, the magnitude of the splitting corresponded well to the excitonic amplitude of the polariton mode. By means of polarization optics (quarter wavelength plate and linear polarizer), the two polarization components of the Zeeman-split polaritons can be spectrally resolved.

We now examine the behavior of trapped polaritons in external magnetic fields of up to 5 T, which were applied to the sample in a superconducting magneto-cryostat in Faraday configuration, and analyze the magnetic response with respect to the excitonic amplitudes to prove a polaritonic nature of the trap emission [10]. For our QW microcavity system, we derive an exciton Zeeman splitting of ≈0.5 meV at 5 T from the experimental value of the planar LP Zeeman splitting of 0.35 meV for $|X|^2(5\ \mathrm{T}) \approx 70\%$. This is in strong agreement with an effective exciton splitting of 0.5 meV that was obtained from dispersion fits to $\sigma^{\pm}$ planar modes at given field and compares well with literature values for QWs with similar material composition and thickness [20,28].

Polaritons confined in 2-μm traps show a well-separated and discrete mode pattern due to a stronger confinement than for 10-μm traps [cf. figure 2 and Ref. 10], on the order of a few meV. In magnetic fields of up to 5 T, we observed both Zeeman splitting and a diamagnetic shift for these modes. We extracted line spectra from FF resolved spectra representing ground-state emission and assigned two observed modes in this spectral range to trap LP and planar LP, respectively, with the planar mode emitting at higher energy. Figure 3a shows $\sigma^{\pm}$ polarized spectra (solid/dashed) in a

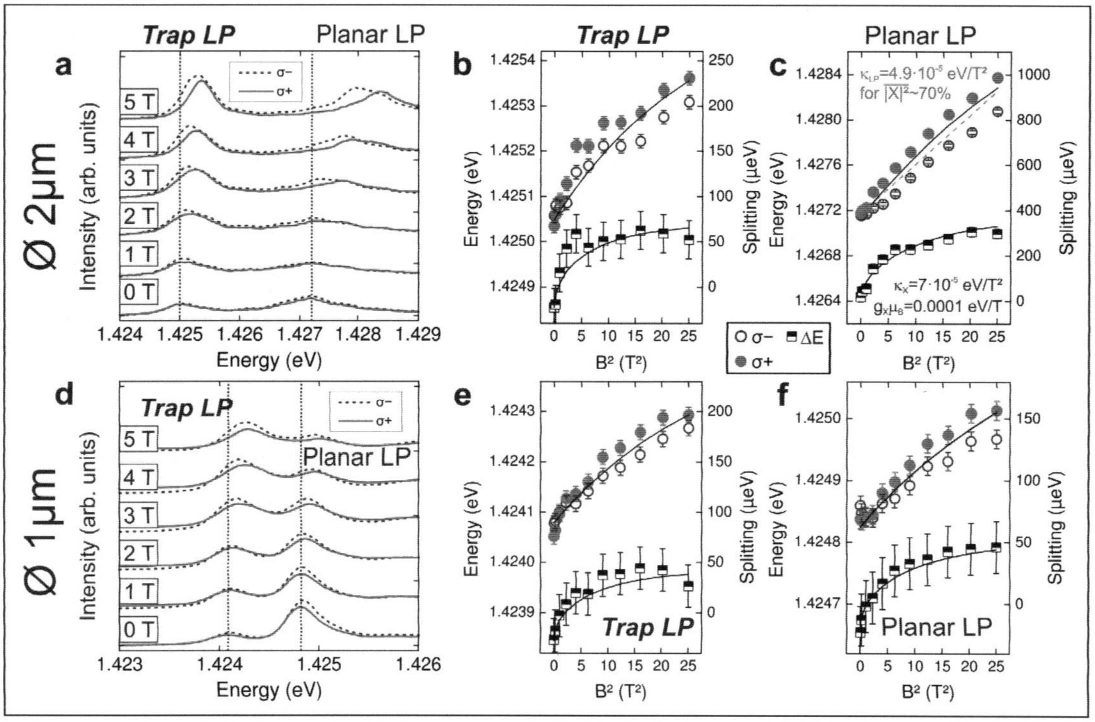

Figure 3: (a) Waterfall presentation of k=0 line spectra for a 2-μm trap mode. $\sigma^{\pm}$ polarized spectra of a magnetic field series are displayed from 0 to 5 T, while dotted vertical lines indicate mode energies at 0 T for guidance. $\sigma^{\pm}$ trap-mode and planar LP energies are plotted vs B^2 (left scale) in (b) and (c), respectively, with the experimental splittings of the modes (squares) shown in the same graphs (right scale) [10]. (d)–(f) represent the same information as in (a)–(c), respectively, for 1-μm trap μPL. Here, solid-line curves represent theoretical estimates, which match well with the experiment, taking into account the Hopfield coefficients at given B.

Journal of Physics: Conference Series **456** (2013) 012033 doi:10.1088/1742-6596/456/1/012033

waterfall diagram for a magnetic field variation from 0 and 5 T, indicating a pronounced energy shift of 0.3/1.0 meV (trap/planar mode) with respect to the 0-T signal. Moreover, a Zeeman splitting of ~0.06/0.35 meV (trap/planar mode) at 5 T can be resolved.

Respective $\sigma^{\pm}$-mode energies are summarized in figures 3b and 3c. In these graphs, mode energies of the trap/planar signal are plotted vs B^2 to reveal the behavior of the shift in comparison to the expectations according to Eq. (1) weighted by corresponding $|X|^2(B)$. In figure 3c, the spectral shift of the planar LP with high excitonic content ($|X|^2(B) \approx 70\%$) shows the expected dependence as a function of the magnetic field. A green dashed line represents a mode shift according to Eq. (1) with diamagnetic coefficient of $\kappa_{LP} = 4.9 \times 10^{-5}$ eV/T^2. However, strong deviations from this characteristic are observed for the trapped LP in figure 3b. This feature can be attributed to a change of the cavity-exciton detuning as a function of the magnetic field, if assuming a linear shift of the QW exciton vs B^2. For LPs with $\Delta \geq 0$ meV, this effect is negligible, but becomes more significant for the LP at the system's transition from resonance ($\Delta \approx 0$ meV) to red detuning ($\Delta < -\hbar\Omega$) [10]. In between, the mixed state of photon and exciton asymptotically approaches the uncoupled photon energy which results in a change of the slope regarding the energy shift.

At 0 T, the detuning of the trap polariton amounts to −1.5meV, for which $|X|^2 \approx 30\%$, while at 5 T, the uncoupled exciton is blue shifted by ≈ 1.7−1.8 meV (with $\kappa_X \approx 7 \times 10^{-5}$ eV/T^2), resulting in a reduction of the excitonic fraction $|X|^2(5\text{ T}) \approx 19\%$. For theoretical estimations, a magnetic-field induced increase of the Rabi-splitting ($E_{\text{Rabi}}(0\text{ T}) = 3.5$ meV) of ≈ 0.5−0.6 meV is also taken into account. Theoretical shifts and splittings for the 2-μm trap polariton of 0.30 and 0.08 meV, respectively, strongly agree with experimental data. Figure 3b displays the splitting $\Delta E_{ZS} = E_{\sigma+} - E_{\sigma-}$ of the polarized modes in the same plot with reference to the right vertical scale. Here, we obtained an initial value of −0.02 meV (at 0 T^2), since the photonic character of the trapped polariton mode adds a polarization imposed splitting (here with negative sign) [10]. An applied magnetic field of ≈1 T compensated this feature by the Zeeman splitting. The expected splitting value at 5 T is thus in strong agreement, if taking this into account, with a thereby determined effective splitting of 0.08 meV.

Due to a detuning change, discussed above for the diamagnetic shift, the Zeeman splitting deviates from the expected linear dependence on the magnetic field with saturation at large fields, while the highly excitonic emission from the planar LP features a linear increase of the splitting according to Eq. (2). Our experimental results indicate that estimates for the diamagnetic shift and Zeeman splitting values need to be determined with respect to the effective Hopfield coefficients corresponding to the respective field B. Taking $|X|^2(B)$ into account, we achieved good congruence of our data with expected values (solid lines) for both energy shift and splitting, as shown in figures 3b and 3c.

In analogy to the 2-μm trap, we present a diamagnetic shift and Zeeman splitting for our 1-μm trap luminescence in figures 3d−3f, providing evidence of 0D polaritons. Energies of the polarized modes presented in figure 3e clearly reveal a diamagnetic shift and Zeeman splitting for the trapped LP with $|X|^2 \approx 15\%/11\%$ (0 T/5 T). Moreover, a good matching of model values with experimental data is again achieved in this study, when considering $|X|^2(B)$. Thus, the experiment proves that strong confinement of ~7.5 meV in polariton traps is feasible by gentle modulation of the cavity height in a planar Fabry-Pérot cavity system, without changing the basic features of exciton-polaritons.

5. Conclusions

In conclusion, we have probed exciton-polariton emission from photonic quantum boxes in a planar, single InGaAs QW microcavity for diameters of 1–10 μm with a strong confinement on the scale of a few meV. Moreover, we have investigated the emission properties of the system subject to an external magnetic field. A dependency of the trap-polaritons' diamagnetic shift and Zeeman splitting on their magnetic-field dependent excitonic fraction has clearly been observed in a magnetic field of up to 5 T, while a simple Hopfield coefficient model has allowed for good theoretical estimates for both Zeeman splitting and diamagnetic shift. Thereby, we have shown that the interaction of polaritons with an external magnetic field serves as a reliable tool to verify the polaritonic character of emission modes under strong confinement.

Acknowledgments
The authors acknowledge financial support by the State of Bavaria and thank M. Emmerling for expert sample processing.

References

* Present address: Institut für Festkörperphysik, Technische Universität Berlin, Hardenbergstrasse 36, 10623 Berlin, Germany

[1] Weisbuch C, Nishioka M, Ishikawa A and Arakawa Y 1992, *Phys. Rev. Lett.* **69**, 3314
[2] Deng H, Weihs G, Santori C, Bloch J and Yamamoto Y 2002, *Science* **298**, 199
[3] Kasprzak J, Richard M, Kundermann S, Baas A, Jeambrun P, Keeling J M J, Marchetti F M, Szymanska M H, Andre R, Staehli J L, Savona V, Littlewood P B, Deveaud B and Dang L S 2006, *Nature* **443**, 409
[4] Balili R, Hartwell V, Snoke D, Pfeiffer L and West K 2007, *Science* **316**, 1007
[5] Lai C W, Kim N Y, Utsunomiya S, Roumpos G, Deng H, Fraser M D, Byrnes T, Recher P, Kumada N, Fujisawa T and Yamamoto Y. 2007, *Nature* **450**, 529
[6] Hohenberg P C 1967, *Phys. Rev.* **158**, 383
[7] El Daïf O, Baas A, Guillet T, Brantut J-P, Kaitouni R I, Staehli J L, Morier-Genoud F and Deveaud B 2006, *Appl. Phys. Lett.* **88**, 061105
[8] Gutbrod T, Bayer M, Forchel A, Reithmaier J P, Reinecke T L, Rudin S and Knipp P A 1998, *Phys. Rev.* B **57**, 9950
[9] Utsunomiya S, Tian L, Roumpos G, Lai C W, Kumada N, Fujisawa T, Kuwata-Gonokami M, Löffler A, Höfling S, Forchel A and Yamamoto Y 2008, *Nat. Phys.* **4**, 700
[10] Rahimi-Iman A, Schneider C, Fischer J, Holzinger S, Amthor M, Höfling S, Reitzenstein S, Worschech L, Kamp M and Forchel A 2011, *Phys. Rev.* B **84**, 165325
[11] Na N and Yamamoto Y 2010, *New J. Phys.* **12**, 123001
[12] Verger A, Ciuti C and Carusotto I 2006, *Phys. Rev.* B **73**, 193306
[13] Armitage A, Fisher T A, Skolnick M S, Whittaker D M, Kinsler P and Roberts J S 1997, *Phys. Rev.* B **55**, 16395
[14] Muller A, Shih C-K, Ahn J, Lu D, Gazula D and Deppe D G 2006, *Appl. Phys. Lett.* **88**, 031107
[15] Hopfield J J 1958, *Phys. Rev.* **112**, 1555
[16] Deng H, Haug H and Yamamoto Y 2010, *Rev. Mod. Phys.* **82**, 1489
[17] Tignon J, Voisin P, Delalande C, Voos M, Houdré R, Oesterle U and Stanley R P 1995, *Phys. Rev. Lett.* **74**, 3967
[18] Whittaker D M, Fisher T A, Afshar A M, Skolnick M S, Kinsler P, Roberts J, Hill G and Pate M 1995, *Il Nuovo Cimento* D **17**, 1781
[19] Tanaka T, Zhang Z, Nishioka M and Arakawa Y 1996, *Appl. Phys. Lett.* **69**, 887
[20] Fisher T A, Afshar A M, Skolnick M S Whittaker D M 1996, *Phys. Rev.* B **53**, R10469
[21] Berger J D, Lyngnes O, Gibbs H M, Khitrova G, Nelson T R, Lindmark E K, Kavokin A V, Kaliteevski M A and Zapasskii V V 1996, *Phys. Rev.* B **54**, 1975
[22] Larionov A V, Kulakovskii V D, Höfling S, Schneider C, Worschech L and Forchel A 2010, *Phys. Rev. Lett.* **105**, 256401
[23] Walker P, Liew T C H, Sarkar D, Durska M, Love A P D, Skolnick M S, Roberts J S, Shelykh I A, Kavokin A V and Krizhanovskii D N 2011, *Phys. Rev. Lett.* **106**, 257401
[24] Reitzenstein S, Münch S, Franeck P, Rahimi-Iman A, Löffler A, Höfling S, Worschech L and Forchel A 2009, *Phys. Rev. Lett.* **103**, 127401
[25] Reitzenstein S, Münch S, Franeck P, Löffler A, Höfling S, Worschech L, Forchel A, Ponomarev I V and Reinecke T L 2010, *Phys. Rev.* B **82**, 121306(R)
[26] Nash K J, Skolnick M S, Claxton P A and Roberts J S 1989, *Phys. Rev.* B **39**, 10943
[27] Walck S N and Reinecke T L 1998, Phys. Rev. B **57**, 9088
[28] Wimbauer T, Oettinger K, Efros A L, Meyer B K and Brugger H 1994, *Phys. Rev.* B **50**, 8889

Journal of Physics: Conference Series **456** (2013) 012034 doi:10.1088/1742-6596/456/1/012034

Magnetic-field-independent superconductivity of ultrathin Pb films on cleaved GaAs surface

Takayuki Sekihara, Ryuichi Masutomi, and Tohru Okamoto

Department of Physics, University of Tokyo, 7-3-1 Hongo, Bunkyo-ku, Tokyo 113-0033, Japan

E-mail: sekihara@dolphin.phys.s.u-tokyo.ac.jp

Abstract. We performed magnetotransport measurements on ultathin Pb films deposited onto cleaved GaAs surface and observed two-dimensional superconductivity for an amorphous 2.2-Å-thick film, which is below one monolayer. The superconducting transition is almost independent of parallel magnetic field as high as 14 T. This means that the superconducting state has much larger critical magnetic field than Pauli paramagnetic limit. We consider two different mechanism relating Rashba spin splitting.

1. Introduction

Two-dimensional (2D) superconductivity in ultrathin films is attractive. The onset of two-dimensional superconductivity have intensely studied by tuning film thickness or applying magnetic fields *etc.* in granular or amorphous thin films [1–3], and recently in crystalline layers [4–10]. Markedly in even a single-atomic layer of Pb and In grown epitaxially on silicon substrate, and superconducting energy gaps were observed by scanning tunneling microscopy (STM) measurements [11]. Futhermore macroscopic zero resistance state in one-atomic layer was confirmed by electric transport measurements [12].

In this article, we examined the magnetic-filed effect of two-dimensional superconductivity in amorphous ultrathin Pb films on cleaved GaAs surface. We used GaAs cleaved surface as a substrate and observed 2D superconductivity even for one monolayer (ML) notwithstanding amorphous films. While the superconducting state was fragile to the perpendicular magnetic field, it was robust to the parallel magnetic filed as high as 14 T. The critical magnetic filed in parallel direction much larger than 14 T cannot be explained by Pauli paramagnetic limit.

2. Experimental Setups

The experimental setups were based on the technique of the previous works on InAs inversion layeres [13, 14]. The substrate was a non-doped semi-insulating GaAs single crystal. It was cleaved at liquid helium temperatures in an ultrahigh vacuum chamber with ^{3}He cryostat. Amorphous Pb films were made by quench-condensation. Magnetotransport data on the (110) cleaved surface were taken in a Hall bar geometry (4 mm $\times$ 0.35 mm) using the standard four probe lock-in technique at 13.1 Hz with two current electrodes and four voltage electrodes on non-cleaved surfaces prepared by deposition of gold at room temperature. The resistance of GaAs substrate was higher than 100 MΩ at 4.2 K. The sample was mounted on a rotatory stage to control the magnetic field direction precisely.

3. Experimental Results

Figure 1 shows the sheet resistance in the normal state and the critical temperature (T_c) as a function of film thickness. For calibrating film thickness we used quartz oscillator assuming the bulk density of Pb. The critical temperature was defined as the value of the half of the sheet resistance in the normal state. We observed superconductivity from a thickness of 2.2 Å, which is below 1 ML. A thickness of 2.2 Å corresponds to a Pb atom coverage of 7.2×10^{14} cm^{-2}, and it is smaller than 8.9×10^{14} cm^{-2} of the atomic density of GaAs (110) surface. In the 2.2-Å-thick film, T_c was 0.90 K and the sheet resistance in the normal state was 4.35 kΩ, which is below the quantum resistance for pairs, $h/4e^2 = 6.45$ kΩ, the threshold value of Superconductor-Insulator transition [1].

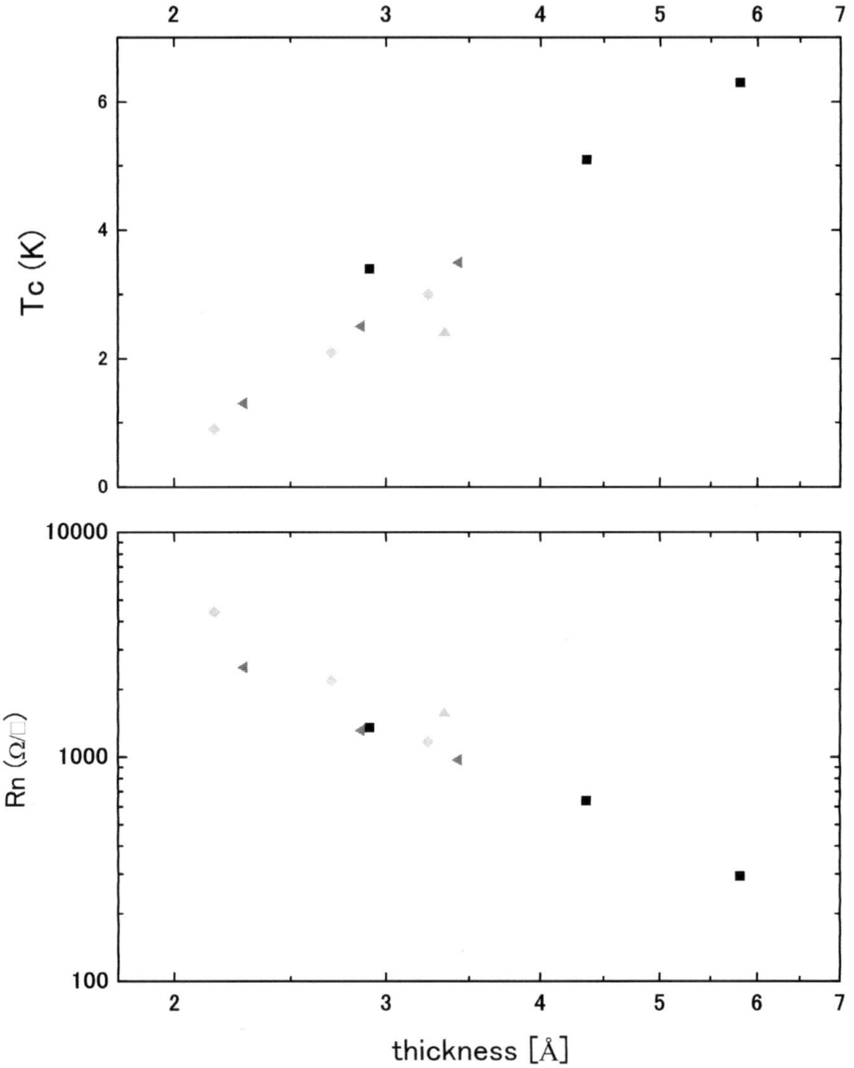

Figure 1. Plot of the sheet resistance in normal state (R_n) and the superconducting critical temperature (T_c) versus film thickness for several runs. 1 ML corresponds to 2.7 Å.

Figure 2 shows the temperature dependence of the sheet resistance for different parallel magnetic field. The results are almost independent of parallel magnetic field. This is surprising because the upper critical filed in parallel direction to the 2D systems is normally limited by Pauli paramagnetic effect. Since at Pauli paramagnetic limit the Zeeman energy arising from spin splitting of Cooper pairs exceeds the superconducting condensation energy, Pauli paramagnetic limit is expressed as $H_P = 1.84T_c$ for an isotropic BCS superconductor [15]. In the 2.9-Å-thick film T_c is 2.6 K and H_P is 4.8 T. In the 3.4-Å-thick film T_c is 3.3 K and H_P is 6.1 T. Though Pauli paramagnetic limit calculated from the expression is much lower than 14 T, the 2D superconductivity is robust to parallel magnetic fields as high as 14 T.

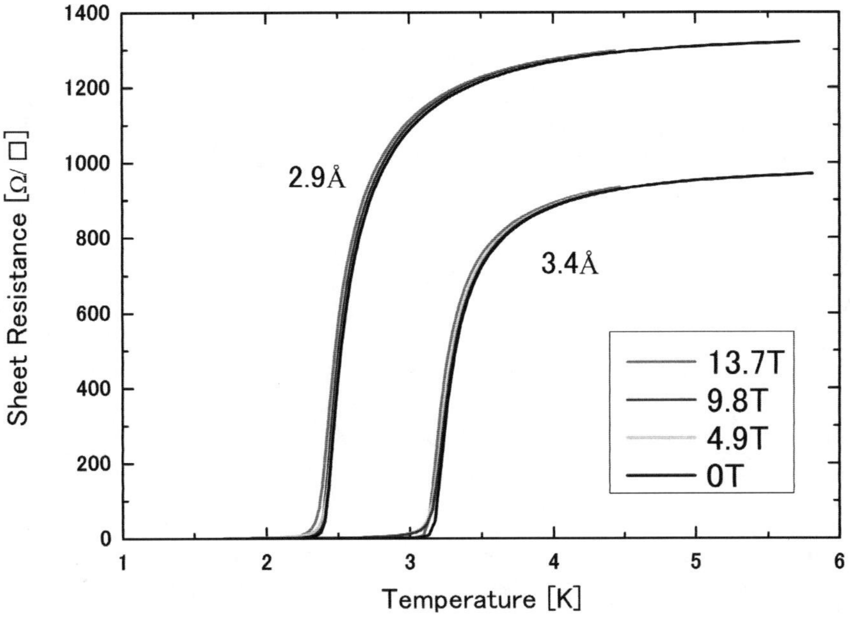

Figure 2. The temperature dependence of the sheet resistance for different parallel magnetic field. The film thickness is 2.9 Å and 3.4 Å. H_P is 4.8 T and 6.1 T.

4. Discussion

We consider two different mechanism to explain upper critical field far above Pauli paramagnetic limit.

The first mechanism is based on spin-orbit scattering. According to the theory including the effect of spin-orbit scattering [16, 17], the upper critical field is given by solving the equation,

$$\ln\left(\frac{T}{T_c}\right) + \psi\left(\frac{1}{2} + \frac{3(\mu_B H_{c2})^2}{4\pi k_B T\hbar\tau_{SO}^{-1}}\right) - \psi\left(\frac{1}{2}\right) = 0 \tag{1}$$

where τ_{SO} is spin-orbit scattering time, μ_B is Bohr magnetron, $\gamma = 1.78$, and $\psi(z)$ is the digamma function. In the limit of $T \to 0$, the upper critical field is given by

$$H_{c2}(0) = \left(\frac{\pi k_B T_c \hbar\tau_{SO}^{-1}}{3\gamma\mu_B^2}\right)^{1/2} = 0.602\left(\frac{\hbar\tau_{SO}^{-1}}{k_B T_c}\right)^{1/2} H_P. \tag{2}$$

With strong spin-orbit scattering the upper critical magnetic field can exceed Pauli paramagnetic limit. The strong symmetry-breaking with respect to the direction perpendicular to the surface is expected to cause a large Rashba spin splitting [18]. On the spin-split Fermi surface, momentum scattering involves the spin relaxation via D'yakonov-Perel mechanism if the Rashba splitting is strong enough. It might lead to strong spin-orbit scattering and large critical magnetic field.

The second mechanism is based on Cooper pairing on spin-split Fermi surfaces caused by Rashba effect. The band splits into two spin-dependent subbands as shown in Figure 3. The pair wave function is a mixture of both singlet and triplet states as

$$|+\vec{k},\uparrow\rangle|-\vec{k},\downarrow\rangle = \frac{1}{2}(|+\vec{k},\uparrow\rangle|-\vec{k},\downarrow\rangle - |+\vec{k},\downarrow\rangle|-\vec{k},\uparrow\rangle) + \frac{1}{2}(|+\vec{k},\uparrow\rangle|-\vec{k},\downarrow\rangle + |+\vec{k},\downarrow\rangle|-\vec{k},\uparrow\rangle) \quad (3)$$

at 0T [19] . In a magnetic field along the x axis, the centers of the normal state Fermi surfaces of two subbands are shifted in opposite y directions while the Fermi surfaces remain circular. If the superconducting state is formed on the basis of the new Fermi surfaces, the spin susceptibility is independent of the superconducting transition and the critical magnetic field can be very large.

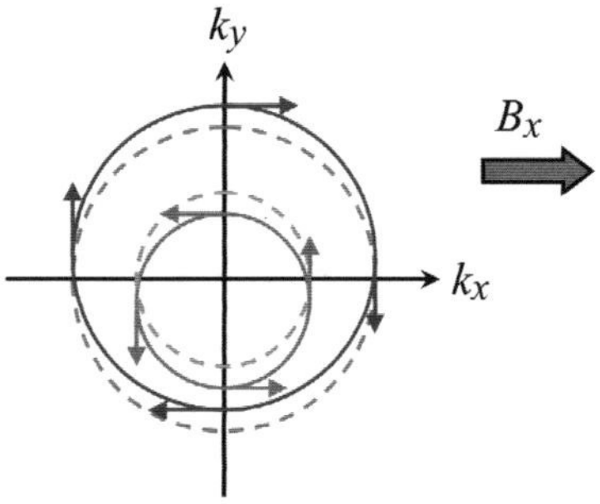

Figure 3. Schematic of 2D free electron Fermi surfaces with the Rashba term. Fermi surfaces at $B = 0$ (dotted lines) shifts in opposite y directions (solid lines) by applying magnetic field along the x axis.

In summary, we have observed 2D superconductivity below 1 ML of Pb films made by quench condensation onto GaAs cleaved surfaces. The 2D superconducting state is almost independent of parallel magnetic field as high as 14 T. We consider two different mechanism to explain it. Rashba effect might have something to do with 2D superconducting state.

Acknowledgments
This work has partly supported by Grain-in-Aid for Scientific Research(A)(No.21244047) from MEXT, Japan.

References

[1] Haviland D B, Liu Y and Goldman A M 1989 *Physical Review Letters* **62** 2180–2183
[2] Yazdani A and Kapitulnik A 1995 *Physical Review Letters* **74** 3037–3040
[3] Goldman A M and Markovic N 1998 *Physics Today* **51** 39–44
[4] Guo Y, Zhang Y F, Bao X Y, Han T Z, Tang Z, Zhang L X, Zhu W G, Wang E G, Niu Q, Qiu Z Q, Jia J F, Zhao Z X and Xue Q K 2004 *Science* **306** 1915–17
[5] Eom D, Qin S, Chou M Y and Shih C K 2006 *Physical Review Letters* **96** 027005
[6] Özer M M, Thompson J R and Weitering H H 2006 *Nature Physics* **2** 173–176
[7] Öezer M M, Jia Y, Zhang Z, Thompson J R and Weitering H H 2007 *Science* **316** 1594–1597
[8] Nishio T, An T, Nomura A, Miyachi K, Eguchi T, Sakata H, Lin S, Hayashi N, Nakai N, Machida M and Hasegawa Y 2008 *Physical Review Letters* **101** 167001
[9] Brun C, Hong I P, Patthey F, Sklyadneva I Y, Heid R, Echenique P M, Bohnen K P, Chulkov E V and Schneider W D 2009 *Physical Review Letters* **102** 207002
[10] Qin S, Kim J, Niu Q and Shih C K 2009 *Science* **324** 1314–1317
[11] Zhang T, Cheng P, Li W J, Sun Y J, Wang G, Zhu X G, He K, Wang L L, Ma X C, Chen X, Wang Y Y, Liu Y, Lin H Q, Jia J F and Xue Q K 2010 *Nature Physics* **6** 104–108
[12] Uchihashi T, Mishra P, Aono M and Nakayama T 2011 *Physical Review Letters* **107** 207001
[13] Tsuji Y, Mochizuki T and Okamoto T 2005 *Applied Physics Letters* **87** 062103
[14] Mochizuki T, Masutomi R and Okamoto T 2008 *Physical Review Letters* **101** 267204
[15] Clogston A M 1962 *Physical Review Letters* **9** 266–267
[16] Maki K 1966 *Physical Review* **148** 362–369
[17] Klemm R A, Luther A and Beasley M R 1975 *Physical Review B* **12** 877–891
[18] In metallic surface-state band, the Rashba spin splitting is reported to be of the order of 0.1 eV, e.g., Yaji K, Ohtsubo Y, Hatta S, Okuyama H, Miyamoto K, Okuda T, Kimura A, Namatame H, Taniguchi M and Aruga T 2010 *Nature Communications* **1** 1–5 and its references.
[19] Gor'kov L P and Rashba E I 2001 *Physical Review Letters* **87** 037004

Journal of Physics: Conference Series **456** (2013) 012035 doi:10.1088/1742-6596/456/1/012035

Electron g-factor in GaAs/AlGaAs quantum wells of different width and barrier Al concentrations.

A. V. Shchepetilnikov[1,2], **Yu. A. Nefyodov**[1], **I. V. Kukushkin**[1], **and W. Dietsche**[3]

[1]Institute of Solid State Physics RAS, 142432 Chernogolovka, Moscow district, Russia
[2]Moscow Institute of Physics and Technology, 141700, Dolgoprudny, Moscow district, Russia
[3]Max-Planck-Institut für Festkörperforschung, Heisenbergstr. 1, 70569 Stuttgart, Germany

E-mail: shchepetilnikov@issp.ac.ru

Abstract. Using electrically detected electron spin resonance technique, three components of electron g-factor tensor are studied in a number of asymmetrically doped GaAs/AlGaAs quantum wells grown in the [001] direction. The investigated quantum wells are characterized by different well widths and barrier Al concentrations. The dependence of g_{zz} component along the growth direction on the heterostructure band gap was shown to be universal. The band gap was measured with the aid of photoluminescence. The two-fold in-plane g-factor anisotropy with the [110] and [$1\bar{1}0$] principal axes turned out to vanish in narrow quantum wells. Three nonzero components of $\hat{a}$ tensor of the linear in the magnetic field corrections to the g-factor are measured for a number of wells of different parameters. The $\hat{a}$ tensor components are strongly dependent on the well width.

The energy of spin-splitting is usually described by the Landé g-factor. In bulk zincblende semiconductors spin splitting is isotropic and the g-factor is reduced to a scalar. The value of the g-factor is strongly dependent on the band gap energy of the crystal [1, 2]. In systems with low symmetry the electron g-factor turns out to be a tensor with several independent components. Thus, in symmetric GaAs/AlGaAs quantum wells grown along the [001] crystallographic direction the g-factor component perpendicular to the plane of the two-dimensional electron gas differs from the one parallel to this plane. Moreover, the strong two-fold in-plane anisotropy of the electron g-factor was observed in asymmetric quantum wells grown in the same direction [3, 4, 5, 6, 7].

As was shown elsewhere [3, 4, 8, 9], the g-factor in asymmetric wells is dependent on the orientation, as well as on the magnitude of the magnetic field. These dependences may be described by the tensor $\hat{a}$ of the first-order corrections in the magnetic field. In our previous work [3, 4], this tensor was shown to have at least three nonzero components, namely a_{xxz}, a_{yyz}, a_{zzz}, where the Ox, Oy and Oz axes correspond to the [110], [$1\bar{1}0$] and [001] directions respectively. All components of the $\hat{g}$ and $\hat{a}$ tensors are strongly dependent on the quantum well parameters, such as well widths and barrier Al concentration. The present paper aims to investigate the in-plane and out-of-plane g-factor anisotropies and to determine the influence of heterostructure parameters on both tensors discussed.

Studies of electron g-factor in GaAs quantum wells have been performed with the aid of several experimental methods: quantum beating spectroscopy, Kerr rotation and electrically detected electron spin resonance techniques. Let us note that optical techniques [10, 11] were mainly

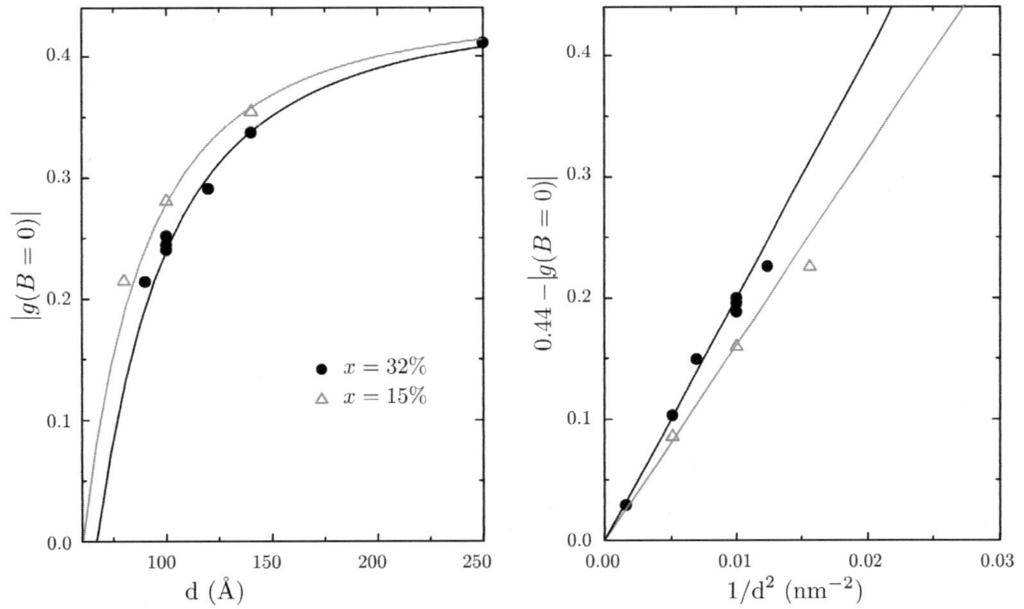

Figure 1. On the left panel the value of the electron g-factor perpendicular to the plane (g_{zz}) of the two-dimensional electron gas measured in two series of quantum wells is plotted vs the width. In each of the series the width of the quantum well was varied, whereas the barrier Al concentration was approximately the same. On the right panel the difference between the bulk GaAs g-factor and the measured value of g_{zz} is depicted vs the inverse well width squared. Solid lines represent the simple model described in the text.

used for the investigation of nominally undoped quantum wells. The g-factor tensor in such wells was theoretically calculated in [12, 13]. In a recent paper [14] the universal dependence of the out-of-plane g-factor component on the quantum well energy gap was revealed in symmetric, nominally undoped, quantum wells. On the contrary, asymmetrically doped GaAs/AlGaAs heterostructures are under investigation in the present paper. The electrically detected electron spin resonance (ESR) technique is particularly suitable for precise measurements of electron g-factor in such wells. For example, this method was successfully applied for the investigation of ESR at subunity filling factors [15] and the spin-exciton relaxation rate [16] in the two-dimensional Hall ferromagnet. In the previous papers we have demonstrated that ESR technique can be utilized for precise measurements of the g-factor and the first-order terms of the g-factor dependence on the magnetic field [3, 4].

The conventional ESR technique cannot be successfully applied to 2D electron systems [17] due to the low number of spins. However, the magnetoresistance of the 2DEG was shown [18] to be very sensitive to spin resonance when the Fermi level is located between spin-split states of a given Landau level. Having measured the fraction δR_{xx} of the magnetoresistance R_{xx}, arising due to the RF absorption, it is possible to observe spin resonance as a peak in $R_{xx}(B)$ at a fixed RF frequency. According to the Larmour theorem, since the RF wave vectors are much smaller than the reciprocal magnetic length, the excitonic effects do not significantly affect spin splitting and, thus, the experimentally measured g-factor corresponds to the single-particle limit.

Several $Al_xGa_{1-x}As/GaAs$ quantum wells of different width and barrier Al concentration were studied. All of them were grown in the [001] direction by MBE. All quantum wells were asymmetrically δ-doped with Si. The electron density varied from $1.2 \times 10^{11} cm^{-2}$ to

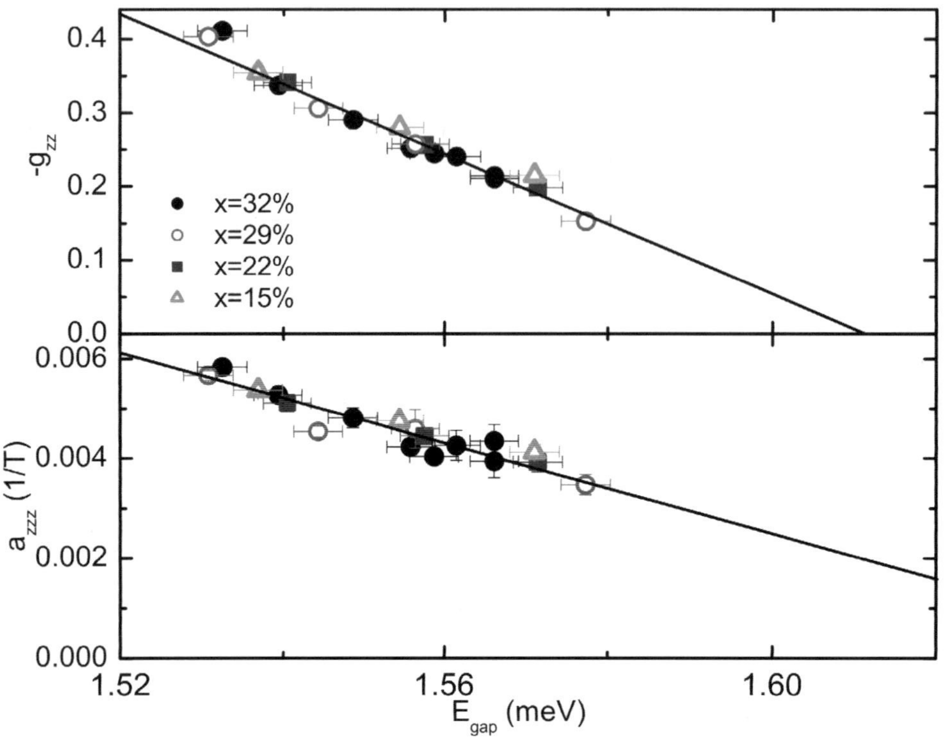

Figure 2. The absolute values of g_{zz} and a_{zzz} (the term representing the dependence of g_{zz} on the magnetic field perpendicular to the plane of two-dimensional electron gas) are plotted vs the quantum well band gap. The band gap energies were measured with the aid of photoluminescence.

$4.4 \times 10^{11} cm^{-2}$ and the mobilities ranged from 0.3×10^5 $V/(cm^2s)$ up to 10×10^5 $V/(cm^2s)$. On all of the samples usual Hall bar mesas were prepared. The barrier Al concentration and the quantum well band gap were checked with the aid of photoluminescence.

An ac probe current of 1μA at the frequency of ~1 kHz was applied from source to drain of the sample under investigation. A lock-in amplifier monitored the channel resistance R_{xx} through two sense contacts along the channel. The sample was illuminated by 100% amplitude modulated radiation at the frequency of f_{mod} ~30 Hz; RF power was delivered from the generator through a coaxial cable terminated by a loop antenna. The second lock-in amplifier, synchronized at the frequency of f_{mod}, was connected to the demodulated output of the first one and, thus, measured the change δR_{xx} in the magnetoresistance, caused by microwave irradiation.

Experiments were carried out at temperatures of 1.3 – 4.2 K in magnetic fields up to 10 T. In our experiments, we fixed the microwave frequency and swept the magnetic field. In [16], we showed that the results of ESR measurements using frequency and magnetic field sweeps do coincide, but the magnetic field sweep is much more convenient in the experiment. From the magnetic field dependence of the ESR-frequency we extracted the g-factor value and its

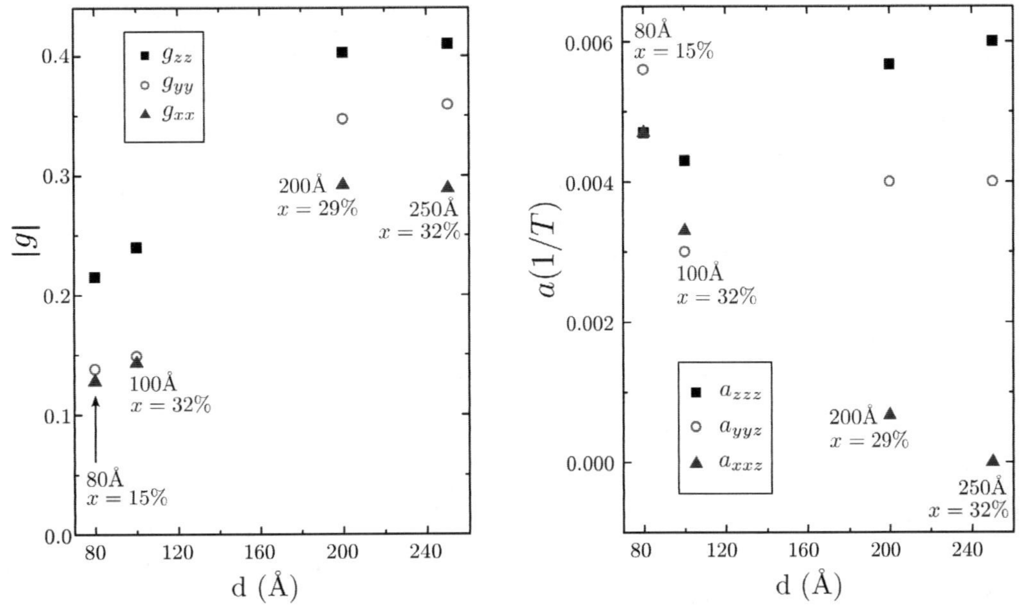

Figure 3. The magnitudes of all three diagonal components of the $\hat{g}$ tensor and three nonzero components of the $\hat{a}$ tensor measured in a number of quantum wells characterized by different well widths and barrier Al concentrations.

dependence on the magnetic field. The linear extrapolation of $g(B)$ dependence to zero magnetic field gives us the $g(B = 0)$, that is the g-factor for the subband bottom. The justification of this procedure can be found in our previous papers, e.g. [3]. The results of the measurements of the out-of-plane component g_{zz} in two series of samples characterized by various well widths are presented in Fig. 1. The barrier Al concentration was approximately the same in each of the series of samples, namely $x = 15\%$ (open triangles) and $x = 32\%$ (solid circles). The accuracy of the $g(B = 0)$ absolute value was typically 0.002, whereas the accuracy in the well-width corresponds to one monolayer. Thus, the errors in Fig. 1 are smaller than the symbols size. Three circles at the well width $d = 100$Å in Fig. 1 correspond to three samples independently grown and having small variation of the Al-content (about 1%) and the well width (one monolayer) one from another, as shown with the aid of photoluminescence technique.

The values of g_{zz} turned out to depend strongly on well widths. This can be easily explained by the dependence of quantum well band gap energy on the well width d. This phenomenon can be taken into account by utilizing a simple model: the electron g-factor depends on the band gap energy of the material [1]; the band gap energy of the heterostructure in turn is equal to the sum of the GaAs bulk band gap energy and a term proportional to $1/d^2$ (in the approximation of a rectangular well with infinite walls); if this term is small enough, the g-factor dependence on the band gap energy of the heterostructure is close to linear. The difference between the bulk $g = -0.44$ and the measured g-factor is depicted versus $1/d^2$ in the right panel of Fig. 1. Solid lines in Fig. 1 represent this model. As can be seen, the model is valid for wide wells but fails in narrow structures. In fact, the band gap energy depends on the heterostructure parameters in a more complicated manner than assumed in this simplified model. Moreover, the g_{zz} dependences on the quantum well width do not coincide in these two series of samples. This may be ascribed to the difference of the band gap energies in two wells of the same width, if the Al concentration is varied.

All the facts mentioned above suggest to measure directly the quantum well band gap and to plot the g-factor against this energy value. The results are presented in Fig. 2. As can be seen, the dependence of g_{zz} on the band gap energy turned out to be universal. What is more surprising, the dependence of a_{zzz} (the term describing the correction to g_{zz} linear in B_z) is also universal. The g_{zz} value reaches zero at some value of energy, whereas a_{zzz} remains finite. This result suggests that in some quantum wells with nonzero g_{zz} at zero magnetic field g_{zz} can reach zero and even change its sign when a moderate magnetic field is applied. The magnitude of such a field diminishes with the increase of quantum well band gap energy. This effect may be utilized as a technique of spin manipulation. In quantum wells the barrier region contributes to the total g-factor. The amount of the wave function penetrating into the barrier and the g-factor in the barrier depends on the potential barrier height, and, thus, the total g-factor value does depend on the Al concentration in the barrier region. Moreover, if the electron density is varied, the well potential profile and the amount of wave function penetrating into the barrier changes. The difference in such well parameters might be responsible for the slight deviations of the g_{zz} and of the a_{zzz} values from universal behavior.

In Fig. 3 the results of $\hat{g}$ and $\hat{a}$ measurements in a number of samples are presented. Parameters of the structures are shown on the plot. As can be seen, the in-plane g-factor anisotropy is strong in wide wells and practically vanishes in narrow ones, whereas the difference between the in-plane and out-of-plane components of the g-factor remains prominent in all of the samples measured. Similarly, the difference between the components of $\hat{a}$ tensor decreases in narrower wells. As can be seen from Fig. 3, the a_{zzz}, g_{zz} and g_{xx} components diminish with well narrowing, whereas a_{xxz} does not. Thus, the magnetic field at which $g_{xx} = 0$ is lower than the field required to make $g_{zz} = 0$. In narrower wells the electron wave function along the growth direction tends to be more symmetric than in wider ones. This might be responsible for the observed reduction of the in-plane anisotropy of the g-factor, since this anisotropy vanishes in symmetric wells [5].

In conclusion, three components of the electron g-factor tensor were measured in a number of asymmetrically doped GaAs/AlGaAs quantum wells grown in the [001] direction. The quantum wells were characterized by different well widths and barrier Al concentrations. The dependences of g_{zz} and a_{zzz} components along the growth direction on the heterostructure band gap were shown to be universal. The two-fold in-plane g-factor anisotropy was shown to vanish in narrow quantum wells. Three nonzero components of $\hat{a}$ tensor of the linear in the magnetic field corrections to the g-factor are measured in several samples. The $\hat{a}$ tensor components are strongly dependent on the well width. The magnetic field required for g_{zz} to reach zero and to change its sign is lower in quantum wells with higher band gap energy. In narrow wells g_{xx} goes to zero at smaller magnetic field than g_{zz}. These findings may facilitate electron spin manipulation with the aid of the magnetic field.

This work was partially supported by the Russian Foundation for Basic Research (RFBR).

[1] Roth L M, Lax B and Zwerdling S 1959 *Phys. Rev.* **114** 90
[2] Weisbuch C and Herrman C 1977 *Phys. Rev.* B **15** 816
[3] Nefyodov Yu A, Shchepetilnikov A V, Kukushkin I V, Dietsche W and Schmult S 2011 *Phys. Rev.* B **83** 041307(R)
[4] Nefyodov Yu A, Shchepetilnikov A V, Kukushkin I V, Dietsche W and Schmult S 2011 *Phys Rev.* B **84** 233302
[5] Kalevich V K and Korenev V L 1993 *Pis'ma v ZhETF* **57** 557 [*JETP Lett.* **57** 571]
[6] Hallstein S, Oestreich M, Ruhle W W and Kohler K 1996 *Proceedings of 12th International Conference High Magnetic Fields in the Physics of Semiconductors* p. 593
[7] Eldridge P S, Hübner J, Oertel S, Harley R T, Henini M, and Oestreich M 2011 *Phys. Rev.* B **83** 041301(R)
[8] Dobers M, Klitzing K v and Weimann G 1998 *Phys. Rev.* B **38** 5453
[9] Lommer G, Malcher F and Rössler U 1985 *Phys. Rev.* B **32** 6965
[10] Malinowski A and Harley R T 2000 *Phys. Rev.* B **62** 2051
[11] Heberle A P, Ruhle W W and Ploog K 1994 *Phys. Rev. Lett.* **72** 3887
[12] Ivchenko E L and Kiselev A A 1992 *Sov. Phys. Semicond.* **26** 827
[13] Pfeffer P and Zawadzki W 2006*Phys. Rev.* B **74** 233303

[14] Yugova I A, Greilich A, Yakovlev D R, Kiselev A A, Bayer M, Petrov V V, Dolgikh Yu K, Reuter D and Wieck A D 2007 *Phys.Rev.* B **75** 245302
[15] Meisels R, Kulac I, Kuchar F and Kriechbaum M 2000 *Phys.Rev.* B **61** 5637
[16] Nefyodov Yu A, Fortunatov A A, Shchepetilnikov A V and Kukushkin I V 2010 *Pis'ma v ZhETF* **91** 385 [*JETP Lett.* **91** 357]
[17] Nestle N, Denninger G, Vidal M, Weinzierl C, Brunner K, Eberl K and Klitzing K v 1997 *Phys. Rev.* B **56** R4359
[18] Stein D, Klitzing K v and Weimann G 1983 *Phys. Rev. Lett.* **51** 130

Journal of Physics: Conference Series **456** (2013) 012036 doi:10.1088/1742-6596/456/1/012036

Fractional quantum Hall states in charge-imbalanced bilayer systems

N. Thiebaut[1], N. Regnault[2,3] and M.O. Goerbig[1]

[1] Laboratoire de Physique des Solides, CNRS UMR 8502, Univ. Paris-Sud, F-91405 Orsay cedex, France

[2] Laboratoire Pierre Aigrain, ENS and CNRS, 24 rue Lhomond, 75005 Paris, France

[3] Department of Physics, Princeton University, Princeton, NJ 08544

E-mail: thiebaut@lps.u-psud.fr

Abstract.
We study the fractional quantum Hall effect in a bilayer with charge-distribution imbalance induced, for instance, by a bias gate voltage. The bilayer can either be intrinsic or it can be formed spontaneously in wide quantum wells, due to the Coulomb repulsion between electrons. We focus on fractional quantum Hall effect in asymmetric bilayer systems at filling factor $\nu = 4/11$ and show that an asymmetric Halperin-like trial wavefunction gives a valid description of the ground state of the system.

1. Introduction

In a two-dimensional electron gas subjected to a high magnetic field repulsively interacting electrons may collectively form a gapped state insensitive to small perturbations. Such states are called fractional quantum Hall states (FQHS) and show up for partial (fractional) fillings of the highest partially occupied Landau level, i.e. for rational values of the filling factor ν, which is the ratio of the number of electrons N and the number of magnetic flux quanta N_ϕ threading the system. Since the discovery of the fractional quantum Hall effect (FQHE) in 1982 [1] new experimental setups nourished our understanding of this multifaceted phenomenon.

The bilayer system, which consists of two parallel 2D electron systems in a semiconductor heterostructure exhibits FQHE at even-denominator filling factors (such as $\nu = 1/2$ [2]) is one of them. Though it is more common to consider symmetric layers, for which charges are equally spread between the two layers ($\nu_{\text{tot}} = 2/n$), we are interested in a bilayer system with charge imbalance, which is obtained by the application of an external gate voltage (fig. 1). The strength of correlations between electrons that belong to different layers depends on their distance d and the energy difference between the two layers Δ is proportional to the external gate voltage V_G.

Beside its own richness, the FQHE in a bilayer system provides a model for its counterpart in a wide quantum well [3, 4, 5], where even-denominator states have also been observed [3, 4, 6]. In the latter, density functional theory calculations indicate that the Coulomb repulsion favors an electron concentration at the borders of the quantum well, which may thus be viewed as a spontaneously formed bilayer system [7]. Charge-imbalanced bilayers can thus depict wide quantum wells in which electrons split in two layers and populate one of them preferentially, either spontaneously or under the influence of external gate electrodes.

Journal of Physics: Conference Series **456** (2013) 012036 doi:10.1088/1742-6596/456/1/012036

Here, we investigate the conditions for the existence of FQHS (at $\nu = 4/11$) in an imbalanced bilayer system by means of exact diagonalization. We introduce asymmetric Halperin states and discuss their relevance for the description of the ground state of the system.

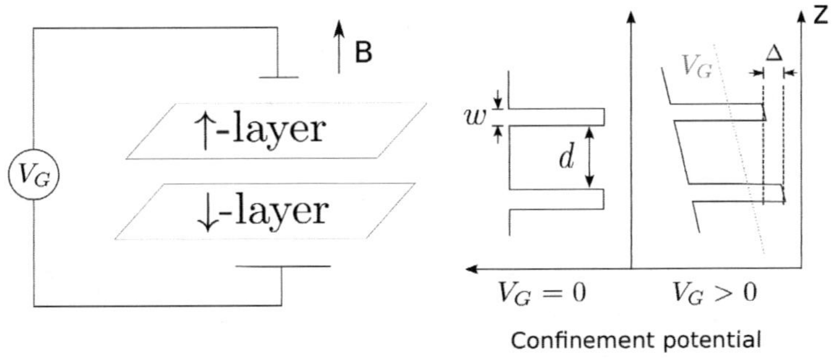

Figure 1: Schematic view of the effect of a gate voltage on a bilayer system. *(Left)* Setup. *(Right)* Effect of the gate voltage on the confinement potential. We consider the limit where the width w of each quantum well is small as compared to the layer separation d, $w \ll d$.

2. Model

We restrict the electron dynamics to the lowest Landau level, which amounts to considering the limit in which adjacent Landau levels are separated by a large cyclotron gap, as compared to the characteristic Coulomb energy scale, $e^2/4\pi l_B$. FQHS are often fully polarized due to the Coulomb repulsion between electrons in a flat Landau energy level, and we consider such states, that is we neglect the electron spin from now on. The effective Hamiltonian of the bilayer system reads

$$\hat{H} = \frac{1}{2}\sum_{\{\sigma_i\}}\sum_{\{m_i\}} V^{\{\sigma_i\}}_{\{m_i\}} c^\dagger_{m_1,\sigma_1} c^\dagger_{m_2,\sigma_2} c_{m_4,\sigma_4} c_{m_3,\sigma_3} + \Delta \sum_m (c^\dagger_{m,\downarrow} c_{m,\downarrow} - c^\dagger_{m,\uparrow} c_{m,\uparrow}) \quad , \tag{1}$$

where $\sigma = \uparrow, \downarrow$ is the layer index and m_i the angular momentum which labels the single-electron states in the symmetric gauge[1] [8]; Δ is the energy difference between the two layers, controlled by the gate voltage V_G in fig. 1. The first term on the right hand side of eq. 1 is the usual interaction term, while the second one is a Zeeman-like term proportional to the density imbalance. In real space the interaction potential depends on the layer indices

$$V^{\{\sigma_i\}}(r) = \begin{cases} \frac{e^2}{4\pi r} & \text{if electrons belong to the same layer (i.e. if } \sigma_1 = \sigma_2 = \sigma_3 = \sigma_4) \\ \frac{e^2}{4\pi\sqrt{r^2+d^2}} & \text{if electrons belong to different layers (i.e. if } \sigma_1 = \sigma_3 \neq \sigma_2 = \sigma_4) \\ 0 & \text{elsewise} \end{cases} \tag{2}$$

[1] Here we assume that the confinement potential is sufficiently narrow (compared to the layer separation, $w \ll d$) to avoid electrons from populating the first excited state of the confinement potential. This permits to neglect the wavefunction extent in the confinement direction, which assumption simplifies the single particle Hilbert space as the layer index does not couple to the orbital motion.

3. Halperin's wavefunctions

A large variety of FQHS in bilayer systems may be described with the help of a generalization of Laughlin's wavefunctions [9] to the case of electrons with internal degrees of freedom, initially introduced by Halperin [10]. Those wavefunctions have three integer parameters that characterise the interaction strength between electrons in the $\uparrow$ and $\downarrow$ layers ($m_\uparrow$ and $m_\downarrow$), and electrons in different layers (n). In terms of the complex positions $z_i^\sigma = x_i^\sigma - iy_i^\sigma$, for a system with N_σ electrons in the σ-layer, Halperin's wavefunctions read

$$\Psi_{(m_\uparrow,m_\downarrow,n)}(\{z_i^\uparrow\},\{z_i^\downarrow\}) = \prod_{i<j\le N_\uparrow} (z_i^\uparrow - z_j^\uparrow)^{m_\uparrow} \prod_{i<j\le N_\downarrow} (z_i^\downarrow - z_j^\downarrow)^{m_\downarrow} \prod_{i\le N_\uparrow, j\le N_\downarrow} (z_i^\uparrow - z_j^\downarrow)^{n} \quad , \tag{3}$$

where the normalization and usual Gaussian factors have been absorbed in a redefinition of the Hilbert space measure $d\mu(\{z_i\}) = \prod_i dx_i dy_i \exp(-|z_i|^2/2l_B^2)/(2\pi l_B^2)$, following Girvin and Jach [11]. Alternatively, we denote the state represented by the Halperin wavefunction (3) the $(m_\uparrow, m_\downarrow, n)$ state. Notice that not all Halperin state (3) are capable of describing physically possible FQHS – it has been shown in Ref. [12] within a generalization of Laughlin's plasma analogy [9] that the exponents must satisfy the condition

$$m_\uparrow m_\downarrow \ge n^2 \tag{4}$$

in order to avoid phase separation of the electron species.

To compute the filling factor associated with a particular Halperin state, one uses the fact that the maximum value of individual orbital momenta is given by the number of flux quanta N_ϕ threading the system. The polynomial expansion of eq. (3) yields a maximal exponent $(N_\sigma - 1)\ m_\sigma + N_{-\sigma}\ n$ for the position of the i-th electron in the σ-layer that must equal N_ϕ. This can be written in matrix form as

$$\begin{pmatrix} N_\phi + m_\uparrow \\ N_\phi + m_\downarrow \end{pmatrix} = \begin{pmatrix} m_\uparrow & n \\ n & m_{-\downarrow} \end{pmatrix} \begin{pmatrix} N_\uparrow \\ N_\downarrow \end{pmatrix} \quad ,$$

and after inverting the matrix, one deduces the relationship between the total number of electrons N and N_ϕ,

$$N = N_\uparrow + N_\downarrow = \frac{1}{m_\uparrow m_\downarrow - n^2}\Big[(m_\uparrow + m_\downarrow - 2n)N_\phi + 2m_\uparrow m_\downarrow - n(m_\uparrow + m_\downarrow)\Big] \quad . \tag{5}$$

The filling factor ν is defined in the thermodynamic limit, for which we obtain

$$\nu = \lim_{N,N_\phi\to\infty} \frac{N}{N_\phi} = \frac{m_\uparrow + m_\downarrow - 2n}{m_\uparrow m_\downarrow - n^2} \quad . \tag{6}$$

We define the polarization as the population difference between the layers

$$P_z = \frac{S_z}{N/2} \tag{7}$$

in terms of the z-component of the total layer pseudospin

$$S_z = \frac{N_\uparrow - N_\downarrow}{2} = \frac{(m_\downarrow - m_\uparrow)(N_\phi - n)}{2(m_\uparrow m_\downarrow - n^2)} \quad , \tag{8}$$

which is zero for symmetric states (i.e. if $m_\uparrow = m_\downarrow$).

Now we focus on asymmetric Halperin states. More precisely we are interested in the $(5,3,2)$ state, which satisfies the condition (4) since it is the simplest (i.e. lowest correlation factors) asymmetric Halperin state for which the filling factor $\nu_{532} = 4/11$ does not belong to the usual composite-fermions sequence $\nu_{\rm CF} = p/(2sp \pm 1)$ [13]. It may thus describe states the existence of which relies on layer asymmetry. Nevertheless it should be mentioned that $\nu = 4/11$ is also the filling factor of a second-generation composite fermion state in monolayer systems [14, 15, 16, 17] generated by interacting composite fermions that has been proposed as a possible explanation for an experimentally observed FQHS at this filling [14].

4. Exact diagonalization

In order to check the physical relevance of the $(5,3,2)$ state we calculate, with the help of exact diagonalization [18], two quantities for the Coulomb interaction (2) in the lowest Landau level: the total angular momentum of the ground state and its polarization as a function of the distance d and of the gate-induced energy difference Δ between the two layers. The calculations are perfomed on the sphere geometry [19]. For the $(5,3,2)$ state, the relation between N_Φ and N is given by $N_\Phi = \frac{11}{4}N - \frac{7}{2}$. Finite-size systems induce a constraint on the number of electrons [see eq. (5)]. Only sizes of the form $N = 2 + 4n$, $n \in \mathbb{N}$ can satisfy this constraint.

Here we will focus on $N = 6$ and $N = 10$ which are the only accessible sizes. FQHS correspond to values of d and Δ for which the gap is finite and the total angular momentum equals zero. Finally we compute the overlap between the $(5,3,2)$ state and the exact ground state for values of d and Δ which correspond to a FQHS.

The polarization of the ground state is shown in fig. 2. For sufficiently high values of Δ the system is fully polarized, i.e all electrons reside in a single layer and one thus obtains the monolayer system with $\nu = 4/11$, studied in Refs. [15, 16, 17]. Conversely for large distances d the polarization tends to be zero. In this situation the potential $V(r) = (r^2 + d^2)^{-1/2}$ between two electrons in different layers is small and the system behaves as two uncorrelated layers with individual filling factors $\nu = 2/11$. It is likely that for such a low filling factor the system tends to form a Wigner crystal [20, 21, 22], and FQHS may therefore be ruled out.

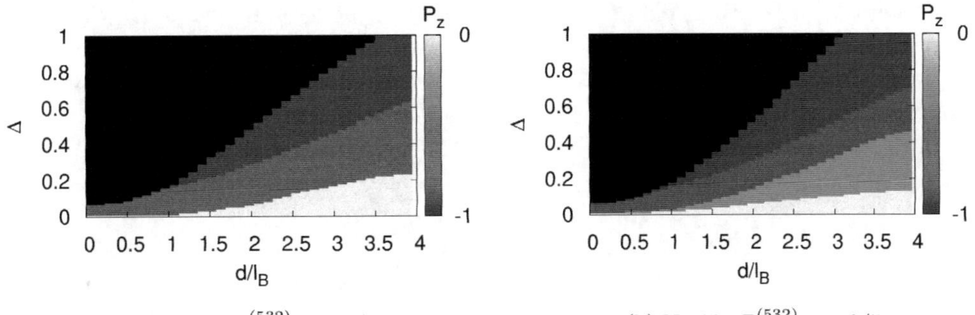

(a) N=6: $P_z^{(532)} = -1/3$ (b) N=10: $P_z^{(532)} = -2/5$

Figure 2: Polarization of the bilayer system as a function of the distance d/l_B and energy shift Δ (energies are given in units of $e^2/4\pi l_B$).

The polarization of the $(5,3,2)$ state can be computed directly from eq. (7) and eq. (8)

$$P_z^{(532)} = -\frac{1}{2} + \frac{1}{N} \tag{9}$$

Journal of Physics: Conference Series **456** (2013) 012036 doi:10.1088/1742-6596/456/1/012036

and may be compared with that obtained from exact diagonalization for $N = 6$ and 10 particles, as a function of the layer separation d/l_B and $\Delta/(e^2/4\pi l_B)$ (see fig. 2). Indeed, one finds that in intermediate regimes the polarization of the ground state is the one of the $(5,3,2)$ state (orange regions).

We now turn to the characterization of the exact ground state. Figure 3 shows the total angular momentum L of the ground state. This latest is rotationally invariant when $L = 0$, as required for an incompressible state. It is zero in the region where the ground-state polarization matches that of the $(5,3,2)$ state, except for large layer separations. As already mentioned in the discussion above, for large distances inter-layer correlations are too weak to allow the system to be in an incompressible state, and an inhomogeneous density state is favored, such as one would for example expect for a Wigner crystal ($d \gtrsim 3.8 l_B$).

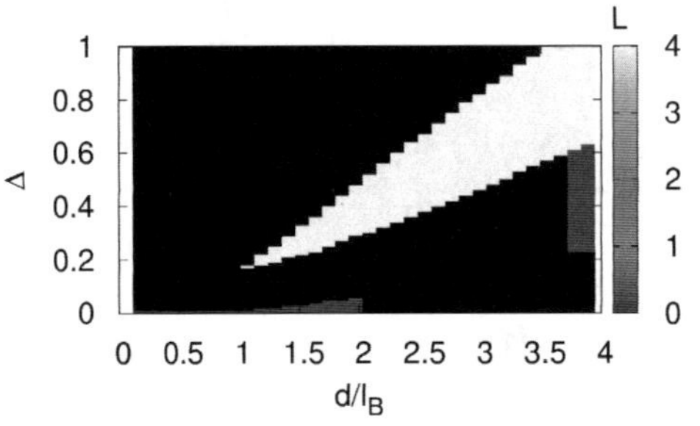

Figure 3: Total orbital momentum of the bilayer system for $N = 6$ electrons as a function of the distance d/l_B and energy shift Δ, in units of $e^2/4\pi l_B$.

The results presented in figs. 2 and 3 indicate that in a certain range of parameters d/l_B and $\Delta/(e^2/4\pi l_B)$, exact diagonalization provides a ground state that matches some physical properties of the $(5,3,2)$ state. To corroborate the relevance of this state, we have calculated the overlap between the exact ground state obtained for the interaction (2) and the $(5,3,2)$ state. In order to obtain the $(5,3,2)$ state numerically, we have used exact diagonalization for a model interaction given in terms of the appropriate pseudopotentials [19]. The results are presented in fig. 4 as a function of d/l_B. The overlap turns out to be high ($> 85\%$) for $d \lesssim 2l_B$, such that in regions of matching polarizations and for sufficiently short interlayer distances the $(5,3,2)$ state gives a good description of the ground state.

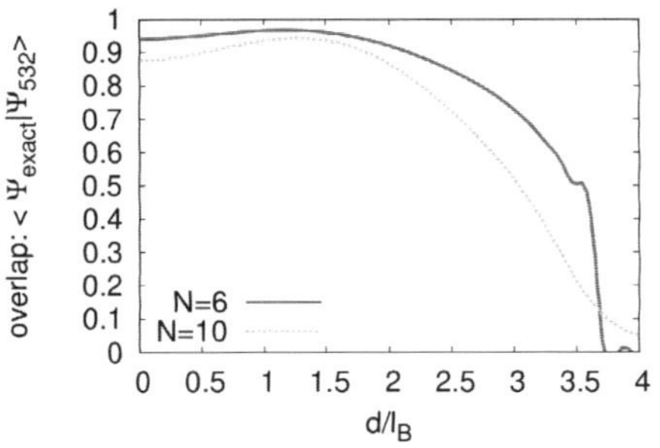

Figure 4: Overlap of the $(5,3,2)$ state with the exact ground state as a function of d/l_B, in the corresponding polarization sector (9).

Journal of Physics: Conference Series **456** (2013) 012036 doi:10.1088/1742-6596/456/1/012036

5. Conclusion

We have performed an exact-diagonalization study of the bilayer system for various distances between the layers and energy shifts, i.e. the energy difference between electrons in the two layers. The ground state is incompressible over a certain range of parameters, which corresponds to an interlayer distance of roughly three times the magnetic length. Asymmetry between layers does not necessarily destroy the FQHS, and the FQHE at exotic fractions may thus be observed in an asymmetric bilayer system. Asymmetric Halperin's wavefunctions are good candidates for the accurate description of those states. Also, using the bilayer modeling of a wide quantum well, a similar behavior could be expected in asymmetric wide quantum wells, the asymmetry being induced, for instance, by a back gate voltage that enables one to tune the electronic density of the system. Experimentalists reported a FQHE at $\nu = 4/11$ in a wide quantum well ($w \sim 3\ l_B$) in 2003 [14]. Since this filling factor corresponds to the one of the $(5, 3, 2)$ state, one may wonder about its relevance for the explanation of this experimental observation. Within this picture this FQHS would be induced by the shape of the confinement potential, in contrast to the picture of second-generation composite fermions that have been proposed as an alternative explanation for the 4/11 state [14, 15, 16, 17].

[1] Tsui D C, Stormer H L and Gossard A C 1982 *Phys. Rev. Lett.* **48**(22) 1559–1562
[2] Eisenstein J P, Boebinger G S, Pfeiffer L N, West K W and He S 1992 *Phys. Rev. Lett.* **68**(9) 1383–1386
[3] Suen Y W, Santos M B and Shayegan M 1992 *Phys. Rev. Lett.* **69**(24) 3551–3554
[4] Luhman D R, Pan W, Tsui D C, Pfeiffer L N, Baldwin K W and West K W 2008 *Phys. Rev. Lett.* **101**(26) 266804
[5] Papić Z, Möller G, Milovanović M V, Regnault N and Goerbig M O 2009 *Phys. Rev. B* **79**(24) 245325
[6] Shabani J, Gokmen T, Chiu Y T and Shayegan M 2009 *Phys. Rev. Lett.* **103**(25) 256802
[7] Nuebler J, Friess B, Umansky V, Rosenow B, Heiblum M, von Klitzing K and Smet J 2012 *Phys. Rev. Lett.* **108**(4) 046804
[8] Girvin S M 1984 *Phys. Rev. B* **29**(10) 6012–6014
[9] Laughlin R B 1983 *Phys. Rev. Lett.* **50**(18) 1395–1398
[10] Halperin B I 1983 *Helv. Phys. Acta* **56** 75
[11] Girvin S M and Jach T 1984 *Phys. Rev. B* **29**(10) 5617–5625
[12] de Gail R, Regnault N and Goerbig M O 2008 *Phys. Rev. B* **77**(16) 165310
[13] Jain J K 1989 *Phys. Rev. Lett.* **63**(2) 199–202
[14] Pan W, Stormer H L, Tsui D C, Pfeiffer L N, Baldwin K W and West K W 2003 *Phys. Rev. Lett.* **90**(1) 016801
[15] Goerbig M O, Lederer P and Smith C M 2004 *Phys. Rev. B* **69**(15) 155324
[16] López A and Fradkin E 2004 *Phys. Rev. B* **69**(15) 155322
[17] Chang C C and Jain J K 2004 *Phys. Rev. Lett.* **92**(19) 196806
[18] Diagham: http://nick-ux.lpa.ens.fr/diagham/wiki/
[19] Haldane F D M 1983 *Phys. Rev. Lett.* **51**(7) 605–608
[20] Lam P K and Girvin S M 1984 *Phys. Rev. B* **30**(1) 473–475
[21] Jiang H W, Willett R L, Stormer H L, Tsui D C, Pfeiffer L N and West K W 1990 *Phys. Rev. Lett.* **65**(5) 633–636
[22] Jiang H W, Stormer H L, Tsui D C, Pfeiffer L N and West K W 1991 *Phys. Rev. B* **44**(15) 8107–8114

Journal of Physics: Conference Series **456** (2013) 012037 doi:10.1088/1742-6596/456/1/012037

Anisotropy of dipolar-broadened nuclear spin resonance

Oleksandr Tsyplyatyev

Department of Physics and Astronomy, University of Sheffield, Sheffield S3 7RH, United Kingdom.

E-mail: o.tsyplyatyev@sheffield.ac.uk

Yevgeniya Avotina

Abstract. We study anisotropy of the homogeneously broadened nuclear spin resonance in a crystal. Using a combinatorial method to restrict the nuclear states to a fixed polarisation, we obtain the first four cumulants of the full line-shape as a function of polarisation of the nuclei and angles between the external magnetic field and the crystalline axes. We demonstrate that the line width is relatively isotropic but the other cumulants such as line shift and higher then the second cumulants are strongly anisotropic. Also the angular dependencies of the first three cumulants are found to be independent of the polarisation but the fourth cumulant changes its angular dependence with a change of the nuclear polarisation. This result shows that the orientation of the external magnetic field is an essential parameter in a study of polarised nuclei.

1. Introduction

Recent interest in studying nuclear spins was fuelled [1, 2, 3, 4, 5, 6, 7, 8] by the dephasing effect in GaAs quantum dots which is induced by fluctuations of the nuclear (Overhauser) field and blocks prospective application of electron spins for spintronics [9] and quantum computing [10]. Among different approaches to this problem, there is a new tool that combines radio frequency excitation and dynamical nuclear pumping techniques to control the Overhauser field and simultaneously to detect its state [11]. In particular, this method allows to access directly some of the local properties of the nuclear bath.

At infinite temperature, a connection between the line width of the nuclear magnetic resonance and the dipole interaction between nuclei in a crystal was established in [12]. Later, it was found that the line shape changes at low spin temperatures [13]. A detailed understanding of the effect of a non-equilibrium polarisation, as occurs in dynamical pumping experiments, may give a way to measure directly the degree of polarisation reached under dynamical pumping and provide further insights in the effect of dipole interaction which plays an important role in the intrinsic dynamics of the nuclear bath in unstrained structures.

In present paper we address this problem by analysing the long wavelength spectrum of polarised nuclei using a cumulant expansion of the full line shape. We employ a combinatorial method to restrict the nuclear states to a fixed polarisation and obtain the first four moments of the full line shape as a function of the nuclear polarisation and angles between the external magnetic field and axes of a crystal. We find that the polarisation and the angular dependencies factorise for the first three moments but are mixed for the fourth moment. For a Zince-Blende lattice of GaAs we find that the angular part of the second cumulant is almost a constant but the other cumulants change by an order of magnitude as the angles change. This result shows that the orientation of the magnetic field is an essential parameter when one wants to study the dipolar interaction between nuclei at a finite polarisation.

The paper is organised as follows. In Section II, we approximate the model of dipole interaction between nuclear spins for a large external magnetic field. In Section III, we analyse the shape the absorption line to obtain its polarisation and angular dependencies. In Section IV, the angular dependencies of the cumulants are studied for a crystall with the Zinc-Blende symmetries.

2. Dipole interaction

The Hamiltonian that describes N nuclear spins located at the sites of a regular lattice, their dipole interaction with the strength g, and an external magnetic field B is given by $H = \mu B \sum_j I_j^z + g\sum_{i<j}\left(3\left(\mathbf{e}_{ij}\cdot\mathbf{I}_i\right)\left(\mathbf{e}_{ij}\cdot\mathbf{I}_j\right) - \mathbf{I}_i\cdot\mathbf{I}_j\right)/r_{ij}^3$, where μ is the nuclear magneton, $I_j^z, I_j^{\pm} = I_j^x \pm iI_j^y$ are the spin-1/2 operators, index j labels the sites on the lattice, r_{ij} is the distance between two sites, and $\mathbf{e}_{ij} = \mathbf{r}_{ij}/r_{ij}$ is the unit vector that connects sites i and j. We consider a d-dimensional crystal with $N = L^d$ nuclear spins and assume periodic boundary conditions, $\mathbf{I}_{j+L} = \mathbf{I}_j$, to exclude edge states from the analysis.

In a strong magnetic field the Hilbert space is partitioned by a large Zeeman energy into a set of subspaces labelled by the total z-projection of all spins, S_z is an eigenvalue of $\sum_j I_j^z$. In each subspace the many spin states that have the same Zeeman energy are split by a part of the dipole interaction that conserves S_z,

$$H_0 = \mu B \sum_j I_j^z + \sum_{i<j} \frac{g\left(3\gamma_{ij}^2 - 1\right)}{r_{ij}^3}\left(I_j^z I_i^z - \frac{I_j^+ I_i^-}{2}\right). \tag{1}$$

The matrix elements of the S_z non-conserving remainder, $H_1 = \sum_{ij} 3g\Big[\left(\alpha_{ij} + i\beta_{ij}\right)^2 I_j^+ I_i^+ + 4\left(\alpha_{ij} + i\beta_{ij}\right)\gamma_{ij} I_i^+ I_j^z + \left(\alpha_{ij} - i\beta_{ij}\right)^2 I_j^- I_i^- + 4\left(\alpha_{ij} - i\beta_{ij}\right)\gamma_{ij} I_i^- I_j^z\Big] / \left(8r_{ij}^3\right)$, connect subspaces with different Zeeman energies and can be treated perturbatively when the magnetic field is large. Here $(\alpha_{ij}, \beta_{ij}, \gamma_{ij})$ are coordinates of the vector $\mathbf{e}_{ij}$ in a reference frame with z-axis parallel to the external B-field.

3. Absorption line shape and cumulant expansion

The absorption of electromagnetic radiation in a strong magnetic field is dominated by a flip of a single nuclear spin at an energy of μB, neglecting the dipole interaction. The dipole interaction of a single spin with other nuclear spins on the lattice makes the width of the resonance finite. Following [12] we study moments of this line shape given by the transition energies $E_f - E_{f'}$ between pairs of the eigenstates $|f\rangle$ and $|f'\rangle$,

$$\left\langle m^k \right\rangle = \frac{\sum_{ff'}\left(E_{f'} - E_f\right)^k \left|\langle f'|I^+|f\rangle\right|^2}{\sum_{ff'}\left|\langle f'|I^+|f\rangle\right|^2}, \tag{2}$$

where $I^+ = \sum_j I_j^+$ is the operator that corresponds to a long wave-length photon and k is the order of the moment. To account for a finite polarisation, the initial states f are restricted to a subspace of fixed S_z. In our study of the dipolar broadening we will use only the truncated part $E_f|f\rangle = H_0|f\rangle$ of the dipole Hamiltonian H. A perturbative treatment of H_1 produces extra peaks at $2\mu B$ and $3\mu B$ with small amplitudes $\sim g/B$, which are neglected.

Table 1. The first four cumulants of the line shape of the dipolar-broadened nuclear resonance $\langle \nu^k \rangle$ and corresponding moments of the crystalline field F_i.

$\langle \nu \rangle = \mu B + \frac{3}{2} F p$	$\langle \nu^3 \rangle = \frac{9}{8} \left(F_{3c} - F_2 F - 5 F_3 \right) p \left(1 - p^2 \right)$
$\langle \nu^2 \rangle = \frac{9}{4} F_2 \left(1 - p^2 \right)$	$\langle \nu^4 \rangle = \frac{9}{16} \left(F_2^2 \left(-6 + 14p^2 - 8p^4 \right) + \left(F_{4q} - 2F_4 \right) \left(6 - 25p^2 + 19p^4 \right) \right)$
	$+ \frac{9}{16} \left(F^2 F_2 + 10 F F_3 - 2 F F_{3c} + F_{4c} \right) p^2 \left(1 - p^2 \right)$
$F = \frac{g \sum_j \left(3\gamma_{ij}^2 - 1 \right)}{r_{ij}^3}$	$F_2 = \frac{g^2 \sum_j \left(3\gamma_{ij}^2 - 1 \right)^2}{r_{ij}^6}$
$F_3 = \frac{g^3 \sum_j \left(3\gamma_{ij}^2 - 1 \right)^3}{r_{ij}^9}$	$F_{3c} = \frac{g^3 \sum_{ij} \left(3\gamma_{lj}^2 - 1 \right) \left(3\gamma_{ij}^2 - 1 \right) \left(3\gamma_{li}^2 - 1 \right)}{r_{lj}^3 r_{ij}^3 r_{il}^3}$
$F_4 = \frac{g^4 \sum_j \left(3z_{ij}^2 - 1 \right)^4}{r_{ij}^{12}}$	$F_{4c} = \frac{g^4 \sum_{ijl} \left(3z_{ij}^2 - 1 \right) \left(3z_{jl}^2 - 1 \right) \left(3z_{lk}^2 - 1 \right) \left(3z_{ki}^2 - 1 \right)}{r_{ij}^3 r_{jl}^3 r_{lk}^3 r_{ki}^3}$
	$F_{4q} = \frac{g^4 \sum_{ij} \left(3z_{ij}^2 - 1 \right)^2 \left(3z_{jl}^2 - 1 \right) \left(3z_{li}^2 - 1 \right)}{r_{ij}^6 r_{jl}^3 r_{li}^3}$

The sums in Eq. (2) can be transformed to traces by means of general quantum mechanical identities. In the denominator, the sum over the eigenstates $|f'\rangle$ gives a unit matrix, $\sum_{f'} |f'\rangle\langle f'| = \hat{1}$, in the subspace of $S^z + 1$ as the operator I^+ changes S_z by 1 only. Then, the remaining sum over a complete set of the eigenstates is $\sum_f \langle f|I^- I^+|f\rangle = \mathrm{Tr}\left(I^- I^+\right)$ where the trace is restricted to a subspace of fixed S^z. Representing the eigenenergies as $E_f = \langle f|H_0|f\rangle$ the numerator can be transformed in a similar way. Explicit expressions for the first three moments are

$$\langle m \rangle = \frac{-\mathrm{Tr}\left(\left[H_0, I^-\right] I^+\right)}{\mathrm{Tr}\left(I^- I^+\right)}, \qquad \left\langle m^2 \right\rangle = \frac{-\mathrm{Tr}\left(\left[H_0, I^-\right]\left[H_0, I^+\right]\right)}{\mathrm{Tr}\left(I^- I^+\right)}, \tag{3}$$

$$\left\langle m^3 \right\rangle = \frac{\mathrm{Tr}\left(\left[H_0, \left[H_0, I^-\right]\right]\left[H_0, I^+\right]\right)}{\mathrm{Tr}\left(I^- I^+\right)}, \quad \left\langle m^4 \right\rangle = \frac{\mathrm{Tr}\left(\left[H_0, \left[H_0, I^-\right]\right]\left[H_0, \left[H_0, I^+\right]\right]\right)}{\mathrm{Tr}\left(I^- I^+\right)}. \tag{4}$$

where $[A, B] = AB - BA$ is the operator commutator. Note, for calculating the absorption spectrum, that I^+has to appear on the right of I^- in the traces.

Unlike the methods of exact diagonalisation [14, 15], in this paper we will use the approach of [12] to evaluate the quantum mechanical averages. As a trace is invariant with respect to a change of basis we will calculate Eq. (3) in the basis of non-interacting spins that are quantised individually along the external magnetic field. In the basis of non interacting states, the denominator is a single sum of traces of individual spin operators $\sum_{ij} \mathrm{Tr}\left(I_i^- I_j^+\right) = \sum_i \mathrm{Tr}\left(I_i^- I_i^+\right)$. The full set of the basis states in a subspace of fixed S^z can be divided in two groups: one has the ith spin in the state $I_i^z = 1/2$ and the other in the state $I_i^z = -1/2$. For each state form the first group the expectation value of $I_i^- I_i^+$ is 0 and for the second group it is 1. Then, the trace of every single spin operator is independent of i, $\mathrm{Tr}\left(I_i^- I_i^+\right) = 0 \cdot C_{N-1}^{n-1} + 1 \cdot C_{N-1}^{n}$, where the binomial factors C_{N-1}^{n-1} and C_{N-1}^{n} give the numbers of states in each group. Here n is the number of 1/2-spins pointing up with respect to the fully polarised state, i. e. $S^z = -N/2 + n$. The sum over all N single spin traces gives the denominator as $\mathrm{Tr}\left(I^- I^+\right) = N!/\left(n!\left(N - n - 1\right)!\right)$.

The numerator of the first moment $k = 1$ requires only one first order commutator, $\left[H_0, I^-\right] = -\mu B \sum_{j=1}^{N} I_j^- - \sum_{i<j} \frac{g}{r_{ij}^3} 3\left(3\gamma_{ij}^2 - 1\right) I_j^z I_i^-$. The trace of a two spin operator, that is involved in Eq. (3), is evaluated analogously to the denominator. We calculate averages over

all possible two spin states accounting for the remaining $N-2$ spin states by binomial factors,

$$\mathrm{Tr}\left(I_j^z I_i^- I_i^+\right) = \frac{(1-\delta_{ij})}{2} \frac{(N-2)!\,(N-2n-1)}{n!\,(N-n-1)!}. \tag{5}$$

Substituting the commutator $[H_0, I^-]$ into Eq. (3) and using the last expression we obtain the line shift, $\langle\nu\rangle = \langle m\rangle$, that corresponds to the first moment $\langle\nu\rangle = \mu B - F3\,(N-2n-1)\,/\,(2N-2)$. Evaluating the limit $N \to \infty$ of this result and identifying the polarisation as $p = -1 + 2n/N$ we obtain the expression in Table. 1. The first moment of the crystalline field F is given in the same table.

The numerator of the second moment $k = 2$ requires an extra first order commutator $[H_0, I^+] = B\sum_j I_j^+ + \sum_{i<j} \frac{g}{r_{ij}^3} 3\left(3z_{ij}^2 - 1\right) I_j^+ I_i^z$. The trace of a three spin operator, which is needed for $\langle m^2\rangle$, is obtained in a similar way,

$$\mathrm{Tr}\left(I_j^z I_i^- I_i^+ I_{i'}^z\right) = \frac{(1-\delta_{ij})\,(1-\delta_{ii'})\,(N-3)!\,((2n-N+1)(2n-N+2) - \delta_{ii'}2n(2n-2N+3))}{4n!\,(N-n-1)!}. \tag{6}$$

Substituting the product of the two commutators $[H, I^-]\,[H, I^+]$ into Eq. (3), and using the above expression, we obtain the second cumulant $\langle\nu^2\rangle = \langle m^2\rangle - \langle m\rangle^2$ that corresponds to the line width, $\langle\nu^2\rangle = F_2 3n\,(2N-2n-3)\,/\,(2\,(N-1)\,(N-2))$. Evaluating the $N \to \infty$ limit of this expression we obtain the result in Table. 1.

The third and the fourth moments contain an admixture of moments of a lower order when the polarisation is finite, $p \neq 0$. Hence, we evaluate the third and the fourth irreducible moments as $\langle\nu^3\rangle = \langle m^3\rangle - 3\langle\nu^2\rangle\langle\nu\rangle - \langle\nu\rangle^3$ and $\langle\nu^4\rangle = \langle m^4\rangle - 4\langle\nu^3\rangle\langle\nu\rangle - 3\langle\nu^2\rangle^2 - 6\langle\nu\rangle^2\langle\nu^2\rangle - \langle\nu\rangle^4$, using the same procedure as for the moments above. In the limit $N \gg 1$, we obtain the results in Table. 1. The dependencies of $\langle\nu^k\rangle$ on the polarisation p and the moments of the crystalline field F_i factorise for $k \leq 3$. This in turn implies that the dependence of the model Eq. (1) on the orientation of the external magnetic field manifests itself equivalently in spectroscopy at different polarisations of the nuclei. However, for $k = 4$ such factorisation does not occur thus the anisotropy changes with a change of the polarisation.

4. Crystalline anisotropy

The moments of the crystalline field F_i depend on the orientation of the external magnetic field. As an example we evaluate the prefactors in Table. 1 numerically for the zinc-blende crystals; GaAs belongs to this class.

These crystals contain two species of atoms. In a typical spectroscopy experiment [4] all nuclei are polarised first then they are probed at a selected frequency to get a signal from a single sublattice while the other sublattice, at a different frequency, stays polarised that makes contribution of the other sublattice small [16]. Having in mind such a setup we will study only one sub-lattice.

The angular dependence of the second moment $\sqrt{F_2}$ is presented in Figure 1b. The shape is relatively isotropic, the biggest change in magnitude of $\sqrt{F_2}$ is ~20% between the directions [001] and [111]. The remaining cumulants are strongly anisotropic. The first moment is the biggest along the direction [111] and in the plane (111), see Figure 1a. In all other directions it is an order of magnitude smaller. The third cumulant is maximum along some twelve directions and is zero along some other twelve directons with a smooth cross-over in-between, see Figure. 1c. The angular dependence of the fourth cumulant at $p = 0$ is present in Figure 1d and the second part of the cumulant that becomes non-zero at $p \neq 0$ is plotted in Figure 4e.

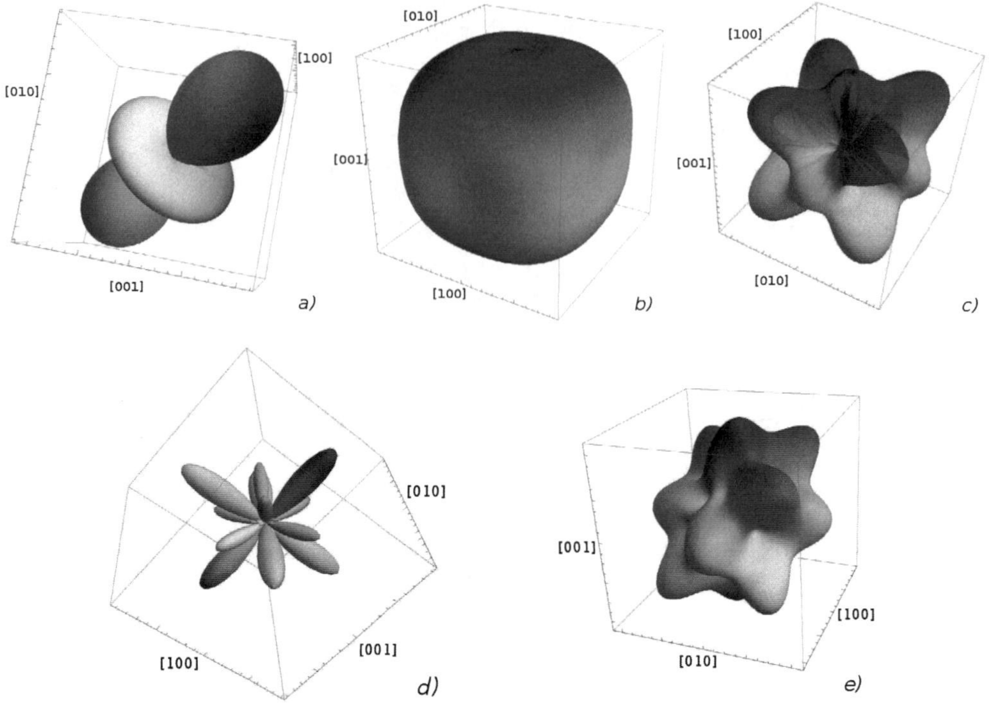

Figure 1. Plots of a) F, b) $\sqrt{F_2}$, c) $\sqrt[3]{F_{3c} - F_2 F - F_3}$, d) $F_{4q} - 2F_4 - F_2^2$, and e) $F^2 F_2 + 10FF_3 - 2FF_{3c} + F_{4c}$ as a function of angles between $\mathbf{B}$ and the axes of a Zince-Blende crystal.

5. Conclusions

We have studied anisotropy of the homogeneously broadened nuclear spin resonance at a finite polarisation in a crystal with respect to mutual orientation of the external magnetic field and axis of the crystal. We have demonstrate that the line width is relatively isotropic but the other cumulants, line shift and higher then the second cumulants, are strongly anisotropic. This result shows that the orientation of the magnetic field is an essential parameter in a study of polarised nuclei.

Acknowledgements

Financial support by the EPSRC (UK) EP/G001642 is acknowledged.

References

[1] J. R. Petta *et. al.*, Science **309**, 2180 (2005).
[2] D. J. Reilly *et. al.*, Science **321**, 817 (2008).
[3] H. Bluhm, S. Foletti, D. Mahalu, V. Umansky, and A. Yacoby, Phys. Rev. Lett. **105**, 216803 (2010).
[4] R. Takahashi, K. Kono, S. Tarucha, and K. Ono, Phys. Rev. Lett. 107, 026602 (2011).
[5] A. V. Khaetskii, D. Loss, and L. Glazman, Phys. Rev. Lett. **88**, 186802 (2002).
[6] D. Stepanenko, G. Burkard, G. Giedke, and A. Imamoglu, Phys. Rev. Lett. **96**, 136401 (2006).
[7] O. Tsyplyatyev and D. Loss, Phys. Rev. Lett. **106**, 106803 (2011).
[8] Daniel Loss, Fabio L. Pedrocchi, and Anthony J. Leggett, Phys. Rev. Lett. **107**, 107201 (2011).
[9] I. Zutic, J. Fabian, and S. Das Sarma, Rev. Mod. Phys. **76**, 323 (2004).
[10] D. Loss and D. P. DiVincenzo, Phys. Rev. A **57**, 120 (1998).
[11] D. Gammon, S. W. Brown, E. S. Snow, T. A. Kennedy, D. S. Katzer and D. Park, Science **277**, 85 (1997).
[12] J. H. Van Vleck, Phys. Rev. **74**, 1168 (1948).
[13] A. Abragam, M. Chapellier, J. F. Jacquinot, and M. Goldman, J. Magn. Reson. **10**, 322 (1973).
[14] O. Tsyplyatyev and D. Loss, Phys. Rev. B **82**, 024305 (2010).
[15] O. Tsyplyatyev, J. von Delft, and D. Loss, Phys. Rev. B **82**, 092203 (2010).
[16] O. Tsyplyatyev and D. M. Whittaker, Phys. Rev. **B** 85, 125123 (2012).

HMF-20 IOP Publishing
Journal of Physics: Conference Series **456** (2013) 012038 doi:10.1088/1742-6596/456/1/012038

Magnetotransport in graphene on silicon side of SiC

P. Vašek[1], L. Smrčka[1], P. Svoboda[1], M. Ledínský[1], V. Jurka[1], M. Orlita[2,3], D. K. Maude[3], W. Strupinski[4], R. Stepniewski[5], R. Yakimova[6]

[1]Institute of Physics ASCR, v.v.i., Cukrovarnická 10, 162 53 Praha 6, Czech Republic
[2]Faculty of Mathematics and Physics, Charles University, Ke Karlovu 3, 121 16 Praha 2, Czech Republic
[3]Laboratoire National des Champs Magnétiques Intenses, CNRS-UJF-UPS-INSA, 25, avenue des Martyrs, 38042 Grenoble, France
[4]Institute of Electronic Materials Technology, Wolczynska 133, Warszawa, 01-919, Poland
[5]Institute of Experimental Physics, Warsaw University, Hoza 69 00-681 Warszawa, Poland
[6]Linköping University of Technology, S-581 83 Linköping, Sweden

E-mail: vasek@fzu.cz

Abstract. We have studied the transport properties of graphene grown on silicon side of SiC. Samples under study have been prepared by two different growth methods in two different laboratories. Magnetoresistance and Hall resistance have been measured at temperatures between 4 and 100 K in resistive magnet in magnetic fields up to 22 T. In spite of differences in sample preparation, the field dependence of resistances measured on both sets of samples exhibits two periods of magneto-oscillations indicating two different parallel conducting channels with different concentrations of carriers. The semi-quantitative agreement with the model calculation allows for conclusion that channels are formed by high-density and low-density Dirac carriers. The coexistence of two different groups of carriers on the silicon side of SiC was not reported before.

1. Introduction

Epitaxial growth of graphene layers on the surface of single-crystalline SiC, reported first in a seminal paper [1], is a well established way to the large scale production of graphene for a future graphene based electronics. Many modifications of this technology have been reported. In addition to standard thermal decomposition of the upper layers of the SiC substrate at temperatures above 1200°C in various atmospheres, chemical vapour deposition (CVD) on the SiC surface can also be used. In the former case graphene arises by rearrangement of carbon atoms left on the SiC surface after Si atoms sublimed out, in the latter carbon atoms have to be added from outside by thermal decomposition of suitable hydrocarbon gas on the surface. Growth details can be expected to influence transport properties of graphene even on SiC substrates of the same crystallographic orientation, due to various types of active defects and/or character of the interface between the graphene and the substrate. In the present paper we report on the experimental study of the and Hall resistances on two series of samples, prepared from wafers epitaxially grown on Si-side of SiC, using either Si sublimation or CVD methods.

Journal of Physics: Conference Series **456** (2013) 012038 doi:10.1088/1742-6596/456/1/012038

2. Experiments

2.1. Graphene growth

Two series of samples have been employed in this study. One of them, denoted as G141, was produced at the University of Linköping, Sweden, on the semi-insulating 6H-SiC(0001) wafer by the Si-sublimation method in an inductively heated furnace. The growth was carried out under highly isothermal conditions at 2000°C and at an ambient argon pressure of 1 bar [2].

Samples in the second series stem from a wafer 699, supplied by ITME Warsaw, Poland. The graphene growth was performed in a commercially available horizontal CVD hot-wall reactor (Epigress V508), inductively heated by an RF generator. Semi-insulating 4H-SiC (0001) substrates were used and propane gas served as the precursor. The substrates were etched in hydrogen and propane mixtures prior to carbon deposition. Precautions had been taken to prevent sublimation of Si atoms in conditions of high temperature (≈1600°C) and a low Ar pressure [3].

Raman spectroscopy was used to confirm the presence of graphene.

2.2. Lithography

Samples were patterned by standard photolithographic techniques. Hall bars with a 100 micrometers wide conducting channel, two current contacts and six contacts for measurement of the longitudinal magnetoresistance and the Hall voltages were formed by oxygen plasma etching. The separation of potential leads was 300 micrometers. The aspect ratio of the sample (width/length) was either 1/3 or 1/6, depending on the pair of contacts employed. The macroscopic size indicates that the sample area covers a large number of terraces on the SiC surface and inhomogeneity in the graphene layer cannot be excluded. Ohmic TiAu contacts were prepared by the electron beam assisted deposition and lift-off technique.

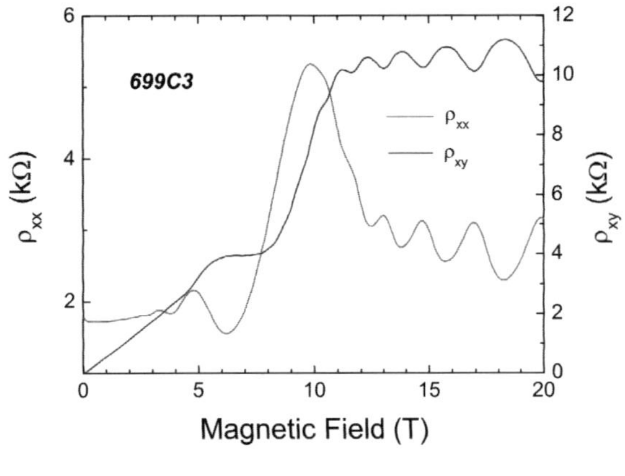

Figure 1. Longitudinal (red curve) and Hall (blue curve) magnetoresistances mesured on a sample prepared in ITME Warsaw, Poland. $N_1 = 9.5 \times 10^{11}$ cm^{-2}, $N_2 = 1.1 \times 10^{13}$ cm^{-2}.

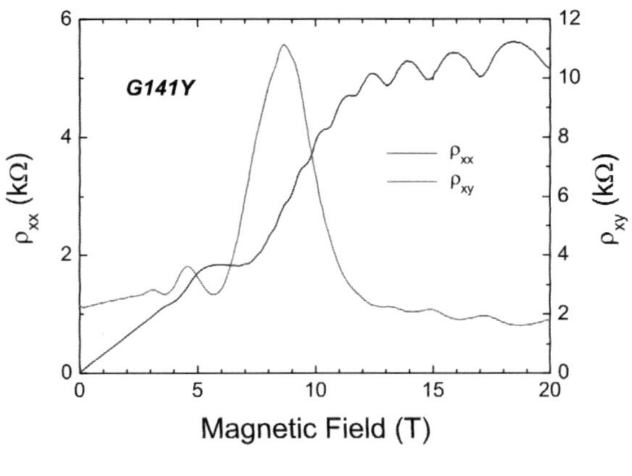

Figure 2. Longitudinal (red curve) and Hall (blue curve) magnetoresistances mesured on a sample prepared in University of Linköping, Sweden. $N_1 = 8.9 \times 10^{11}$ cm^{-2}, $N_2 = 1.4 \times 10^{13}$ cm^{-2}.

2.3. Magnetotransport measurement

Magnetoresistance $\rho_{xx}(B)$ and Hall resistance $\rho_{xy}(B)$ have been simultaneously measured by a 4-probe low frequency ($f = 10.66$ Hz) AC lock-in method. Excitation current up to 100 nA was taken from the output of the lock-in amplifier using 10 MΩ or 100 MΩ load resistors. Most experiments were performed with samples directly immersed in a bath of liquid helium at 4.2 K. For the investigation of the temperature dependence up to 100 K a VTI has been used with a calibrated Cernox thermometer as a temperature sensor. Transverse magnetic fields up to 22 T were generated by a 10 MW Bitter magnet. Admixtures of ρ_{xx} to ρ_{xy} due to contacts misalignment was removed by symmetrization of all traces measured in positive and negative magnetic fields.

All experiments were realized in LNCMI CNRS Grenoble, France.

3. Results and discussion

The dominant feature observed on the magnetoresistance (MR) curves $\rho_{xx}(B)$ shown in Fig. 1 and Fig. 2 is the presence of two types of oscillations, both periodic in $1/B$ but with different periods. Simultaneously, the Hall resistance $\rho_{xy}(B)$ oscillates at fields above 10 T. All these oscillations disappeared after the samples were rotated into parallel configuration with respect to applied magnetic field, which confirms that the oscillations originate in two-dimensional sheets of charge carriers.

Such a behavior cannot stem from a single conducting sheet and therefore two independent channels have to exist. From the periods of oscillations in longitudinal magnetoresistance $\rho_{xx}(B)$ we can estimate charge carriers concentrations in both channels. Denoting the channels responsible for the ,,low field" and ,,high field" oscillations by indices 1 and 2, respectively, we obtained for the sample 699C3 (Fig. 1) the values $N_1 = 9.5\times10^{11}\text{cm}^{-2}$ and $N_2 = 1.1\times10^{13}\ \text{cm}^{-2}$. The same analysis of the data presented in Fig. 2 gives the concentrations $N_1 = 8.9\times10^{11}\text{cm}^{-2}$ and $N_2 = 1.4\times10^{13}\text{cm}^{-2}$.

Apart from the modulation by the oscillations at higher fields, the Hall curves for both samples clearly exhibit QHE plateaux at $\rho_{xy}(B) = (h/\nu e^2)$. Absence of the plateaux at $\nu = 4$ confirms, that the ,,low field" oscillations can be attributed to the single layer graphene. The identification of the second conducting layer with substantially higher carrier concentrations N_2

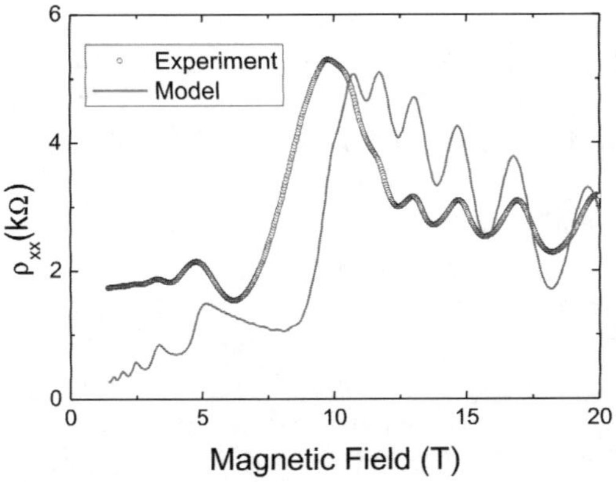

Figure 3. Experimental and model curves of longitudinal magnetoresistance of a sample 699.

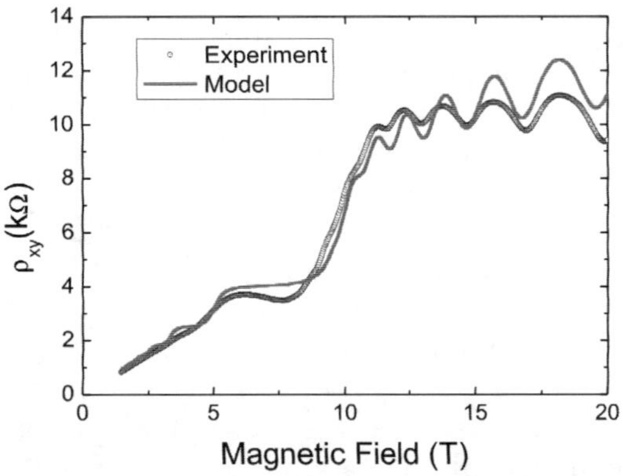

Figure 4. Experimental and model curves of Hall magnetoresistance of a sample 699.

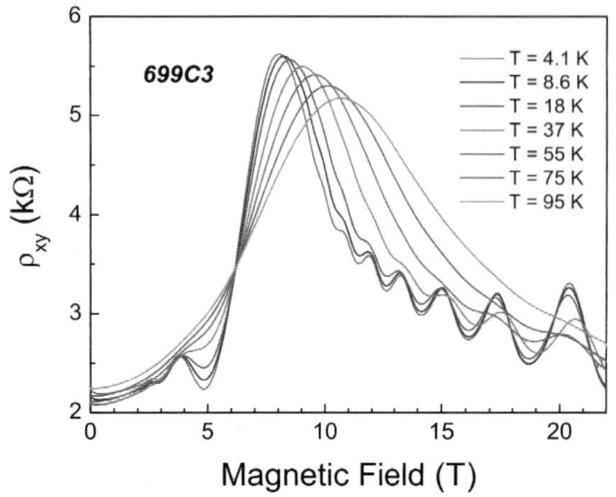

Figure 5. Longitudinal magnetoresistances mesured at various temperatures on sample prepared in ITME Warsaw, Poland.

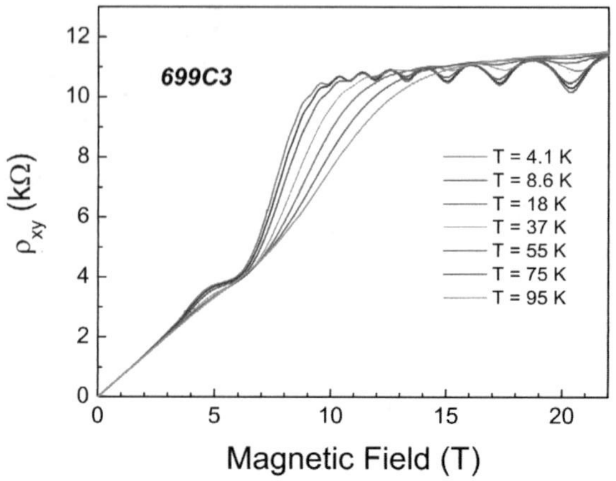

Figure 6. Hall magnetoresistances mesured at various temperatures on sample prepared in ITME Warsaw, Poland.

seems to be less straightforward, the Hall plateaux cannot be reached in magnetic fields at our disposal. Assuming the same character of both channels, we could compare experimental data for the sample 699C3 in Fig. 1 with calculations based on the model described in [4], including the appropriate description of the conductivity tensor for graphene [5], and assuming two parallel conductivity channels in the sample.

In these calculations, concentrations N_1 and N_2, together with the half-widths $\Gamma_1 = 2.9$ meV and $\Gamma_2 = 6$ meV of Landau levels serve as fitting parameters. The results of the best fit are presented in Figs. 3 and 4. The model describes the data reasonably well, in particular those in Fig. 4 for the Hall resistance. Moreover, the best fit provides concentrations $N_1 = 1.2\times10^{12}\text{cm}^{-2}$ and $N_2 = 1.1 \times 10^{13}\text{cm}^{-2}$. These agree quite well with the concentrations derived directly from the periods of both types of MR oscillations observed on the same sample.

In order to further characterize the two conducting channels, we have measured temperature dependence of $\rho_{xx}(B)$ and $\rho_{xy}(B)$ up to about 100 K. The results can be seen in Fig. 3 and Fig. 4. Both types of oscillations persist up to quite high temperatures, which is characteristic property of graphene. Temperature damping of the amplitude of oscillations provides an estimate of the cyclotron effective mass m_c^* of the charge carriers [7]. It implies a reliable subtraction of the monotonous ,,background" part, which is not easy here, in particular for the ,,low field" oscillations. Nevertheless, in highest fields, where the resistivity $\rho_{xx}(B)$ from the channel responsible for the ,,low field" SdH oscillations approaches zero, we got values of m_c^* that agree reasonably well with those derived from the concentrations N_i. The latter stem from the expression $m_c^* = \hbar/v_F$ for the effective mass of the single-layer graphene [7]. Taking $v_F = 1.1 \times 10^6$ m/s [7] and the concentrations N_1 and N_2 given in caption to Fig. 1, we get $m_{c1}^* = 0.018\, m_0$ and $m_{c2}^* = 0.061\, m_0$, respectively. The curves presented in Figs. 5 and 6 were obtained more than a year later than those shown in Fig. 1 (see [6]). This confirms the stability of the sample properties in time.

We can conclude, that our results for the two wafers produced by widely different technologies can be interpreted as a superposition of oscillations from two parallel two-dimensional conducting channels with substantially different carrier concentrations and apparently also mobilities.

It seems to be surprising, since different type of substrates together with different growth technology can be expected to provide quite different morphology of the epitaxial graphene. Detailed study of the graphene formed on SiC under various environments [8] has shown no apparent differences between results for 6H-SiC and 4H-SiC wafers. Substantial differences can however arise between samples produced by the purely sublimation method and the CVD assisted one. In the former case the existence of an interfacial (buffer) layer is well established and it influences the electronic transport properties of graphene [9], most probably through the concentration and mobility of the charge carriers. But due to the bonding of the buffer layer to the Si atoms in the substrate it loses its graphene-like linear band structure near the Fermi level [9] and cannot thus generate SdH oscillations characteristic for the single layer graphene. In the CVD assisted method, the structure of the interface between graphene and the SiC substrate is less understood, but it probably also influences only the concentration and mobility of the charge carriers in uppermost graphene layers. It is important that both technologies produce graphene, that is far from being flat and homogeneous on the scale larger than a few microns. Graphene seems to nucleate on the edges of the terraces and grow further in islands until they fuse into a layer, which is supported by the diffusion of the carbon atoms along the surface [9]. It can be expected, that not only the morphology and thickness [9], but also the transport properties can be different for graphene at the terraces and their edges [10]. Therefore we speculate, that the surface area of our samples is broken into small interconnected regions with either low or high concentration of carriers, which form a network of parallel and/or series resistors. It may be a combination of the monolayer graphene at the terraces and bilayer graphene at their edges [8], but the electron transport measurement itself can hardly reveal the detailed microstructure of the epitaxial graphene on SiC substrate. Explanation of the origin of dual SdH oscillations observed in our samples thus requires further investigation.

Acknowledgments

The support of the European Science Foundation EPIGRAT project (GRA/10/E006, 670 and 671/N-ESF-EPI/2010/0), GACR No. P204/10/1020 and No. P108/11/0894 projects, programme ,,Transnational access" contract No. 228043-EuroMagNET II-Integrated Activities, AVCR research program AVOZ10100521, the Academy of Sciences of the Czech Republic project KAN400100652 are acknowledged.

References

[1] Berger C, Song Z, Li T, Li X, Ogbazghi A Y, Feng R, Dai Z, Marchenkov A N, Conrad E H, First P N and de Heer W A 2004 *J. Phys. Chem.* B **108** 19912
[2] Virojanadara C, Zakharov A A, Yakimova R and Johansson L I 2010 *Surf. Sci.* 604, L4
[3] Strupinski W, Grodecki K, Wysmolek A, Stepniewski R, Szkopek T, Gasskell P E, Grüneis A, Haberer D, Bozek R, Krupka J and Baranowski J M 2011 *Nano Lett.* **11** 1786
[4] van der Burgt M, Karavolas V C, Peeters F M, Singleton J, Nicholas R J, Herlach F, Harris J J, Van Hove M and Borghs G 1995 *Phys. Rev.* B **52** 12218
[5] Zheng Y and Ando T 2002 *Phys. Rev.* B **65** 245420
[6] Orlita M, Maude D K, Goncharuk N A, Jurka V, Vašek P, Svoboda P, Smrčka L, Strupinski W, Yakimova R and Jawad-ul-Hassan 2011 *Annual Report LNCMI Grenoble-Toulouse* eds F Duc and D Maude CARBON ALLOTROPES p.10
[7] Zhenbing Tan, Gangling Tan, Li Ma, Liu G T, Lu L and Yang C L 2011 *Phys. Rev.* B **84** 115429
[8] Shrivastava N, Guowei He, Luxmi, Mende P C, Feenstra R M and Yugang Sun 2012 *J. Phys. D: Applied Phys.* **45** 154001
[9] Hibino H, Tanabe S, Mizuno S and Kageshima H 2012 *J. Phys. D: Applied Phys.* **45** 154008
[10] Grodecki K, Bozek R, Strupinski W, Wysmolek A, Stepniewski R and Baranowski J M 2012 *Appl. Phys. Lett.* **100** 261604

Journal of Physics: Conference Series **456** (2013) 012039 doi:10.1088/1742-6596/456/1/012039

Microwave-induced parametric instability of 2D magnetoplasmons

V A Volkov[1,2] **and A A Zabolotnykh**[2,1]

[1]V. A. Kotelnikov Institute of Radio-engineering and Electronics, Russian Academy of Sciences, Mokhovaya 11-7, 125009, Moscow, Russia

[2]Moscow Institute of Physics and Technology, Institutskii per. 9, Dolgoprudny, 141700, Moscow Region, Russia

Abstract. Recent experiments on microwave-irradiated ultraclean 2DES revealed a colossal narrow photoresistivity peak in the vicinity of the second cyclotron resonance harmonic $2\omega_c$. Here we propose an explanation of this puzzle phenomenon in terms of the parametric instability of 2D magnetoplasmons. We develop the hydrodynamic theory of 2D magnetoplasmons under an inhomogeneous electric field of the MW pumping with frequency Ω. Such a strong inhomogeneous MW field is always induced by lateral metal contacts, which are used to measure photoresistivity. We show that the parametric instability of the 2D magnetoplasmons arises near $\Omega = 2\omega_c$ when the gradient of the electric field of the pumping exceeds a threshold value. The instability leads to heating of 2DES, which in turn leads to the sharp photoresistivity peak.

1. Introduction

Remarkable features of microwave (MW) response of high-mobility two-dimensional electron systems (2DESs) at low temperature in magnetic field have excited much interest the last few years, for review see [1]. The point is that even at low MW power the resistivity of these structures varies dramatically [2, 3]: it oscillates strongly with inverse magnetic field. The period of the oscillations is determined by the relation of MW frequency Ω to cyclotron frequency ω_c. With the increase of the MW power the oscillation minima become deeper and reach zero values [4, 5].

Recently in ultraclean (mobility 3×10^7 cm^2/Vs) 2DESs a new peculiarity of the resistance was discovered. When the MW frequency is near double cyclotron frequency an extremely high and narrow spike of resistivity was observed [6]-[9]. The origin of the resistance spike is not still clear.

The goal of this article is to propose a possible explanation of the effect in terms of classical approach to the description of 2DES. Our explanation is based on the parametric resonance of 2D magnetoplasmons at the MW irradiation. Under conditions of the parametric resonance amplitudes of plasmon velocity and electron charge density increase exponentially with time, i.e. the plasma instability arises in the system. The instability takes place provided the resonance detuning is sufficiently small and the gradient of the MW field acting on 2D electrons is sufficiently large. The instability leads to heating of 2DES, which in turn leads to the resistance spike. Note that the parametric resonance under different conditions was studied theoretically in [10]-[12].

2. Basic equations

Consider 2DES situated in plane $z = 0$ in constant magnetic field $\vec{B}$ along OZ-axis. Electrons are affected by the pumping electric field with the amplitude $\vec{E}_\Omega$ and frequency Ω. Electric field acting on electrons is always inhomogeneous near metal contacts, so we assume that it depends on the coordinates $\vec{E}_\Omega = \vec{E}_\Omega(x)$. We don't consider a uniform part of the pumping field because it doesn't effect the parametric resonance, see below. We assume the electric field of the pumping $\vec{E}_{ext}(x,t) = \vec{E}_\Omega(x)\cos\Omega t$ is directed along x-axis and depends only on coordinate x, and the system is uniform along y-axis.

Plasmons are treated in the frame of the hydrodynamic approach in terms of the hydrodynamic velocity $\vec{V}(x,t)$, the deviation of 2D charge density $\rho(x,t)/e$ from the equilibrium concentration n_0 and the self-consistent potential $\varphi(x,z,t)$.

The hydrodynamic velocity and the self-consistent potential are related by the Euler equation

$$\frac{\partial \vec{V}}{\partial t} + \frac{\vec{V}}{\tau} + (\vec{V},\vec{\nabla})\vec{V} + \frac{e}{m}\vec{\nabla}\varphi = \frac{e}{m}\vec{E}_{ext}(x,t) + \frac{e}{mc}[\vec{V},\vec{B}], \tag{1}$$

where τ is the phenomenological time that describes the relaxation rate in the system, e, m are the electron charge and effective mass, c is the speed of light, $\vec{\nabla} = (\partial_x, \partial_y)$, $\varphi = \varphi(x, z = 0, t) = \varphi(x,t)$ is the self-consistent electron potential. Further we assume that the conditions $\Omega\tau \gg 1, \omega_c\tau \gg 1$ take place, i.e. the dissipation is small; $\omega_c = |eB/(mc)|$.

The self-consistent electron potential $\varphi(x,z,t)$ and fluctuations of the 2D charge density $\rho(x,t)$ in quasi-static approximation are related by the Poisson equation

$$\Delta_{3D}\varphi(x,z,t) = -\frac{4\pi}{\varkappa}\rho(x,t)\delta(z), \tag{2}$$

where $\delta(z)$ is the Dirac delta function, $\varkappa$ is a dackground static dielectric constant, $\Delta_{3D} = (\partial_x^2 + \partial_y^2 + \partial_z^2)$.

Finally, $\vec{V} = \vec{V}(x,t)$ and $\rho = \rho(x,t)$ are related by the continuity equation

$$\partial_t\rho + div(en_0 + \rho)\vec{V} = 0. \tag{3}$$

Thus, we obtain system of Eqs. (1)-(3) for φ, ρ, and $\vec{V}$.

3. Scheme of solution

Suppose that solutions of Eqs. (1)-(3) can be found in the form

$$\begin{aligned} \varphi(x,t) &= \varphi_0(x,t) + \delta\varphi(x,t), \\ \rho(x,t) &= \rho_0(x,t) + \delta\rho(x,t), \\ \vec{V}(x,t) &= \vec{V}_0(x,t) + \delta\vec{V}(x,t), \end{aligned} \tag{4}$$

where $\varphi_0(x,t), \rho_0(x,t)$, and $\vec{V}_0(x,t)$ are the solutions of Eqs. (1)-(3) linearized with respect to φ, ρ, and $\vec{V}$.

In high-mobility 2D systems with $\omega_c\tau \gg 1$, $\omega\tau \gg 1$, the Fourier components of $\vec{V}_0(x,t)$, $\rho_0(x,t)$, $\varphi_0(x,t)$ have the form

$$V_{0x}(q,\omega) = \frac{e}{m}E_{ext}(q,\omega)\frac{-i\omega}{-\omega^2 + \omega_{mp}^2(q)^2}, \qquad V_{0y}(q,\omega) = \frac{\omega_c}{i\omega}V_{0x}(q,\omega), \tag{5}$$

$$\rho_0(q,\omega) = \frac{en_0iq}{i\omega}V_{0x}(q,\omega), \qquad \varphi_0(q,\omega) = \frac{2\pi}{\varkappa|q|}\rho_0(q,\omega).$$

Here $\omega_{mp}^2(q) = \omega_c^2 + \omega_p^2(q)$ is the square magnetoplasma frequency for a given q. One can write down Eqs. (5) for $\vec{E}_{ext}(x,t) = \vec{E}_\Omega(x)\cos\Omega t$ in the space-time representation as

$$V_{0x}(x,t) = V_{sx}(x)\sin(\Omega t), \quad V_{0y}(x,t) = V_{cy}(x)\cos(\Omega t), \quad \rho_0(x,t) = \rho_c(x)\cos(\Omega t). \tag{6}$$

After substitution of Eqs. (4) into Eqs. (1)-(3), linearization with respect to fluctuations $\delta\varphi(x,t)$, $\delta\vec{V}(x,t)$, $\delta\rho(x,t)$ and exclusion of $\delta\varphi(x,t)$, $\delta V_y(x,t)$, one can derive equations for $\delta\rho(x,t)$ and $\delta V_x(x,t)$. We don't give an explicit form of these cumbersome equations. The Eqs. contain coefficients periodic with frequency Ω, therefore, their solutions obey the Floquet-Bloch theorem. According to [13] the solution near the fundamental parametric resonance (when Ω is near $2\omega_c$) can be found in the two-wave approximation in the form

$$\delta V_x = \left(A(x)\cos\left(\frac{\Omega}{2}t\right) + B(x)\sin\left(\frac{\Omega}{2}t\right)\right)e^{st}, \tag{7}$$

$$\delta\rho = \left(A_\rho(x)\cos\left(\frac{\Omega}{2}t\right) + B_\rho(x)\sin\left(\frac{\Omega}{2}t\right)\right)e^{st}, \tag{8}$$

where s is the amplification coefficient. We assume that s is a real constant. Condition $s > 1/\tau$ must be fulfilled to achieve unstable solutions. Equation $s = 1/\tau$ takes place at the threshold.

Substituting Eqs. (7), (8) to the system of equations mentioned above, then equating coefficients at $\sin(\Omega t/2)$ and $\cos(\Omega t/2)$, and neglecting other harmonics, equations for $A(x), B(x), A_\rho(x), B_\rho(x)$ can be derived. Then we exclude $A_\rho(x), B_\rho(x)$ neglecting the terms containing the second or the higher powers of E_Ω, s and τ^{-1}. The resulting system is as follows

$$\begin{pmatrix} \hat{\omega}_p^2/\Omega - \triangle\omega + \hat{U}(x) & s \\ s & -\hat{\omega}_p^2/\Omega + \triangle\omega + \hat{U}(x) \end{pmatrix}\begin{pmatrix} A(x) \\ B(x) \end{pmatrix} = \begin{pmatrix} 0 \\ 0 \end{pmatrix}, \tag{9}$$

where $\triangle\omega = \Omega/2 - \omega_c$ is the resonance detuning. The linear in the external field "perturbation" operator $\hat{U}$ is determined as follows:

$$\hat{U}(x) = \Delta(x) + \frac{\hat{\omega}_p^2}{2en_0\Omega}\rho_c(x) - [\hat{\omega}_p^2, V_{sx}(x)]\frac{\partial_x}{\Omega^2}, \tag{10}$$

where $\Delta(x) = \partial_x\left(V_{sx}(x) + V_{cy}(x)2\omega_c/\Omega\right)/4$, $\hat{\omega}_p^2 = -\left(e^2n_0/m\right)\partial_x^2\hat{G}$ is the operator in the coordinate representation. The operator $\hat{G}$ acts as follows $F_q[\hat{G}f(x)] = G(q)f(q)$, where $F_q[g(x)] = g(q)$ is the Fourier transform of function $g(x)$, $G(q) = 2\pi/(\varkappa|q|)$; $[\hat{\omega}_p^2, V_{sx}(x)] = \hat{\omega}_p^2 V_{sx}(x) - V_{sx}(x)\hat{\omega}_p^2$ is the commutator of operator $\hat{\omega}_p^2$ and function $V_{sx}(x)$. Operator of derivative and $\hat{\omega}_p^2$ act on all the functions behind them.

It should be noted that if $\vec{E}_{ext}(x,t)$ is uniform then $\hat{U} = 0$, and the instability doesn't appear. The system loses stability as electric field acting on electrons becomes more inhomogeneous.

Let us estimate the form of the solutions of Eqs. (9) in simple models.

4. Solution of Eqs. (9) for models

Solution without the self-consistent field. Consider the problem without taking into account the self-consistent field. This is the case when the electron-electron interaction in 2DES is suppressed, for example, via the background dielectric constant $\varkappa$. When $\varkappa$ is sufficiently high one can neglect $\hat{G}$ and $\hat{\omega}_p^2$ in (9). Then $\hat{U}(x) = \Delta(x)$ and the system of eqs. (9) becomes as follows

$$\begin{pmatrix} -\triangle\omega + \Delta(x) & s \\ s & \triangle\omega + \Delta(x) \end{pmatrix}\begin{pmatrix} A(x) \\ B(x) \end{pmatrix} = \begin{pmatrix} 0 \\ 0 \end{pmatrix}. \tag{11}$$

If there is at least one point in which condition $\Delta^2(x) = s^2 + (\triangle\omega)^2$ is satisfied then the system of local equations (11) has nontrivial solutions. After replacement of s by its minimum value τ^{-1} we obtain the conditions for the local derivative of the electric field:

$$\left|\frac{eE'_\Omega(x)}{2m\Omega}\right| = \sqrt{(1/\tau)^2 + (\triangle\omega)^2}. \tag{12}$$

The solutions have local behavior:

$$\begin{pmatrix} A(x) \\ B(x) \end{pmatrix} = \sum_n C_n \begin{pmatrix} 1 \\ \frac{\triangle\omega - \Delta(x_n)}{1/\tau} \end{pmatrix} \delta(x - x_n), \tag{13}$$

where x_n are the positions in which condition (12) is satisfied, C_n are an arbitrary constants.

Assume that the amplitude of the electric field in the sample area $x > 0$ is described as a power law $E_\Omega(x) = E_c l/(x + l)$ or as an exponential law $E_\Omega(x) = E_c \exp(-x/l)$, where from the contact situated at $x = 0$ and l is the length characterising the decrease of the electric field. Using condition (12) the expression for the threshold electric field can be presented in the form

$$\left|\frac{eE_c^{th}}{2m\Omega l}\right| = \sqrt{(1/\tau)^2 + (\triangle\omega)^2}. \tag{14}$$

Obviously the threshold field increases with the increase of the resonance detuning $\triangle\omega$ and with the decrease of the relaxation time τ.

Smooth pumping field. We exclude component $B(x)$ from Eqs. (9) and obtain the equation for $A(x)$:

$$\left(\left(\frac{\hat{\omega}_p^2}{\Omega} - \triangle\omega\right)^2 + s^2 + \frac{1}{\Omega}[\hat{\omega}_p^2, \hat{U}(x)] - \hat{U}^2(x)\right) A(x) = 0. \tag{15}$$

Assume that $V_{sx}(x) = \alpha x$, i.e. that the screened electric field E_Ω^{scr}, which is proportional to velocity $V_{sx}(x)$, is a linear function of x. Then $[\hat{\omega}_p^2, \hat{U}(x)] = 0$ and Eq. (15) after the Fourier transformation is reduced to the algebraic one

$$\left(\frac{\omega_p^4(q)}{\Omega^2}\left(1 - \frac{\alpha^2}{4\Omega^2}\right) - \frac{2\omega_p^2(q)}{\Omega}\left(\triangle\omega - \Delta\frac{\alpha}{2\Omega}\right) + (\triangle\omega)^2 + s^2 - \Delta^2\right) A(q) = 0. \tag{16}$$

Leaving only the largest terms in the coefficients of Eq.(16), we obtain the condition for existence of unstable solutions in this approximation:

$$\begin{aligned} \left|\frac{e(E_\Omega^{scr})'}{2m\Omega}\right| &\geq \sqrt{(1/\tau)^2 + (\triangle\omega)^2}, \qquad \triangle\omega \leq 0, \\ \left|\frac{e(E_\Omega^{scr})'}{2m\Omega}\right| &\geq \frac{1}{\tau}, \qquad \triangle\omega > 0, \end{aligned} \tag{17}$$

where $(E_\Omega^{scr})'$ is the spatial derivative of the screened pumping field. Functions $A(x), B(x)$ oscillate in space. Notice that the condition (17) for $\triangle\omega \leq 0$ is in agreement with the condition (12).

5. Conclusion

We developed the hydrodynamic theory of 2D magnetoplasmons under an inhomogeneous electric MW field oscillating with frequency Ω. In fact, the strong heterogeneity of the MW field is always induced by the lateral metal contacts which are used to measure photoresistivity. We use the linear approximation on the plasmon variables such as the hydrodynamic velocity and the 2D charge density and the self-consistent potential, but take into account nonlinear terms in the MW pumping. It is shown that magnetoplasmon variables obey equations containing inhomogeneous parametric terms with frequency Ω, therefore the Floquet-Bloch theorem is valid for them. If Ω is near $2\omega_c$ then the instability of magnetoplasmons was found to occur, i.e. their amplitudes start to grow exponentially with time with increasing microwave pumping. The parametric instability of the 2D magnetoplasmons arises when the gradient of pumping electrical field exceeds a threshold value. Conditions for the instability are analysed in simple approximations, and the expressions (12), (17) for the threshold are derived. The threshold field decreases with the increase of 2DES mobility and with the decrease of the resonance detuning $\triangle\omega = \Omega/2 - \omega_c$. The instability leads to heating of 2DES, which in turn leads to an increase of the sample resistance. Thus the sharp photoresistivity peak occurs when Ω is near $2\omega_c$.

Let us estimate the threshold field for the following parameters: $\mu = 10^7 cm^2/Vs$, $\Omega = 2\pi 10^{11} rad/s$, $m = 0.066 m_0$, $l = 0.1\mu m$, $\triangle\omega = 0$, here m is the electron effective mass of $GaAs$, $\mu = e\tau/m$. From the expression (14) we get $E_\Omega^{th} \approx 0.03$ V/cm and this is less than the amplitude of electric fields of MW radiation in the experiments [6]-[9].

The work was supported by the Russian Foundation for Basic Research (projects 11-02-01290 and 11-02-12143).

References

[1] Dmitriev I A, Mirlin A D, Polyakov D G, and Zudov M A *Rev. Mod. Phys.*, to be publ.
[2] Zudov M A, Du R R, Simmons J A, and Reno J L 2001 *Phys. Rev.* B **64** 201311(R)
[3] Ye P D, Engel L W, Tsui D C, Simmons J A, Wendt J R, Vawter G A, and Reno J L 2001 *Appl. Phys. Lett.* **79** 2193
[4] Mani R G, Smet J H, von Klitzing K, Narayanamurti V, Johnson W B, and Umansky V 2002 *Nature (London)* **420** 646
[5] Zudov M A, Du R R, Pfeiffer L N, and West K W 2003 *Phys. Rev. Lett.* **90** 046807
[6] Dai Ya, Du R R, Pfeiffer L N, and West K W 2010 *Phys. Rev. Lett.* **105** 246802
[7] Hatke A T, Zudov M A, Pfeiffer L N, and West K W 2011 *Phys. Rev.* B **83** 121301(R)
[8] Hatke A T, Zudov M A, Pfeiffer L N, and West K W 2011 *Phys. Rev.* B **83** 201301(R)
[9] Dai Ya, Stone K, Knez I, Zhang C, Du R R, Yang C, Pfeiffer L N, and West K W 2011 *Phys. Rev.* B **84** 241303(R)
[10] Orgad D, and Levit S 1996 *Phys. Rev. Lett.* **77** 719
[11] Mikhailov S A 1998 *Appl. Phys. Lett.* **73** 1886
[12] Weick G, and Mariani E 2011 *Phys. Rev.* B **84** 125441
[13] Landau L D, and Lifshitz E M 1976 *Mechanics* (Oxford: Butterworth-Heinemann) p 80

Effective mass from microwave photoresistance measurements in GaAs/AlGaAs quantum wells

A. T. Hatke[1], M. A. Zudov[1], J. D. Watson[2,3], M. J. Manfra[2,3], L. N. Pfeiffer[4], and K. W. West[4]

[1]School of Physics and Astronomy, University of Minnesota, Minneapolis, Minnesota 55455, USA
[2]Department of Physics, Purdue University, West Lafayette, Indiana 47907, USA
[3]Birck Nanotechnology Center, School of Materials Engineering and School of Electrical and Computer Engineering, Purdue University, West Lafayette, Indiana 47907, USA
[4]Department of Electrical Engineering, Princeton University, Princeton, New Jersey 08544, USA

Abstract. We have performed microwave photoresistance measurements in high mobility GaAs/AlGaAs quantum wells and investigated the value of the effective mass. The effective mass, obtained from the period of microwave-induced resistance oscillations (MIRO), was found to be about 12 % *lower* than the band mass in GaAs, $m_\mathrm{b}^\star$. In contrast, the measured magnetoplasmon dispersion (MPR) revealed an effective mass which is close to $m_\mathrm{b}^\star$, in accord with previous studies. These findings suggest that, in contrast to MPR, the MIRO dispersion contains corrections due to electron-electron interaction effects.

1. Introduction

The most frequently quoted value of the effective mass $m^\star$ in GaAs/AlGaAs-based two-dimensional electron systems (2DESs) is the value of the band mass of bulk GaAs, $m_\mathrm{b}^\star = 0.067\,m_0$, where m_0 is the free electron mass [1]. One of the common methods to obtain $m^\star$ is based on the temperature damping of Shubnikov-de Haas oscillations (SdHOs) [2], which are controlled by the filling factor, $\nu = 2\varepsilon_F/\hbar\omega_\mathrm{c}$, where $\varepsilon_F = \pi\hbar^2 n_e/m^\star$ is the Fermi energy, n_e is the carrier density, $\hbar\omega_\mathrm{c} = eB/m^\star$ is the cyclotron energy, and B is the magnetic field. Since $m^\star$ does not enter the filling factor, it cannot be obtained from the oscillation period; instead, one has to analyze the temperature damping of the SdHO amplitude.

The SdHO approach applied to 2DESs with $n_e \gtrsim 10^{11}$ cm^{-2} usually yields $m^\star$ values which are close to, or somewhat higher than, $m_\mathrm{b}^\star$ [3,4]. However, there exist studies [5,6] which report values significantly ($\simeq 10$ %) lower than $m_\mathrm{b}^\star$. The disagreement in obtained mass values can, at least in part, be accounted for by a relatively low accuracy of the SdHO approach.[1] There also exist other factors which might affect extracted $m^\star$, even when the procedure seems to work properly, which itself is not always the case [5, 7–9]. According to Ref. [6], the lower values of $m^\star$ might very well be a signal of electron-electron interactions which, in contrast to the case of dilute 2DESs, can actually *reduce* the effective mass at intermediate densities [10–15].

[1] At $n_e \simeq 3\cdot 10^{11}$ cm^{-2}, Ref. [5] (Ref. [6]) obtained $m^\star$ between 0.061 (0.057) and 0.065, including error bars.

Therefore, it is both interesting and important to revisit the issue of low effective mass values using alternative experimental probes.

In addition to SdHOs, several other types of magnetoresistance effects are known to occur in high mobility 2DESs [16–23]. Unlike the filling factor entering SdHOs, the parameters controlling these oscillations *do depend* on $m^\star$, thus making it available directly from the oscillation period. In what follows, we briefly discuss one such oscillation type, microwave-induced resistance oscillations (MIRO) [16], whose period can be measured with high precision.

MIRO appear in magnetoresistivity when a 2DES is irradiated by microwaves. Being a result of electron transitions between Landau levels, owing to photon absorption, MIRO are controlled by ω/ω_c, where $\omega = 2\pi f$ is the radiation frequency. It is well established, both theoretically [24–27] and experimentally [28–30], that MIRO are well described by $-\sin(2\pi\omega/\omega_c)$, provided that $2\pi\omega/\omega_c \gg 1$ and one is in the low power limit [31, 32]. As a result, the higher order ($i = 3, 4, ...$) MIRO maxima are *accurately* described by

$$\omega = \frac{e}{m^\star} B_i (i - \delta)\,, \tag{1}$$

where $\delta \approx 1/4$.[2] Once the value of δ is verified experimentally, one can obtain $m^\star$ using, e.g., the dispersion of the i-th MIRO maximum, $f(B_i)$.[3] Equivalently, the mass can be obtained directly from the oscillation period at a given ω, e.g., from the dependence of i on B_i, $i = \omega m^\star/eB_i + \delta$.

In this work we investigate the effective mass in a high mobility GaAs/AlGaAs quantum well using microwave photoresistance measured over a wide frequency range from 100 to 175 GHz. Remarkably, the effective mass extracted from the MIRO is found to be considerably lower than the band mass value. More specifically, MIRO are found to be well described by Eq. (1) with the effective mass $m^\star \approx 0.059\, m_0$ at *all* frequencies studied. These findings provide strong evidence for electron-electron interactions, which can be probed by microwave photoresistance in very high Landau levels. In contrast, the measured dispersion of the magneto-plasmon resonance (MPR) reveals $m^\star \approx m_b^\star$, in agreement with previous studies.

2. Experimental details

Our sample is a lithographically defined Hall bar of width $w = 50$ μm fabricated from a 300 Å-wide GaAs/$Al_{0.24}Ga_{0.76}As$ quantum well grown by molecular beam epitaxy with a density and mobility of $n \approx 2.7 \times 10^{11}$ cm^{-2} and $\mu \approx 1.3 \times 10^7$ cm^2/Vs, respectively.[4] Microwave radiation, generated by a backward wave oscillator, was delivered to the sample, placed in a ^{3}He cryostat, via a 1/4"(6.35 mm)-diameter light pipe. The magnetoresistivity $\rho_\omega(B)$ was measured under continuous microwave irradiation using a standard low-frequency lock-in technique.

3. Results and discussion

In Fig. 1(a) we present $\rho_\omega(B)$ measured at $T = 0.65$ K under microwave irradiation of $f = 170$ GHz. The data reveal MIRO, zero-resistance states [34–39], sharp spikes near the second harmonic of the cyclotron resonance [40–42], and a peak which corresponds to the lowest mode of the dimensional magneto-plasmon resonance (MPR) [43–45]. One can determine the effective mass entering Eq. (1) by trial and error, namely, by adjusting $m^\star$ until *each* cyclotron resonance harmonic falls symmetrically between maximum and minimum of the same order. Remarkably,

[2] Strictly speaking, $\delta \approx 1/4$ in Eq. (1) should be replaced by δ_i which approaches 1/4 for $i \gtrsim 3$. For instance, according to Ref. [32], $\delta_3 \approx 0.23$ and $\delta_4 \approx 0.24$. However, using 1/4 in Eq. (1) instead of more accurate values is well justified since it will result in less than 1 % error in the mass.

[3] Alternatively, $m^\star$ can also be obtained from the MIRO minima, described by $\omega = \omega_c(i + 1/4)$, or from the zero-response nodes, $\omega = \omega_c \cdot i$, where microwave photoresistance vanishes.

[4] Measurements of MIRO in a Hall bar of width $w = 200$ μm with $n \approx 3.2 \times 10^{11}$ cm^{-2}and $\mu \approx 1.1 \times 10^7$ cm^2/Vs also revealed $m^\star \approx 0.059\, m_0$ [33].

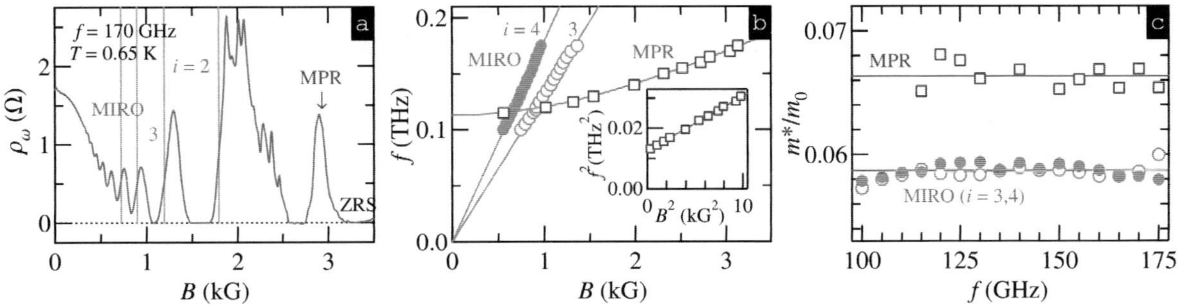

Figure 1. (Color online) (a) Magnetoresistivity $\rho_\omega(B)$ measured at $T = 0.65$ K irradiated with microwaves of $f = 170$ GHz. The vertical lines (marked by i) are drawn at the harmonics of the cyclotron resonance, $\omega/\omega_c = i$, calculated using $m^\star = 0.0590m_0$. (b) Dispersions $f(B)$ of the MIRO maxima for $i = 3$ (open circles), $i = 4$ (filled circles), and of the MPR peak (squares). The lines are fits to the MIRO data, $f = (i - 1/4)eB_i/2\pi m^\star$, with $i = 3, 4$ and the solid curve is given by $f = \sqrt{f_0^2 + (eB/2\pi m^\star)^2}$, where $f_0 = 112.5$ GHz and $m^\star = 0.066\, m_0$. Inset shows the MPR dispersion as f^2 vs B^2 (squares) and a linear fit (solid line), $f^2 = f_0^2 + (eB/2\pi m^\star)^2$. (c) $m^\star$ obtained from the MIRO maxima for $i = 3$ (open circles), $i = 4$ (filled circles) and the MPR peak (squares) vs f. Solid lines represent averages for the $i = 4$ MIRO maxima, $m^\star = 0.0587\, m_0$ and for the MPR peak, $m^\star = 0.0664\, m_0$, respectively.

such a procedure applied to the data in Fig. 1(a) results in $m^\star \approx 0.059\, m_0$, used to calculate the positions of vertical lines drawn at $\omega/\omega_c = 2, 3, 4$, and 5.

We have repeated our measurements at various microwave frequencies, from 100 to 175 GHz, and obtained the magnetic field positions of the MIRO maxima and of the MPR peak. Our findings are presented in Fig. 1(b) showing microwave frequency f as a function of B corresponding to the MIRO maxima at $i = 3$ (open circles), $i = 4$ (filled circles), and to the MPR peak (squares). It is clear that the MIRO maxima follow a linear dispersion relation, which extrapolates to the origin, as expected from Eq. (1). By fitting the MIRO data with Eq. (1), $f = (i - 1/4)eB_i/2\pi m^\star$, we obtain $m^\star = 0.0586\, m_0$ ($m^\star = 0.0587\, m_0$) for $i = 3$ ($i = 4$). Since the obtained values are both very close to each other, we conclude that the effective mass entering the MIRO period is given by $m^\star \approx 0.059\, m_0$.

The MPR peak follows a dispersion [cf. open squares in Fig. 1(b)] characteristic of a magnetoplasmon resonance [46],

$$\omega^2 = \omega_c^2 + \omega_0^2\,, \tag{2}$$

where ω_0 is the frequency of the lowest mode of standing plasmon oscillation (with a wavevector determined by the Hall bar width). As shown in the inset, f^2 is a linear function of B^2, in agreement with Eq. (2). From the slope of the fit to the data, $f^2 = f_0^2 + (eB/2\pi m^\star)^2$ (cf. solid line in the inset), we obtain $m^\star \approx 0.066m \approx m_b^\star$. We also notice that previous MPR experiments obtained $m^\star$ values ranging from 0.067 to 0.071 [43, 44, 47–49].

We summarize our findings in Fig. 1 (c) showing effective mass values, obtained from the two dispersion relations, as a function of microwave frequency. More specifically, $m^\star$ obtained from the MIRO maxima for $i = 3$ (open circles), $i = 4$ (filled circles), and from the MPR (squares). Solid horizontal lines represent the averages of the measured values. Figure 1 (c) further confirms that the masses extracted from our fits accurately describe our experimental data over the entire range of frequencies studied.

As mentioned above, one can also easily obtain $m^\star$ from the MIRO period. This method is based on the scaling of multiple oscillations and does not *a priori* assume $\delta = 1/4$. To illustrate this approach, we present in Fig. 2 microwave photoresistivity $\delta\rho_\omega = \rho_\omega - \rho$ (right axis) as a

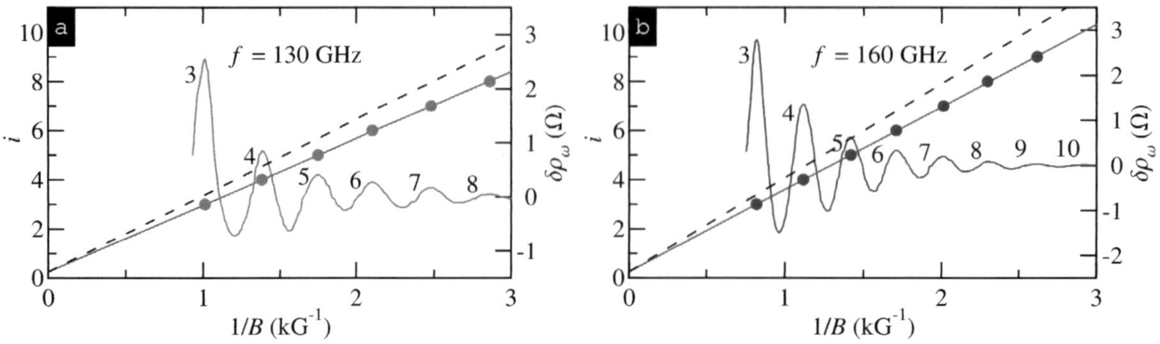

Figure 2. (Color online) Microwave photoresistivity $\delta\rho_\omega$ (right axis, solid curve) and the order of the MIRO maxima i (left axis, circles) vs. $1/B$ measured at (a) $f = 130$ GHz and (b) $f = 160$ GHz. Fits to the data (solid lines) with $i = 2\pi f m^\star/eB + \delta$ yield $\delta \approx 0.25$ and $m^\star \approx 0.0585\, m_0$ ($m^\star \approx 0.0587\, m_0$) for $f = 130$ GHz ($f = 160$ GHz). Dashed lines are calculated using Eq. (1) and $m^\star = m_b^\star = 0.067\, m_0$.

function of $1/B$ measured at (a) $f = 130$ GHz and (b) $f = 160$ GHz. Both data sets exhibit multiple oscillations whose period scales with $1/(fm^\star)$. To extract $m^\star$ from the data, we plot the order of the MIRO maxima i (circles, left axis) as a function of $1/B$ for both frequencies and observe the expected linear dependence. From the slope of the linear fits to the data (solid lines), $i = 2\pi f m^\star/eB + \delta$, we find $m^\star \approx 0.0585\, m_0$ ($m^\star \approx 0.0587\, m_0$) for $f = 130$ GHz ($f = 160$ GHz).[5] These values are in excellent agreement with the $m^\star$ values found from the dispersions of the $i = 3, 4$ MIRO maxima. In addition, we find that both fits intercept the vertical axis at $\delta \approx 0.25$, in agreement with Eq. (1), confirming the equivalence of the two approaches. Finally, to illustrate that our data cannot be described by the band mass we include dashed lines which are calculated using $i = 2\pi f m_b^\star/eB + 0.25$.

4. Summary

In summary, we have investigated microwave photoresistance in a high mobility GaAs/AlGaAs quantum well over a wide range of microwave frequencies. The analysis of the period of the microwave-induced resistance oscillations reveals an effective mass $m^\star \approx 0.059\, m_0$, which is considerably lower than the GaAs band mass $m_b^\star = 0.067\, m_0$. These findings provide strong evidence for electron-electron interactions and for sensitivity of MIRO to these interactions. On the other hand, the measured dispersion of the magneto-plasmon resonance is best described by $m^\star \approx m_b^\star$. It would be interesting to examine if the low value of the effective mass is confirmed in studies of other nonlinear phenomena, such as Hall-field induced resistance oscillations [18,50,51].

Acknowledgments

We thank E. Kolomeitsev and B. Shklovskii for discussions and J. Jaroszynski, J. Krzystek, G. Jones, T. Murphy, and D. Smirnov for technical assistance. This work was supported by the US Department of Energy, Office of Basic Energy Sciences, under Grant Nos. DE-SC002567 (Minnesota) and DE-SC0006671 (Purdue). A portion of this work was performed at the National High Magnetic Field Laboratory (NHMFL), which is supported by NSF Cooperative Agreement No. DMR-0654118, by the State of Florida, and by the DOE. The work at Princeton was partially funded by the Gordon and Betty Moore Foundation and the NSF MRSEC Program through the Princeton Center for Complex Materials (DMR-0819860).

[5] Similar $m^\star$ values have been obtained from the MIRO minima and from the zero-response nodes (not shown).

References

[1] Sze S M 1981 *Physics of Semiconductor Devices* (Wiley, New York) p 850
[2] Pudalov V M 2011 *Low Temp. Phys.* **37** 8
[3] Smrčka L, Vašek P, Koláček J, Jungwirth T and Cukr M 1995 *Phys. Rev. B* **51** 18011
[4] Hang D, Huang C, Zhang Y, Yeh H, Hsiao J and Pang H 2007 *Solid State Commun.* **141** 17
[5] Coleridge P, Hayne M, Zawadzki P and Sachrajda A 1996 *Surf. Sci.* **361-362** 560
[6] Tan Y W, Zhu J, Stormer H L, Pfeiffer L N, Baldwin K W and West K W 2005 *Phys. Rev. Lett.* **94** 016405
[7] Hayne M, Usher A, Harris J J and Foxon C T 1992 *Phys. Rev. B* **46** 9515
[8] Hayne M, Usher A, Harris J J and Foxon C T 1997 *Phys. Rev. B* **56** 10446
[9] Coleridge P T 1991 *Phys. Rev. B* **44** 3793
[10] Smith A P, MacDonald A H and Gumbs G 1992 *Phys. Rev. B* **45** 8829
[11] Kwon Y, Ceperley D M and Martin R M 1994 *Phys. Rev. B* **50** 1684
[12] Zhang Y and Das Sarma S 2005 *Phys. Rev. B* **72** 075308
[13] Asgari R, Davoudi B, Polini M, Giuliani G F, Tosi M P and Vignale G 2005 *Phys. Rev. B* **71** 045323
[14] Asgari R and Tanatar B 2006 *Phys. Rev. B* **74** 075301
[15] Drummond N D and Needs R J 2009 *Phys. Rev. B* **80** 245104
[16] Zudov M A, Du R R, Simmons J A and Reno J L 2001 *Phys. Rev. B* **64** 201311(R)
[17] Zudov M A, Ponomarev I V, Efros A L, Du R R, Simmons J A and Reno J L 2001 *Phys. Rev. Lett.* **86** 3614
[18] Yang C L, Zhang J, Du R R, Simmons J A and Reno J L 2002 *Phys. Rev. Lett.* **89** 076801
[19] Zhang W, Zudov M A, Pfeiffer L N and West K W 2007 *Phys. Rev. Lett.* **98** 106804
[20] Zhang W, Zudov M A, Pfeiffer L N and West K W 2008 *Phys. Rev. Lett.* **100** 036805
[21] Hatke A T, Chiang H S, Zudov M A, Pfeiffer L N and West K W 2008 *Phys. Rev. B* **77** 201304(R)
[22] Hatke A T, Chiang H S, Zudov M A, Pfeiffer L N and West K W 2008 *Phys. Rev. Lett.* **101** 246811
[23] Khodas M, Chiang H S, Hatke A T, Zudov M A, Vavilov M G, Pfeiffer L N and West K W 2010 *Phys. Rev. Lett.* **104** 206801
[24] Durst A C, Sachdev S, Read N and Girvin S M 2003 *Phys. Rev. Lett.* **91** 086803
[25] Vavilov M G and Aleiner I L 2004 *Phys. Rev. B* **69** 035303
[26] Dmitriev I A, Vavilov M G, Aleiner I L, Mirlin A D and Polyakov D G 2005 *Phys. Rev. B* **71** 115316
[27] Dmitriev I A, Mirlin A D, Polyakov D G and Zudov M A 2012 *Rev. Mod. Phys.* **84** 1709
[28] Zudov M A 2004 *Phys. Rev. B* **69** 041304(R)
[29] Mani R G, Smet J H, von Klitzing K, Narayanamurti V, Johnson W B and Umansky V 2004 *Phys. Rev. Lett.* **92** 146801
[30] Hatke A T, Zudov M A, Pfeiffer L N and West K W 2009 *Phys. Rev. Lett.* **102** 066804
[31] Hatke A T, Khodas M, Zudov M A, Pfeiffer L N and West K W 2011 *Phys. Rev. B* **84** 241302(R)
[32] Hatke A T, Zudov M A, Pfeiffer L N and West K W 2011 *Phys. Rev. B* **84** 241304(R)
[33] Hatke A T, Zudov M A, Watson J D, Manfra M J, Pfeiffer L N and West K W 2012 *unpublished*
[34] Mani R G, Smet J H, von Klitzing K, Narayanamurti V, Johnson W B and Umansky V 2002 *Nature (London)* **420** 646
[35] Zudov M A, Du R R, Pfeiffer L N and West K W 2003 *Phys. Rev. Lett.* **90** 046807
[36] Smet J H, Gorshunov B, Jiang C, Pfeiffer L, West K, Umansky V, Dressel M, Meisels R, Kuchar F and von Klitzing K 2005 *Phys. Rev. Lett.* **95** 116804
[37] Zudov M A, Du R R, Pfeiffer L N and West K W 2006 *Phys. Rev. B* **73** 041303(R)
[38] Zudov M A, Du R R, Pfeiffer L N and West K W 2006 *Phys. Rev. Lett.* **96** 236804
[39] Dorozhkin S I, Pfeiffer L, West K, von Klitzing K and Smet J H 2011 *Nature Phys.* **7** 336
[40] Dai Y, Du R R, Pfeiffer L N and West K W 2010 *Phys. Rev. Lett.* **105** 246802
[41] Hatke A T, Zudov M A, Pfeiffer L N and West K W 2011 *Phys. Rev. B* **83** 121301(R)
[42] Hatke A T, Zudov M A, Pfeiffer L N and West K W 2011 *Phys. Rev. B* **83** 201301(R)
[43] Vasiliadou E, Müller G, Heitmann D, Weiss D, von Klitzing K, Nickel H, Schlapp W and Lösch R 1993 *Phys. Rev. B* **48** 17145
[44] Kukushkin I V, Muravev V M, Smet J H, Hauser M, Dietsche W and von Klitzing K 2006 *Phys. Rev. B* **73** 113310
[45] Hatke A T, Zudov M A, Watson J D and Manfra M J 2012 *Phys. Rev. B* **85** 121306(R)
[46] Chaplik A V 1972 *Sov. Phys. JETP* **35** 395
[47] Yang C L, Du R R, Pfeiffer L N and West K W 2006 *Phys. Rev. B* **74** 045315
[48] Muravev V M, Jiang C, Kukushkin I V, Smet J H, Umansky V and von Klitzing K 2007 *Phys. Rev. B* **75**(19) 193307
[49] Dorozhkin S I, Smet J H, von Klitzing K, Pfeiffer L N and West K W 2007 *JETP Lett.* **86** 543
[50] Hatke A T, Zudov M A, Pfeiffer L N and West K W 2009 *Phys. Rev. B* **79** 161308(R)
[51] Hatke A T, Zudov M A, Pfeiffer L N and West K W 2011 *Phys. Rev. B* **83** 081301(R)

Journal of Physics: Conference Series **456** (2013) 012041 doi:10.1088/1742-6596/456/1/012041

Shubnikov-de Haas oscillations in high mobility GaAs quantum wells in tilted magnetic fields

A. T. Hatke[1], **M. A. Zudov**[1], **L. N. Pfeiffer**[2], **and K. W. West**[2]

[1]School of Physics and Astronomy, University of Minnesota, Minneapolis, Minnesota 55455, USA

[2]Department of Electrical Engineering, Princeton University, Princeton, New Jersey 08544, USA

E-mail: zudov@physics.umn.edu

Abstract. We report on quantum magneto-oscillations in a high mobility GaAs/AlGaAs quantum well at very high ($\geq 87^\circ$) tilt angles. Unlike previous studies, we find that the spin and cyclotron splittings become equal over a *continuous range* of angles, but only near certain, angle-dependent filling factors. At high enough tilt angles, Shubnikov-de Haas oscillations reveal a prominent beating pattern, indicative of *consecutive* level crossings, *all* occurring at the same angle. We explain these unusual observations by an in-plane field-induced increase of the carrier mass, which leads to accelerated, filling factor-driven crossings of spin sublevels in tilted magnetic fields.

1. Introduction

When a two-dimensional electron system (2DES) is subject to a perpendicular magnetic field $B_\perp$, its low temperature magnetoresistivity is known to exhibit Shubnikov-de Haas oscillations (SdHO). SdHO originate from the magnetic quantization of the energy spectrum,

$$E_{N,\mp} = \hbar\omega_c (N + 1/2) \pm \Delta_s/2\,, \tag{1}$$

where $\hbar\omega_c = \hbar e B_\perp/m^\star$ is the cyclotron energy, $m^\star$ is the effective mass, N is the Landau level index, and Δ_s is the sum of the single particle Zeeman energy Δ_Z and the exchange energy Δ_X. The resistivity of the 2DES senses the density of electron states at the Fermi level, ε_F, and will show a minimum whenever ε_F falls in the middle between the energy levels, see Eq. (1), i.e., when the filling factor $\nu = 2\varepsilon_F/\hbar\omega_c$ acquires an integer value.

The strength of the SdHO at odd and even ν is determined by the *spin splitting*, $\Delta_{odd} = \Delta_s$, and by the *cyclotron splitting*, $\Delta_{even} = \hbar\omega_c - \Delta_s$, respectively. Furthermore, the exchange contribution to the spin splitting becomes important only when there is an appreciable population difference between $(N,+)$ and $(N,-)$ sublevels. The latter turns off rather abruptly at some critical, disorder-dependent filling factor ν_s. The associated disappearance of the odd-ν SdHO minima is referred to as a critical collapse of the exchange-enhanced spin splitting [1, 2, 3, 4, 5]. At odd $\nu \gg 1$, but considerably lower than ν_s, the exchange energy scales with the cyclotron energy $\Delta_X = \alpha_X \hbar\omega_c$ [1, 6], while at even ν, one can set $\Delta_X \approx 0$ and

Journal of Physics: Conference Series **456** (2013) 012041 doi:10.1088/1742-6596/456/1/012041

$\Delta_s \approx \Delta_Z = \alpha_Z \hbar\omega_c$. Since in GaAs $\alpha_Z \ll 1$, $\alpha_X < 1$[1] and, most importantly, since there are no critically collapsing contributions to $\Delta_{\rm even}$, the SdHO at even-ν persist to filling factors much larger than ν_s.[2]

Adding an in-plane magnetic field will increase the Zeeman energy since the latter scales with the total field. On the other hand, one usually assumes that both the cyclotron and the exchange energies depend only on $B_\perp$ [2, 4].[3] As a result, when the magnetic field is tilted by angle θ away from the sample normal, for any given ν, the cyclotron splitting, $\Delta_{\rm even} \propto 1-\alpha_Z/\cos\theta$, will decrease with θ and eventually approach the increasing spin splitting, $\Delta_{\rm odd} \propto \alpha_X + \alpha_Z/\cos\theta$. This approach forms a basis of the *coincidence method* [7], which was successfully used to study the exchange contribution [2, 8]. The "coincidences", manifested as equally strong even-ν and odd-ν SdHO, will occur when[4] $\Delta_{\rm even} = \Delta_{\rm odd}$ or

$$\gamma_i \equiv \frac{1}{\cos\theta_i} = \frac{i-\alpha_X}{2\alpha_Z}, \quad i = 1, 3, 5, \ldots . \tag{2}$$

Equation (2),[5] dictating that the coincidences should occur *simultaneously* for all ν, was confirmed experimentally [2] for $i = 1$, $\theta_1 \approx 87.2^\circ$ ($\gamma_1 \approx 20$).

In this work we report on spin-resolved SdHO in an ultra-high mobility GaAs/AlGaAs quantum well at very high tilt angles ($\theta \geq 87^\circ$). We find that the evolution of the SdHO waveform depends sensitively not only on θ, as predicted by Eq. (2), but also on the filling factor. More specifically, we find that coincidence conditions, $\Delta_{\rm even} = \Delta_{\rm odd}$, can be satisfied *over a range of tilt angles* but only at certain filling factors. These filling factors monotonically increase with θ and eventually diverge at $\theta = \theta_i$. At $\theta > \theta_1$, SdHO reveal a beating pattern indicating *cosecutive* ($i = 3, 5$), filling factor-driven coincidences, occurring at the same θ. These findings are in contrast with Eq. (2) and previous experiments [2] and can be explained by an in-plane field-induced increase of the effective mass due to the finite width of our 2DES. This increase leads to a non-monotonic dependence of the cyclotron splitting on ν and its eventual collapse resulting in accelerated crossings of spin sublevels.

2. Experimental details

Our sample is a 200 μm-wide Hall bar fabricated from a 29 nm-wide GaAs-$Al_{0.24}Ga_{0.76}As$ quantum well, grown by molecular beam epitaxy, with density $n_e \approx 2.85 \times 10^{11}$ cm^{-2} and mobility $\mu \approx 2.4 \times 10^7$ cm^2/Vs . The magnetoresistivity was recorded using a standard low-frequency (a few Hertz) lock-in technique at temperature $T \approx 0.3$ K. The tilt angle θ, controlled *in situ* by a precision rotator, was held constant during each magnetic field sweep.

3. Results and discussion

In Fig. 1 we present the oscillatory part of the resistivity $\delta\rho$ (left axes), obtained from the total resistivity by subtracting a slowly varying background, versus ν at different tilt angles θ, as marked. The corresponding amplitudes, $\delta\rho_{\rm max}$, at even (circles) and odd (squares) filling factors are shown on the right (logarithmic) axes. These amplitudes were evaluated using $\delta\rho$ values at the minima and the average of the neighboring maxima.

[1] According to Ref. [6], $\alpha_X = \ln(k_F a_B)/\pi k_F a_B$, where k_F is the Fermi wave number, and a_B is the effective Bohr radius. In our 2DES we estimate $\alpha_X \approx 0.234$.

[2] While odd-ν minima disappear abruptly at $\nu = \nu_s$, the even-ν oscillations decay roughly exponentially with ν.

[3] While (fully-enhanced) $\Delta_X \propto B_\perp$, ν_s will increase with $\Delta_Z \propto \gamma$ making the exchange contribution relevant at progressively higher ν with increasing θ (Refs. [2], [4]).

[4] Equation (2) for $i = 1$ coincides with Eq. (11) of Ref. [2] if α_X is replaced by $2\alpha_X$. The extra factor appears because in Ref. [2] Δ_X is included in both $\Delta_{\rm odd}$ and $\Delta_{\rm even}$.

[5] Equation (2) should also be approximately valid at even i, corresponding to level crossings, if one takes the exchange part to be only partially ($\simeq 50\%$) enhanced (Ref. [8]).

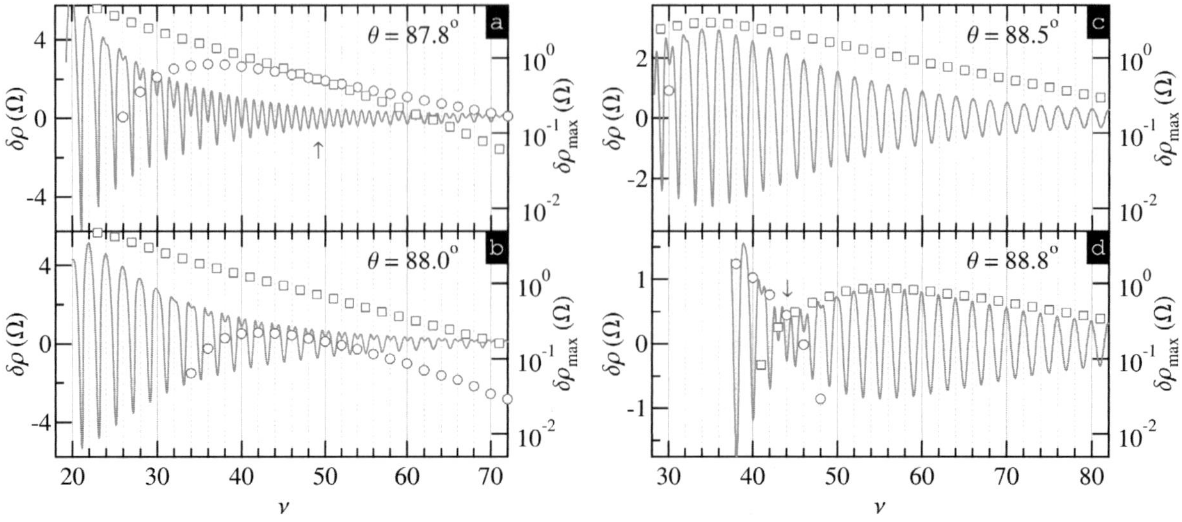

Figure 1. Oscillatory part of the resistivity $\delta\rho$ (left axis) and amplitude $\delta\rho_{\max}$ (right axis) versus the filling factor ν measured at tilt angles (a) $\theta \approx 87.8°$, (b) $\theta \approx 88.0°$, (c) $\theta \approx 88.5°$, and (d) $\theta \approx 88.8°$. Even-ν (circles) and odd-ν (squares) amplitudes become equal at $\nu = \nu_1$ (cf. ↑) and then again at $\nu = \nu_3$ (cf. ↓).

At the lowest θ, Fig. 1(a), we observe that the amplitudes at odd-ν and even-ν become equal at $\nu_1 \approx 49$ (cf. ↑). The data at lower and somewhat higher θ (not shown) reveal that ν_1 increases with θ. At lower filling factors, $\nu < \nu_1$, the odd-ν amplitude continues to grow but the even-ν amplitude first saturates, then starts to *decrease*, and eventually disappears at $\nu = \nu_c \approx 24$.

The observed amplitude crossing indicate that at $\nu = \nu_1$ the spin splitting becomes equal to the cyclotron splitting. On the other hand, the vanishing of the even-ν amplitude at $\nu = \nu_c$ signals a collapse of the cyclotron gap, which is a precursor of the crossing of spin sub-levels from the neighboring Landau levels. Such crossing would occur when $\Delta_{\rm even} = \hbar\omega_c - \Delta_Z = 0$, so the observed collapse indicates that Δ_Z is growing faster than $\hbar\omega_c$ with decreasing ν. We further notice that the exchange energy at odd filling factors should be close to its maximum value, $\alpha_X \hbar\omega_c$ since, at $\nu < \nu_1$, the separation between spin-split maxima is close to unity and the odd-ν amplitude shows excellent exponential dependence. Finally, $\gamma\alpha_Z$ is no longer a small fraction and Δ_Z gives a significant contribution to the spin splitting.

At a higher tilt angle, Fig. 1(b), we find that the even-ν amplitude remains considerably smaller than the odd-ν amplitude over the whole range of ν and the crossing never takes place. At still higher angle, Fig. 1(c), the even-ν SdHO disappear completely indicating that the condition for the first level crossing is approximately satisfied for all filling factors under study. However, closer examination of Fig. 1(c) reveals a maximum of the odd-ν amplitude at $\nu \approx 35$, and the re-emergence of the even-ν SdHO to the left of this maximum, indicating that at these filling factors the crossing of spin sublevels has already occurred.

The data at the next tilt angle, Fig. 1(d), reveal a node which occurs at $\nu = \nu_3$ (cf. ↓). Similar to the "coincidence" at $\nu = \nu_1$, this node moves to progressively higher ν with increasing θ. Passing through the node from high to low filling factors, odd-ν SdHO decay away while the even-ν SdHO set in. As a result, at $\nu \approx \nu_3$ the amplitudes of the odd-ν and even-ν SdHO again become close to each other, reflecting equal gaps, similar to the situation at $\nu \approx \nu_1$ (cf. Fig. 1(a)).[6] While the SdHO shown in Fig. 1(d) differs in detail from those shown in Fig. 1(a)

[6] The data at higher angles [9] reveal a node at $\nu = \nu_5$, which signals the next crossing.

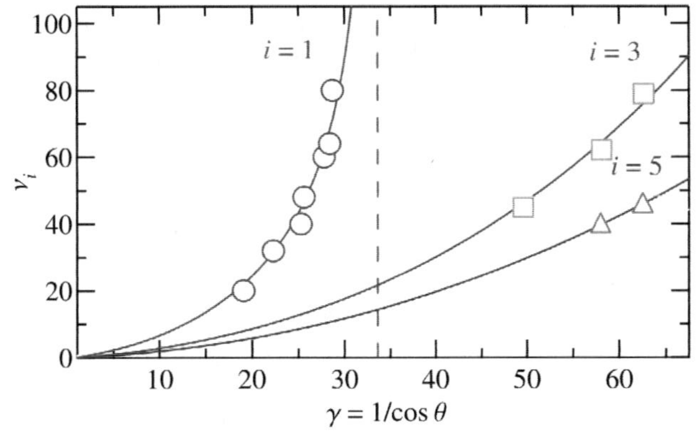

Figure 2. Filling factors ν_i for $i = 1$ (circles), $i = 3$ (squares) and $i = 5$ (triangles) corresponding to crossings of the even-ν and odd-ν amplitudes as functions of $1/\cos\theta$. Solid lines are fits to the data with Eq. (4). The dashed vertical line is the asymptote of Eq. (4) for $i = 1$ coincidence at large ν. This line also corresponds to θ_1 given by Eq. (2). Adapted from Ref. [9].

and Fig. 1(b), as we show next, all observations can be attributed to subsequent level crossings. Clearly, the observed behavior cannot be described by Eq. (2), which is independent of ν.

To explain our findings we consider finite thickness effects which, in combination with $B_{\|}$, can lead to an increase of the effective mass. This increase was studied theoretically [10, 11, 12, 13, 14, 15, 16] and confirmed in experiments examining the temperature damping of the SdHO [17] and the shift of magnetoplasmon resonances [18, 19]. Following Ref. [19], we use the expression for a parabolic potential [10, 16] and substitute the intersubband splitting of our quantum well, ε_{10}, for the intersubband spacing in the parabolic well. At $\gamma \gg 1$, the effective mass increases with respect to its value at $\theta = 0$ by a factor $1/\alpha_\gamma$, which is given by [10, 16]

$$\frac{1}{\alpha_\gamma} = \frac{m^\star_\theta}{m^\star_0} \approx \sqrt{1 + \left(\frac{\beta\gamma}{\nu}\right)^2}, \tag{3}$$

where $\beta = 2\varepsilon_F/\varepsilon_{10}$ and ε_F is evaluated at $\theta = 0$. Since β is of the order of unity,[7] a significant mass increase is expected at $\nu < \gamma$. This increase will result in a sublinear dependence of the cyclotron energy on $1/\nu$ and its eventual saturation at $\hbar\omega_c = \varepsilon_{10}/\gamma \ll \varepsilon_{10}$, in the limit of $\nu \ll \gamma$. As a result, at $\theta \neq 0$, the cyclotron splitting, $\Delta_{\rm even} \equiv \hbar\omega_c - \Delta_Z \propto [\alpha_\gamma(\nu) - \alpha_Z\gamma]/\nu$, is no longer a monotonic function, giving rise to the ν-dependence of the even-ν SdHO amplitude, see Fig. 1. Coincidence conditions, which take into account the mass increase, Eq. (3), can now be obtained by replacing i with $i\alpha_\gamma$ in Eq. (2) and solving for ν, which results in

$$\nu_i \approx \frac{\beta\gamma}{\sqrt{i^2/\left(2\alpha_Z\gamma + \alpha_X\right)^2 - 1}}. \tag{4}$$

Tilting the sample with respect to the magnetic field is also known to affect α_Z due to the anisotropy of the g-factor. Recent experiments on electron-spin resonance have shown that $g_0 = \sqrt{g_\perp^2\cos^2\theta + g_\|^2\sin^2\theta}$, where $g_\perp = 0.414$ and $g_\perp = 0.340$ [20]. At $\gamma \gg 1$, we obtain $g_0 \approx g_\| = 0.340$. It is also known that at high magnetic fields quadratic in B corrections to Δ_Z become important [20, 21, 22]. These terms can be incorporated into the g-factor which acquires linear B dependence, $g = g_0 - aB$. At $\theta \neq 0$, $a \approx (\cos\theta/g)(g_\perp a_\perp \cos^2\theta + g_\| a_\| \sin^2\theta)$, where $a_\perp \sim a_\| \sim 0.01\ \mathrm{T}^{-1}$ [20]. At $\gamma \gg 1$, $a \approx a_\|/\gamma \ll a_\|$, and the correction to the g-factor, aB, remains less than 1% in our experiment even at the highest magnetic fields. We will therefore assume a B- and θ-independent g-factor, $g = g_0 = 0.34$ in our analysis.

To demonstrate that our findings can be described by Eq. (4), we present in Fig. 2 the filling factors ν_1 (circles), ν_3 (squares), and ν_5 (triangles) corresponding to crossings of the even-ν

[7] Using infinite square well approximation for our 2DES gives $E_{10} \approx 225$ K and $\beta \approx 1.1$ ($E_F \approx 125$ K).

and odd-ν amplitudes as functions of γ. Solid lines are fits to the data using Eq. (4) with $\alpha_Z \equiv g_0 m^\star/2m = 0.0114$, $\alpha_X = 0.234$, and β as a fitting parameter. For ν_1 the fit generates $\beta \approx 1.3$, which is close to our estimate of $\beta = 1.1$. The vertical line is drawn at $\theta = \theta_1$, calculated using Eq. (2). For ν_3 and ν_5 the fits produce $\beta \approx 1.8$ and $\beta \approx 2.1$, respectively. These values are somewhat higher than our estimate, which likely reflects the limitations of our model.

4. Summary

In summary, we have shown that in a typical high mobility GaAs/AlGaAs quantum well, the Shubnikov-de Haas oscillations in tilted magnetic fields cannot be described by a simple series of level crossings at a set of angles defined by Eq. (2). Instead, multiple level crossings can be realized for a given angle, but only near some characteristic filling factors, see Eq. (4) and Fig. 2. These findings can be explained by finite thickness effects, which give rise to an effective mass increase induced by an in-plane magnetic field. Finally, we mention that an in-plane magnetic field can also lead to the overall suppression of the Shubnikov-de Haas oscillations [23].

Acknowledgments

We thank I. Dmitriev, R. Haug, M. Khodas, T. Shahbazyan, B. Shklovskii, S. Studenikin, and M. Dyakonov for discussions and G. Jones and T. Murphy for technical assistance. This work was supported by the US Department of Energy, Office of Basic Energy Sciences, under Grant No. DE-SC002567. The work at Princeton was funded by the Gordon and Betty Moore Foundation and by the NSF MRSEC Program through the Princeton Center for Complex Materials (DMR-0819860). A portion of this work was performed at the NHMFL, which is supported by NSF Cooperative Agreement No. DMR-0654118, by the State of Florida, and by the DOE.

References

[1] Fogler M M and Shklovskii B I 1995 *Phys. Rev. B* **52**(24) 17366
[2] Leadley D R, Nicholas R J, Harris J J and Foxon C T 1998 *Phys. Rev. B* **58** 13036
[3] Piot B A, Maude D K, Henini M, Wasilewski Z R, Friedland K J, Hey R, Ploog K H, Toropov A I, Airey R and Hill G 2005 *Phys. Rev. B* **72**(24) 245325
[4] Piot B A, Maude D K, Henini M, Wasilewski Z R, Gupta J A, Friedland K J, Hey R, Ploog K H, Gennser U, Cavanna A, Mailly D, Airey R and Hill G 2007 *Phys. Rev. B* **75**(15) 155332
[5] Pan W, Baldwin K W, West K W, Pfeiffer L N and Tsui D C 2011 *Phys. Rev. B* **84**(16) 161307
[6] Aleiner I L and Glazman L I 1995 *Phys. Rev. B* **52**(15) 11296
[7] Fang F F and Stiles P J 1968 *Phys. Rev.* **174**(3) 823
[8] Nicholas R J, Haug R J, Klitzing K v and Weimann G 1988 *Phys. Rev. B* **37** 1294
[9] Hatke A T, Zudov M A, Pfeiffer L N and West K W 2012 *Phys. Rev. B* **85** 241305(R)
[10] Khaetskii A and Shklovskii B 1983 *Sov. Phys. JETP* **58** 421
[11] Maan J C 1990 *Two Dimensional Systems, Heterostructures, and Superlattices* ed Bauer G, Kuchar F and Heinrich H (Springer-Verlag, Berlin)
[12] Merlin R 1987 *Solid State Commun.* **64** 99
[13] Tang H and Butcher P N 1988 *J. of Phys. C: Solid State Phys.* **21** 3313
[14] Smrcka L 1990 *J. of Phys. Condens. Matter* **2** 8337
[15] Kirpichev V E, Kukushkin I V, Timofeev V B and Fal'ko V I 1990 *JETP Letters* **51** 436
[16] Smrcka L and Jungwirth T 1994 *J. of Phys. Condens. Matter* **6** 55
[17] Smrčka L, Vašek P, Koláček J, Jungwirth T and Cukr M 1995 *Phys. Rev. B* **51** 18011
[18] Batke E and Tu C W 1986 *Phys. Rev. B* **34** 3027
[19] Kozlov V, Gubarev S and Kukushkin I 2011 *JETP Letters* **94** 397
[20] Nefyodov Y A, Shchepetilnikov A V, Kukushkin I V, Dietsche W and Schmult S 2011 *Phys. Rev. B* **84** 233302
[21] Dobers M, von Klitzing K and Weimann G 1988 *Phys. Rev. B* **38** 5453
[22] Nefyodov Y A, Shchepetilnikov A V, Kukushkin I V, Dietsche W and Schmult S 2011 *Phys. Rev. B* **83** 041307
[23] Bogan A, Hatke A T, Studenikin S A, Sachrajda A, Zudov M A, Pfeiffer L N and West K W 2012 *Phys. Rev. B* **86** 235305

AUTHOR INDEX

AUTHOR INDEX